Informatik aktuell

Herausgeber: W. Brauer
im Auftrag der Gesellschaft für Informatik (GI)

Paul Levi

Michael Schanz

Reinhard Lafrenz

Viktor Avrutin (Hrsg.)

Autonome Mobile Systeme 2005

19. Fachgespräch
Stuttgart, 8./9. Dezember 2005

Herausgeber

Paul Levi
Michael Schanz
Reinhard Lafrenz
Viktor Avrutin
Universität Stuttgart
Fakultät Informatik, Elektrotechnik und Informationstechnik
Institut für Parallele und Verteilte Systeme (IPVS), Abt. Bildverstehen
Universitätsstr. 38, 70569 Stuttgart
http://www.ipvs.uni-stuttgart.de/abteilungen/bv/

Bibliographische Information der Deutschen Bibliothek
Die Deutsche Bibliothek verzeichnet diese Publikation in der Deutschen Nationalbibliografie; detaillierte
bibliografische Daten sind im Internet über http://dnb.ddb.de abrufbar.

CR Subject Classification (2001): I.2.9, I.2.10, I.2.11, I.4.7, I.4.8, I.5.4, J.7

ISSN 1431-472-X
ISBN-10 3-540-30291-3 Springer Berlin Heidelberg New York
ISBN-13 978-3-540-30291-9 Springer Berlin Heidelberg New York

Springer Berlin Heidelberg New York
Springer ist ein Unternehmen von Springer Science+Business Media

springer.de

© Springer-Verlag Berlin Heidelberg 2006
Printed in Germany

Satz: Reproduktionsfertige Vorlage vom Autor/Herausgeber
Gedruckt auf säurefreiem Papier   SPIN: 11420231   33/3142-543210

Vorwort

Das 19. Fachgespräch Autonome Mobile Systeme (**AMS 2005**) findet am 8. und 9. Dezember 2005 in Stuttgart statt und wird zum vierten Mal von der Abteilung Bildverstehen des Instituts für Parallele und Verteilte Systeme (IPVS) der Universität Stuttgart organisiert. Ein Ziel dieser Fachgespräche ist es, Wissenschaftlerinnen und Wissenschaftlern aus Forschung und Industrie, die auf dem Gebiet der autonomen mobilen Systeme arbeiten, eine Basis für den Gedankenaustausch zu bieten und wissenschaftliche Diskussionen sowie Kooperationen auf diesem Forschungsgebiet zu fördern beziehungsweise zu initiieren.

Autonome mobile Systeme werden nicht nur in den mittlerweile traditionellen Bereichen wie Konstruktion, Fertigung und Logistik eingesetzt, sondern auch immer mehr im Alltag. Dabei werden sowohl Methoden und Konzepte der Sensorfusion, der Modellierung sowie der künstlichen und verteilten künstlichen Intelligenz in neue Bereiche übertragen, als auch die kognitiven Fähigkeiten der Systeme sukzessive erweitert. Die aktuellen Entwicklungen bei autonomen Fahrzeugen, in der Servicerobotik sowie bei vielen medizinischen Anwendungen und nicht zuletzt auch bei autonomen Spielzeugen zeigen deutlich diesen Trend. Mit zunehmender Miniaturisierung, Performanzsteigerung und einem höheren Grad an Integration und Vernetzung ist zu erwarten, dass immer mehr autonome Systeme sowohl in bereits etablierten als auch neuen Bereichen eingesetzt werden. Die Fachgespräche sollen dabei vor allem zum Austausch von Wissen und Erfahrungen beitragen.

Von den in diesem Jahr eingereichten 52 Kurzfassungen wurden vom Fachgesprächsbeirat 41 Beiträge aufgrund ihrer Qualität und Originalität angenommen. Die thematischen Schwerpunkte bei den Grundlagen liegen in diesem Jahr auf den Gebieten Bildverarbeitung und Lokalisierung, während bei den Anwendungen ein Trend in Richtung Outdoor-Anwendungen zu erkennen ist.

Die Organisatoren danken dem Fachgesprächsbeirat für die Auswahl der Beiträge. Bei den Autoren bedanken wir uns für das Einreichen ihrer Beiträge und die darin investierte wissenschaftliche Arbeit. Unser Dank gilt außerdem Herrn Prof. Dr. Dr. h.c. Brauer, dem Herausgeber der Buchreihe „Informatik Aktuell" sowie dem Springer-Verlag für die Herstellung dieses Bandes. Besonders bedanken möchten wir uns bei Frau Glaunsinger vom Springer-Verlag für die freundliche und effiziente Zusammenarbeit. Allen Teilnehmerinnen und Teilnehmern wünschen wir einen erfolgreichen wissenschaftlichen Gedanken- und Erfahrungsaustausch auf dem 19. Fachgespräch Autonome Mobile Systeme und einen angenehmen Aufenthalt in Stuttgart.

Stuttgart, im Oktober 2005   Paul Levi, Michael Schanz
                                                   Reinhard Lafrenz, Viktor Avrutin

Tagungsorganisation

Abteilung Bildverstehen
Institut für Parallele und Verteilte Systeme (IPVS)
Fakultät Informatik, Elektrotechnik und Informationstechnik
Universität Stuttgart
Universitätsstraße 38, 70569 Stuttgart
http://www.ipvs.uni-stuttgart.de/abteilungen/bv/

Fachgesprächsbeirat

Prof. Dr.-Ing. R. Dillmann	(Universität Karlsruhe)
Prof. Dr.-Ing. G. Färber	(TU München)
Prof. Dr.-Ing. G. Hirzinger	(DLR, Oberpfaffenhofen)
Prof. Dr.-Ing. A. Knoll	(TU München)
Dr. rer. nat. G. Lawitzky	(Siemens AG, München)
Prof. Dr. rer. nat. P. Levi	(Universität Stuttgart)
Prof. Dr.-Ing. Dr.-Ing. E.h. G. Schmidt	(TU München)
Prof. Dr.-Ing. H. Wörn	(Universität Karlsruhe)

Inhaltsverzeichnis

Lokalisierung und Kartographierung

Outdoor-Systeme

Fahrerassistenzsysteme

Kognitive Sensordatenverarbeitung

Architekturen und Anwendungen

Steuerung und Navigation

Kooperative Systeme

Architektur und Komponenten für ein heterogenes Team kooperierender, autonomer humanoider Roboter

Jutta Kiener, Sebastian Petters, Dirk Thomas, Martin Friedmann, Oskar von Stryk

Fachgebiet Simulation und Systemoptimierung, Fachbereich Informatik,
Technische Universität Darmstadt, D-64289 Darmstadt,

E-mail: {kiener,petters,dthomas,friedmann,stryk}@sim.tu-darmstadt.de

Zusammenfassung. Für ein kooperierendes Team autonomer, humanoider Roboter, das derzeit aus insgesamt vier unterschiedlichen, ca. 37–68 cm großen Robotertypen besteht, werden eine plattformübergreifende, modulare Softwarearchitektur sowie plattformübergreifende und individuelle Module zur Sensordatenverarbeitung, Planung und Bewegungssteuerung entwickelt.Das entwickelte funktionale Framework ermöglicht die Kommunikation der Softwaremodule, d.h. Algorithmen für die unterschiedlichen Aufgaben innerhalb der Architektur untereinander, sowie die Kommunikation per WLAN zwischen verschiedenen Rechnern und Robotern. Als Anwendungsszenario für die Teamkooperation in einer dynamischen und strukturierten Umgebung wird Roboterfußball untersucht. Die entwickelten Methoden wurden im Juli 2005 von den Darmstadt Dribblers beim RoboCup in Osaka bei der Premiere von Teamspielen in der Humanoid Robot League eingesetzt. Daneben werden Kooperationsszenarien von heterogenen Robotersystemen bestehend aus vierbeinigen und humanoiden Robotern untersucht.

1 Einleitung

Motivation und Zielsetzung. Kooperierende, autonome Mehrrobotersysteme werden derzeit für unterschiedliche Anwendungen, z.B. den kooperativen Transport einer Last durch mehrere fahrende oder fliegende Roboter, die Überwachung und Aufklärung eines Katastrophengebiets oder beim Roboterfußball in einer besonders dynamischen Umgebung untersucht.

Zur effektiven, flexiblen und robusten Entwicklung eines heterogenen Mehrroboterteams werden neben *Modulen* zur Lösung der grundlegenden Fragen der Sensordatenverarbeitung, Planung und Bewegungssteuerung *Softwarearchitekturen* sowie ein effizientes *Framework* benötigt. Anforderungen sind dabei *Modularität* für die Integration plattformunterschiedlicher oder konkurrierender Module zur Sensor-, insbesondere Kameradatenverarbeitung, Lokalisierung, Bewegungssteuerung, Verhaltenssteuerung sowie *Flexibilität* zur Anpassung an wechselnde Hardware wie unterschiedliche Prozessoren, Kamerasysteme oder Bewegungsapparate. Ebenso unterstützt werden muss die *Kommunikation* zwischen einzelnen Modulen und verschiedenen Robotern und Rechnern in unterschiedlichen Phasen von Entwicklung und der Betrieb der Mehrrobotersysteme unter Berücksichtigung der besonderen Anforderungen laufender, zwei- und vierbeiniger Roboter, die u.a. in der besonderen Schwierigkeit und Vielfalt der Bewegungsmöglichkeiten sowie der Auge-Bein Koordination liegen.

Stand der Forschung. In den letzten Jahren wurden wesentliche Fortschritte bei humanoiden Robotern erzielt. Dennoch enthält die robuste und schnelle Fortbewegung beim zweibeinigen Laufen sowie die autonome Navigation mit Auge-Bein-Koordination noch viele ungelöste Fragen. Die Herausforderungen beim Fußballspielen mit autonomen humanoiden Robotern liegen unter anderem in der Ausführung möglichst schneller, zielorientierter und situationsabhängiger Bewegungen unter Berücksichtigung von Bewegungsstabilität und Echtzeitanforderungen.

Bisherige Ansätze für Roboterarchitekturen für zielorientiert kooperierende Mehrrobotersysteme erfüllen die vorstehend ausgeführten Anforderungen für die hier betrachteten Humanoid-Roboter nur bedingt. Beispielsweise ist das für rollende Mehrrobotersysteme entwickelte auf CORBA basierende Miro [1] für leistungsfähige Mehrprozessorsysteme optimiert. Auf stromsparenden, leistungsschwächeren Ein- oder Mehr–prozessorsystemen wie auf den hier betrachteten Humanoid-Robotern bringt die Verwendung von CORBA jedoch einen Effizienzverlust mit sich, da nur wenige Vorteile dieser Middleware ausgenutzt werden können.

Saphira [2,3] ist die Softwareumgebung für die verbreiteten, radgetriebenen Pioneer-Roboter und ermöglicht eine direkte Anwendungsprogrammierung. Diese ist nicht transparent und im Quelltext nicht zugänglich, so dass die Integration nicht vom Hersteller vorgesehener Sensor- oder Aktuator-Hardware nicht auf einfachem Wege möglich ist. Auch ermöglicht es keine genaue Zeitstempelung von Sensordaten. CLARAty [4] ist als ein Framework für generische und wiederverwendbare Roboterkomponenten entwickelt worden, das zwar auf unterschiedliche Plattformen angepaßt werden kann, aber kooperative Mehrrobotersysteme nicht direkt unterstützt. Darüber hinaus sind sämtliche dieser Architekturen für radgetriebene und nicht für die besonderen Anforderungen bei laufenden, vier- oder zweibeinigen Robotern ausgerichtet.

Ansätze für funktionale Architekturen bei autonomen, humanoiden Robotern wurden erst in jüngster Zeit vorgestellt, ohne jedoch die Lösung komplexer Aufgaben durch Teamkooperation zu berücksichtigen. Beispielsweise die für den japanischen HRP2 Humanoid-Roboter jüngst vorgestellte Systemsoftware [5] ist nicht für den kompletten Austausch einzelner Module, eine Portierung auf andere Humanoid-Roboter-Plattformen oder zur Roboterkooperation vorgesehen. Eine modulare, verteilte Steuerungsarchitektur für einen humanoiden Oberkörper auf Rädern zur Anwendung bei der Mensch-Roboter Kooperation wird in [6] beschrieben, wobei Aspekte der zielorientierten Mehrroboterkooperation und der Bewegungen auf zwei Beinen keine Rolle spielen.

Die in der Four-Legged Robot League des RoboCup im GermanTeam (HU Berlin, U Bremen, TU Darmstadt, U Dortmund) entwickelte Architektur [7], die in aktueller Version [8] im Oktober 2004 veröffentlicht und seither von mindestens einem Dutzend Teams weltweit eingesetzt wird, berücksichtigt einen Teil der genannten Anforderungen für ein homogenes Team vierbeiniger Roboter. Diese enthält jedoch fix verbundene Schnittstellen zwischen den einzelnen Modulen, so dass die Struktur nicht flexibel geändert werden kann, beispielsweise für das Testen oder den Austausch neuer Module mit modifizierten Ein- und Ausgaben. Zudem ist die Anwendung nicht auf anderen Plattformen als dem vierbeinigen Roboter getestet, die graphischen Oberflächen zum

Abb. 1. Die vier untersuchten, ca. 37-68 cm großen Humanoid-Roboter-Prototypen (links) und kinematische Struktur von Mr. DD (rechts).

Debuggen der Anwendung sind aufgrund von Windows-spezifischen Bibliotheken nur auf diesem Betriebssystem lauffähig.

Das hier vorgestellte, neu entwickelte plattformunabhängige Framework *Robo-Frame* bietet effiziente Mechanismen zum Datenaustausch auf Ein- oder Mehrprozessor-systemen, wie sie auf Humanoid-Robotern eingesetzt werden, sowie einfach zu benutzende Schnittstellen für den Austausch zwischen den Modulen. Zum Debuggen wird eine erweiterbare und plattformunabhängige graphische Benutzeroberfläche verwendet.

2 Humanoide Roboterplattformen

Derzeit wird an vier unterschiedlichen Prototypen von Humanoid-Robotern geforscht. Der Bewegungsapparat des 68 cm große Roboters Mr. DD (24 DoF, ganz links in Abb. 1) ist ein von der japanischen Firma iXs hergestelltes Unikat. Die Bewegungsapparate der beiden kleineren, ca. 37 cm großen Roboter Mr. DD junior 1 (zweiter von links) und 2 (ganz rechts) mit je 21 DoF sind als Bausätze KHR1 und YDH der japanischen Firma Kondo verfügbar, die jeweils um ein Drehgelenk in den Beinen erweitert wurden. Der vierte Prototyp (dritter von links in Abb. 1) wurde in Zusammenarbeit mit der deutschen Firma PM-Solutions entwickelt. Alle Roboter sind aus Servomotoren aufgebaut und besitzen jeweils sechs nicht-redundante Bewegungsfreiheitsgrade pro Bein, jedoch in unterschiedlichen kinematischen Anordnungen. Die Bewegungsapparate wurden jeweils um ein Stereo- bzw. Monokamerasystem, um WLAN sowie Berechnungskapazitäten durch einen Pocket PC erweitert, wobei Mr. DD über ein Controller-Board mit 64 Bit MIPS Prozessor verfügt. Darüber hinaus wurden teilweise weitere Sensoren wie Gyroskop im Oberkörper, Beschleunigungssensor und Bodenkontaktsensoren in den Füßen integriert. Weitere Informationen zum Aufbau der Roboter stehen auf der Team-Homepage www.dribblers.de zur Verfügung.

3 Softwarearchitektur

Die der Architektur zugrunde liegende objektorientierte Software ist modular aufgebaut. Sie gliedert sich in die einzelnen Module für die verschiedenen Aufgaben innerhalb der

Bildverarbeitung, Objekterkennung, Lokalisierung, Bewegungs- und Verhaltenssteuerung und ein Framework, das die Kommunikation zwischen den Modulen abwickelt. Da das Framework nicht nur bei unterschiedlichen HumanoidRobotern, sondern mittelfristig auch bei heterogenen Teams von Humanoid-Robotern mit vierbeinigen oder radgetriebenen Robotern eingesetzt werden soll, muss es plattformunabhängig, d.h. auch unabhängig von spezifischer Hardware oder Betriebssystemen, gehalten werden. Unterstützte Betriebssysteme sind Linux, Unix, Windows und Windows CE. Auch für spätere Weiterentwicklungen muss die Erweiterbarkeit des Frameworks gegeben sein.

Das Framework stellt die Interprozess-Kommunikation zwischen den Modulen, die in Threads zusammengefaßt oder auf mehrere Threads verteilt werden können, her. Die Threads können dabei asynchron auf einem Rechner oder verteilt auf mehreren Rechnern ausgeführt werden. Für den zeitoptimierten Datenaustausch sind zwei verschiedene Kommunikationsmechanismen installiert: SharedMemory für große auf einem Rechner vorgehaltene Daten sowie Nachrichten in Ringpuffern für den allgemeinen Datenaustausch auf einem Rechner sowie zwischen verschiedenen Rechnern, die mit einer exakten Zeitstempelung versehen werden. Zur Kommunikation kann je nach Anwendungsfall TCP als zuverlässiges verbindungsorientiertes oder UDP als minimales, verbindungsloses Protokoll verwendet werden. Für eine Roboter-Roboter-Kommunikation wird in der Regel UDP gewählt, um auch für die Inter-Robot-Kommunikation die Echtzeitanforderungen erfüllen zu können. Die zentralen Betriebssystemfunktionen wie Synchronisation und Netzwerkzugriff sind gekapselt und durch eine Abstraktionsschicht plattformübergreifend verfügbar.

Die Architektur selbst wird über die Applikation aufgebaut. Dort werden benutzerdefiniert die in einem oder mehreren Threads verwendeten Module angegeben, ggf. mit unterschiedlichen Taktraten für jeden Thread. Da der Datenaustausch pufferbasiert oder mittels SharedMemory erfolgt, müssen keine festen Schnittstellen zum Datenaustausch in den einzelnen Modulen angegeben werden. Der zeitliche Ablauf der Module regelt sich über das Vorhandensein der von den Modulen angeforderten Daten in den jeweiligen Puffern. Über einen Log-Mechanismus können Modul-Informationen zum textbasierten Debugging auf die Konsole oder in Log-Dateien ausgegeben werden.

Eine graphische Benutzeroberfläche (GUI) innerhalb des Frameworks ermöglicht über eine zuverlässiges TCP-Verbindung zusätzlich die schnelle Auswertung der programmierten Algorithmen sowie deren graphisches Debugging und das Arbeiten mit aufgezeichnete Nachrichten. Für Module können eigene Visualisierungsroutinen implementiert und in die GUI eingebunden werden. Desweiteren sind Standardmechanismen wie zum Beispiel das Aufbauen einer Netzwerkverbindung in der GUI steuerbar.

4 Module

Die einzelnen Software-Module für Sensorik, Planung und Aktorik können nicht nur in einem hierachischen, deliberativen Ablauf aufgerufen werden, sondern es sind beliebige Steuerungsstrukturen implementierbar. Sensorwerte und deren weiterverarbeitete Werte stehen in den oben beschriebenen Ringpuffern oder im SharedMemory bereit und werden erst auf Anfrage von anderen Software-Modulen zur Verfügung gestellt. Die Daten sind dabei von Frameworkseite automatisch mit dem zur Generierungszeit erstellten Zeitstempel versehen. Durch das dedizierte Anfragen der Daten durch die Module wird unnötiges Generieren und Umkopieren der Daten vermieden. Es stehen immer die aktuellsten Werte in einer zuvor definierten Anzahl zur Verfügung. Das Weiterleiten der Daten im Puffer zwischen einzelnen Modulen, Threads oder Rechnern wird komplett vom Framework übernommen. Somit beschränkt sich die Datenabfrage und -ausgabe innerhalb der Module auf jeweils einen Funktionsaufruf, wodurch eine hohe Benutzerfreundlichkeit und Lesbarkeit des Code garantiert wird.

Die einzelnen Software-Module gliedern sich im Wesentlichen in die Bereiche Bildverarbeitung, Verhalten, Sensordatenaufnahme und Generierung einer Laufbewegung (vgl. Abb. 2). Die nachfolgend beschriebenen Module kommen auf den Humanoid-Robotern zum Einsatz.

Sensor-Aktuator-Schnittstelle. Die externen (Kamera, Abstandssensoren) und internen (Winkelencoder, Gyroskop, Beschleunigungssensor) Sensordaten werden über eine USB-, SDIO- oder serielle Schnittstelle eingelesen und mit einem eindeutigen Zeitstempel versehen. Die berechneten Gelenkwinkel werden über eine USB- oder serielle Schnittstelle an die Motoren übergeben.

Bewegungen. Die Art der Laufbewegung (nach vorne, zur Seite, rückwärts) wird zur Laufzeit durch die Verhaltenssteuerung festgelegt und derzeit online mit Hilfe einer roboterspezifischen inversen Beinkinematik als statisch stabile Bewegung bestimmt.

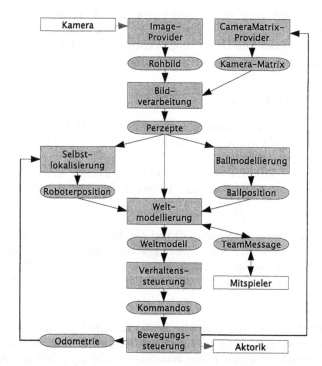

Abb. 2. Übersicht über die verwendeten Module (Rechtecke) und ausgetauschte Daten (Ellipsen) der realisierten Architektur für die Anwendung Roboterfußball.

Neben den Laufbewegungen sind Sonderbewegungen wie auf den Boden legen, Aufstehen aus der Bauch- oder Rückenlage sowie Schussbewegungen und Übergänge zwischen Elementarbewegungen auswählbar.

Bildverarbeitung. In der Humanoid League des RoboCup gibt es ein definiertes Spielfeld. Daher können aufgrund der Größe bekannter Objekte wie der Tore und des Balls auch mit einem Monokamerasystem Abstände berechnet werden. Aufbauend auf einer Farbsegmentierung werden derzeit die auftretenden Objekte wie Tor, Ball, Feldlinien, Mitspieler und Gegner klassifiziert und lokalisiert.

Selbstlokalisierung. Im Unterschied zur Four-Legged Robot League stehen keine Landmarken als zusätzliche Orientierungshilfen zur Verfügung. Die Selbstlokalisierung muß daher allein auf Basis der erkannten Objekte (Tore, Linien) und ggf. auf über WLAN kommunizierten Informationen der Mitspieler beruhen. Zu diesem Zweck wird derzeit ein kombiniertes Markov-Partikel-Verfahren entwickelt.

Weltmodellierung. Die Umwelt des Roboters wird anhand der erkannten Objekte (Tore, Linien, Ball, Roboter) sowie der über WLAN kommunizierten Informationen der Mitspieler modelliert. Dabei wird unterschieden, ob eine bekannte Umgebung (wie beim RoboCup) anhand der lokalisierten Objekte bestimmt wird oder ob eine unbekannte Umwelt in einer potentiellen anderen Anwendung exploriert wird.

Verhalten. Zur Verhaltenssteuerung beim Roboterfußball werden hierarchische Zustandsautomaten zur Steuerung von Kopf-, Bein-, Armbewegungen sowie von komplexeren Verhaltensweisen von Torwart und Feldspielern entwickelt, wie sie bereits im GermanTeam [8,9] in der Four-Legged Robot League erfolgreich eingesetzt wurden.

5 Ergebnisse und Ausblick

Die entwickelten Humanoid-Roboter, Module und Architektur werden u.a. im Rahmen der Wettbewerbe des RoboCup evaluiert.

RoboCup 2004. Mr. DD war der erste Humanoid-Roboter einer deutschen Forschungseinrichtung, der in der Humanoid League in der Disziplin Penalty Kick teilgenommen hat.

RoboCup German Open 2005. Mr. DD und Mr. DD jr. 1 führten gemeinsam mit dem Freiburger Team Penalty Kick Demonstrationen durch. Das Torwartverhalten ermöglichte das Fallen zur Seite, beim Stürmerverhalten konnte der Ball lokalisiert, Schritte und Positionierung zum Ball geplant und mit einem Schuß aufs Tor abgeschlossen werden.

RoboCup 2005 Osaka. An den erstmalig durchgeführten 2–2 Spielen sowie am Elfmeterschießen konnte erfolgreich teilgenommen werden. Darüber hinaus fand eine Demonstration eines gemischten Teams autonomer vierbeiniger Roboter (Aibos) und humanoider Roboter statt.

Kooperatives Verhalten. Durch kooperatives Verhalten von homogenen oder heterogenen Robotern kann ein ausfallsicherees Gesamtsystem erzeugt werden, da Sensorausfälle auf einem Robotersysteme durch Daten der anderen Roboter kompensiert werden können. Die entsprechende Verhaltenssteuerungsarchitektur ist für die Kommunikation der Roboter untereinander so ausgelegt, dass die von den anderen Robotern verschickten Daten in die eigenen Module eingelesen werden können, bei einem Übertragungsausfall der anderen Roboter die Module innerhalb der Architektur aber nicht auf die Fremddaten für das Erzeugen eines Verhaltens angewiesen ist. Auf Modulebene kann der Zugriff auf die Fremddaten benutzerfreundlich analog wie auf die robotereigenen Pufferdaten erfolgen. Ein kooperatives Verhalten wird für zwei verschiedene Konstellationen demonstriert, zwischen zwei Humanoid-Robotern und einem vierbeinigen und einem Humanoid-Roboter.

In einem vereinfachten RoboCup-Szenario befinden sich zwei Humanoid-Roboter mit einem Ball auf einem Spielfeld. Position und Orientierung der Roboter ist vorgegeben. Die Roboter tauschen untereinander aus, an welcher Stelle der Ball von ihnen bzgl. eines jeweils festen Roboterkoordinatensystems gesehen worden ist. Die von jedem Roboter erkannte Position wird mit der vom anderen Roboter verglichen, aus beiden zusammen wird die endgültige Ballposition berechnet. Fällt ein Kamerasystem bei einem der beiden Humanoid-Roboter aus, so kann die Ballposition trotzdem noch aus den Daten des anderen Robotersystems rekonstruiert werden, wodurch einen robusteres Gesamtsystem erreicht wird. Weiter soll die Position der Roboter beliebig in einer definierten Umgebung sein. Mittels Landmarken können die Roboter ihre eigene Position und Orientierung innerhalb der Umgebung bestimmen. Ausgetauscht werden dann die Daten des robotereigenen Weltmodells zur Verbesserung der jeweiligen Weltmodelle.

Als zweites Szenario wird die Kooperation zweier heterogener Robotersysteme, eines vierbeinigen AIBO ERS-7 Roboters mit 64 Bit MIPS-Prozessor unter Aperios/OPEN-R, auf dem eine entsprechende Verhaltenssteuerung auf Basis des entwickelten Frameworks implementiert wurde, und des Humanoid-Roboters Mr. DD jr. 2, gezeigt. Der vierbeinige Roboter erkennt mit seinem Kamerasystem einen aus Sicht des Humanoid-Roboters verdeckten Ball und übermittelt diesem die Koordinaten des Balls. Daraufhin gibt der Humanoid-Roboter durch Heben des jeweiligen Arms die Lage des vom vierbeinigen Roboters erkannten Balls (links, rechts) zu erkennen. Liegt kein Ball im Sichtfeld des vierbeinigen Roboters, reagiert der Humanoid-Roboter nicht, vgl. Abb. 3.

Ausblick. Während die genannten Szenarien für sich allein auch mit erheblich weniger Architekturaufwand direkt realisierbar sind, dienen diese als erste, erfolgreiche Tests für die beschriebenen Entwicklungen und damit als Ausgangsbasis für die Untersuchung komplexerer Mehrroboterkooperationen. Bis zum RoboCup 2006 soll aufbauend auf einer auf dem Framework aufsetzenden Steuerungsarchitektur sowie einer Weiterentwicklung der verwendeten Module und Hardware ein erfolgreiches Team humanoider Roboter für 3–3 Spiele entwickelt werden. Für Anwendungen heterogener Roboterteams außerhalb des Roboterfußballs werden außer humanoiden und vierbeinigen Robotern auch radgetriebene Systeme wie ein Pioneer 2DX integriert.

Abb. 3. Der hinter einem Sichtschutz stehende Humanoid-Roboter reagiert auf die vom vier-beinigen Roboter gesendeten Nachrichten bzgl. der erkannten Ballposition. Links: Kein Ball erkennbar, keine Reaktion. Mitte/Rechts: Ball links bzw. rechts, Humanoid-Roboter winkt mit rechtem bzw. linkem Arm.

Literaturverzeichnis

1. H. Utz, S. Sablatnög, S. Enderle, G.K. Kraetzschmar: Miro – middleware for mobile robot applications. *IEEE Trans. on Robotics and Automation*, Vol. 18, 2002, 493-497.
2. K. Konolige: Saphira robot control architecture version 8.1.0. *SRI International*, April, 2002.
3. A. Orebäck, H.I. Christensen: Evaluation of architectures for mobile robotics. *Autonomous Robots*, 14, 2003, 33-49.
4. I. Nesnas, A. Wright, M. Bajracharya, R. Simmons, T. Estlin, W.S. Kim: CLARAty: An architecture for reusable robotic software. *SPIE Aerosense Conference*, Orlando, Florida, April 2003.
5. K. Okada, T. Ogura, A. Haneda, D. Kousaka, H. Nakai, M. Inaba, H. Inoue: Integrated system software for HRP2 humanoid. *Proc. 2004 IEEE ICRA*, New Orleans, April, 2004, 3207-3212.
6. D.N. Ly, K. Regenstein, T. Asfour, R. Dillmann: A modular and distributed embedded architecture for humanoid robots. *Proc. IEEE/RSJ IROS*, Japan, Sep. 28–Oct. 2, 2004.
7. T. Röfer: An architecture for a national RoboCup team. In G.A. Kaminka, P.U. Lima, R. Rojas (eds.): *RoboCup 2002: Robot Soccer World Cup VI*, Lecture Notes in Artificial Intelligence, Springer, 2003, 417-425.
8. GermanTeam. http://www.germanteam.org
9. M. Lötzsch, J. Bach, H.-D. Burkhard, M. Jüngel: Designing agent behavior with the extensible agent behavior specification language XABSL. In *th International RoboCup Symposium 2003 (Robot World Cup Soccer Games and Conferences)*, Padovy, Italy, Lecture Notes in Artificial Intelligence, Springer, 2004, to appear.

Dynamic Task Assignment in a Team of Agents

Michael Schanz, Jens Starke, Reinhard Lafrenz, Oliver Zweigle, Mohamed Oubbati,
Hamid Rajaie, Frank Schreiber, Thorsten Buchheim, Uwe-Philipp Käppeler,
Paul Levi

Institute of Parallel and Distributed Systems, University of Stuttgart, Universitätsstraße 38,
70569 Stuttgart, Germany
Institute of Applied Mathematics, University of Heidelberg, Im Neuenheimer Feld 294,
69120 Heidelberg, Germany

E-mail: robocup@informatik.uni-stuttgart.de, jens.starke@iwr.uni-heidelberg.de

Abstract. In dynamic and complex multi-robot scenarios, the task assignment
is a challenging and crucial topic, because it has to be very flexible and robust
or in some sense fault-tolerant to achieve a certain degree of redundancy which
is often required in these scenarios. To cope with these requirements, a dynamic
task assignment approach based on self-organization principles adopted from
nature is presented. In this work, a team of soccer playing robots is used to validate the suggested self-organized dynamic task assignment in dynamic domains
with real-time constraints.

1 Introduction

One of the most critical parts in multi-agent applications is the assignment of tasks to
robots or more general agents, especially in highly dynamical scenarios. For realistic
scenarios, for instance in service robotics or in the RoboCup [9], usually the number
of tasks is much larger than the number of agents. Hence, the goal is firstly to select
relevant tasks which have highest priority and secondly to assign them to the agents in
such a way, that the required flexibility is guaranteed and the overall performance is
optimized. There are several techniques to cope with this problem, ranging from static
mapping to complex utility functions. The performance of a system in highly dynamical environments depends critically on its ability to react on changes, disturbances and
faults in a flexible way. In some applications it is possible, that the utility of the different agents for a specific task or action varies. This implies, that the system must be
able to re-assign tasks or actions and to choose a reasonable subset of them for the
currently available agents within the team.

In this paper, we present an approach for a dynamic assignment of agents to tasks,
where break-downs of single agents, changing requests or (environmental) conditions
are handled. The aimed practical scenarios are on the one hand the soccer playing
robot team CoPS Stuttgart and on the other hand the specific service robotics application occurring in the Nexus project [12,6]. In both scenarios an agent is represented
by a robot. The presented approach is based on our previous work presented
in [17,1,10,11].

The basic concept of this mechanism relies on coupled selection equations [13], which are inspired by self-organization principles [7,3,2] derived from natural systems. In contrast to our previous work, we have applied two important enhancements introduced in [15]. The first one is the dynamical assignment instead of re-calculation and the second one is the ability to activate or deactivate specific tasks or actions and therefore to control the overall system behavior. The practicability is demonstrated in RoboCup experiments by simulations as well as with real robots.

2 State of the Art

To perform any kind of cooperative work with a team of robots, the execution of tasks or actions must be distributed. In this paper we focus on the assignment of specific actions for situations related to the RoboCup scenario. Currently, many research work in the field of RoboCup copes with continuous re-assignments in highly dynamic situations. In almost all approaches, the handling of this highly dynamic situations results in complex decision trees.

The naive approach is to write a sequence of if .. then .. else or case statements. This approach is very fast, but hard to implement and to debug and almost impossible to extend or to adapt to new strategies or team behaviors. Especially if the system is required to be fault-tolerant and robust, many special cases have to be taken into consideration.

An improved approach to assign the actions to robots is to specify the decisions in form of rule-based systems. One successful way to implement such a rule-based system uses XABSL (Extensible Agent Behavior Specification Language) [4], which is a powerful language for modeling decision systems in robotics. However, for complex frameworks, also this code becomes quickly unreadable and is difficult to maintain or to extend.

To overcome these problems, we developed a graphical modeling tool to design and edit decision networks in the form of modified Petri nets [18]. In these nets, the places correspond to actions, which are executed if the preconditions, modeled in form of transition conditions, are satisfied. Such a decision net can be structured by defining a hierarchy of subsets, where places can recursively be expanded to subnets. From this model, XABSL code can be generated [8,18]. Anyhow, in many situations, it is not obvious how to model a behavior for a specific situation, which means to assign actions to robots and how this assignment scheme has to be worked out by a human expert. Therefore, we decided to introduce self-organization to optimize the assignment of actions to robots.

3 Evolution of Preferences to Decisions

For a self-organized assignment of actions to robots, we need to define time-dependent utility functions for individual robots and actions, and an assignment matrix, which also considers global interdependencies.

The preferences for each possible decision of a robot i to serve a strategic target j or perform an action j is mathematically described by a time-dependent variable $\xi_{ij} \in \mathbb{R}$ whose temporal evolution is defined by a selection process described by coupled selection equations [13]. These equations ensure that the solution tends to a state, which is a clear decision. In the case with n robots and n targets, the matrix (ξ_{ij}) finally approximates a permutation matrix. If (ξ_{ij}) is a non-square matrix, the set (targets or robots) with the smaller number of elements, assigns each of its elements to one and only one of the possible choices. As an example the decision making between two defense strategies under environmental changes is presented in Section 5.

In contrast to the original coupled selection equations [5,13,14,16] the extended coupled selection equations [15] used in this work, have additional situation dependent utility parameters α_{ij} and activation parameters λ_{ij} for the dynamic activation or deactivation of targets or actions. The extended coupled selection equations with the time scaling κ are defined by:

$$\frac{d}{dt}\xi_{ij} = \kappa\,\xi_{ij}\left(\alpha_{ij}\left(\lambda_{ij} - \xi_{ij}^{\,2}\right) + 2\beta\,\xi_{ij}^{\,2} - \beta\left(\sum_{i'}\xi_{i'j}^{\,2} + \sum_{j'}\xi_{ij'}^{\,2}\right)\right) \qquad (1)$$

Note, that in the extended coupled selection equations (1), not only the state variables, i.e., the preferences ξ_{ij} for the assignment of robot i to target j depend on time, but also the utility parameters α_{ij} and the activation parameters λ_{ij}. In contrast to the original selection equations used in [5] for the control of mobile robots, the time dependent parameters α_{ij} in (1) are introduced to cope with the dynamically changing environment defined by the movement of other robots. Sudden changes caused by the antagonistic robots or the moving ball are considered instantaneously through the changing utility functions, whereby larger values of α_{ij} represent greater utility. Furthermore, targets or actions can be activated or deactivated by setting the parameters λ_{ij} to the values +1 for the activated or −1 for the deactivated state according to a desired strategic plan adequate for the current situation. This will be demonstrated in Section 5 in a switching scenario between two defense strategies.

4 Team Behavior Modeling by Role Assignment

To improve the overall performance, it is useful to define certain roles and behaviors to support the cooperative teamwork. Such a role, for instance *defender* or *forward* defines the general behavior of a robot and is more or less statically assigned. However, the behavior in detail is determined dynamically depending on the current situation. For instance a robot may have the role of a *defender*. But this robot can also take the behavior of an *attacker*, if it is in the current situation the most suitable one. Hence, in our RoboCup implementation, every robot has an assigned role and a dynamically assigned behavior.

Designing the behavior patterns and the corresponding decisions manually is a sophisticated and complex task. To reduce the complexity we introduce two levels of abstraction in the decision tree. The top level deals with external events caused by referee signals or game rules. The decisions on this level are still manually modeled using the above mentioned graphical tool. The decisions on the lower level are mainly influenced by geometrical relations between the objects on the soccer field and are highly dynamic. To model the decision rules on this level is very difficult even for humans because many possible situations have to be considered. Therefore we applied the self-organization approach described in Section 3. In [1,10] we already showed, that self-organized assignment techniques perform well in multi-robot applications and are hence promising in the RoboCup scenario. In contrast to these papers, the control and therefore the positions of the robots in the field are not directly related to the ξ_{ij}. Only an abstract, symbolic behavior is selected, which leads to the execution of a plan for this behavior. This enables the implementation of complex control strategies, considering task-specific restrictions, e.g., dribbling of a ball, where the robot is subject to constraints like maximum curvature or acceleration without losing the ball.

To demonstrate the applicability of the above described method, we show the assignment of robots to different behaviors, where we have more possible behaviors than robots. Therefore it is necessary to select the most useful behaviors for the current game situation. This selection is governed by the parameters λ_{ij} and α_{ij} of the extended coupled selection equations (1). Hereby, the activation parameters λ_{ij} activate or deactivate specific behaviors j for the robot i, whereas the utility parameters α_{ij} express the corresponding utility, i.e. larger utility results in a larger α_{ij}. The α_{ij} are calculated dynamically based on the robot's world model. In the current implementation, especially geometrical and topological relations are considered. However, this is not the only choice and one can think about more sophisticated relations improving the overall performance even further.

The λ_{ij} can be used in special situations, for example after referee events, e.g. in case of a goal, all robots should go to their homing positions. In such cases, complete sets of behaviors are deactivated and a set of special behaviors is activated by setting the λ_{ij} to ±1. After the situation is over, the system switches back to the standard setting.

The extended coupled selection equations (1) assign the robots to the most useful behaviors which are available in the set of activated behaviors in a dynamic way. The property of these equations guarantees that each robot is assigned to one and only one behavior in the long time perspective. However, if another assignment becomes more useful in the course of time, the changing values of the α_{ij} cause a changing of the ξ_{ij} and thus a switching of the assignment, which optimizes the overall performance of the system and guarantees flexibility and fault-tolerance.

In order to perform real RoboCup experiments, the self-organized behavior selection was integrated into the XABSL-framework as one element of the lower level of the decision process.

5 Experimental Results: Example of a Defense Scenario

In simulation experiments, we have shown that the system is able to assign the robots dynamically to the currently most important tasks. Hereby, the tasks correspond to the robot behaviors forward, defender, and attacker. The forward and defender roles base on zone defense, that means that the robots act only in a local area defined by the specific behavior. In case of the defender, we have three such areas, for forward behavior only two. The special behavior of the attacker is not locally restricted, it follows the action sequence get Ball, dribble, shoot.

The α_{ij} represent the utility of the assignment of the i-th robot to the j-th behavior. The values α_{ij} are determined as shown in equation (2).

$$\alpha_{ij}(t) = \begin{cases} a_0 \dfrac{\min_{i'} d(rob_{i'}, ball)}{d(rob_i, ball)} & \text{if } j = 0 \\[4mm] a_1 \dfrac{\min_{j'} d(rob_i, pos_{j'})}{d(rob_i, pos_j)} + a_2 \dfrac{\min_{j'} d(ball, pos_{j'})}{d(ball, pos_j)} & \text{else} \end{cases} \tag{2}$$

Hereby, rob_k, $ball$ and pos_l are the positions of robot k and the ball and the reference positions of target l (see Fig. 1). Furthermore, $d(\cdot, \cdot)$ denotes the Euclidean distance between two positions and a_1, a_2 and a_3 are tuning parameters. The target $l = 0$ represents the behavior of the attacker, which always tries to get the ball. The rest of the behaviors correspond to the specific roles defender or forward, for instance left defender or right forward. This example scenario represents a typical case whereby there are more possible behaviors than available robots. Therefore, only the most important behaviors, according to predefined priorities, are assigned. In equation (2), $a_0 \geq 1$ is a gain factor, to ensure that the behavior of the active defender is always assigned. In ambiguous situations, it is better that more than one robot try to get this behavior and therefore to go to the ball, than having no attacker. To calculate the utility of the other behaviors $j \neq 0$, two geometric relations are used: First, the Euclidean distance between the particular robot and the center of the behavior's area and secondly the Euclidean distance of the ball to that area. These two relations are weighted by the factors a_1 and a_2. In a similar way, the utility function can be extended by additional parameters a_k taking other than pure geometrical information into account. An example of a such a non-geometrical information is the remaining number of shots for a specific robot, depending on the pressure left in the pressure tank of it's pneumatic kicking device. Due to the intrinsic properties of the extended coupled selection equations (1), the dynamics leads to unambiguous assignments in the course of time.

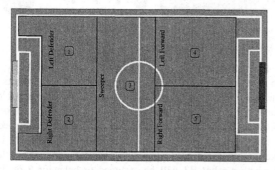

Fig. 1. Reference positions of the behaviors. Every numbered position is assigned to the corresponding behavior and serves among others as the reference position for calculating the α_{ij} factors.

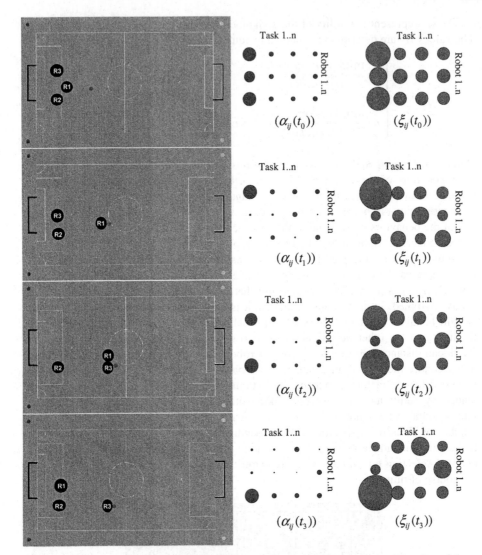

Fig 2. Simulation at four points in time $(t_0 \ldots t_3)$ with matrices (α_{ij}) and (ξ_{ij}). The size of the circles represents the relative values of the corresponding matrix components. In each matrix, the rows represent the three robots and the columns correspond to four possible behaviors. The most left column stands for the attacker, the others for the defender and forward behaviors. Four snapshots are shown: t_0 is the initial situation, where all robots tend to become an attacker; t_1 is an intermediate state on the way to a clear assignment, where robot R1 is already clearly assigned to the attacker behavior, robot R2 is more or less clearly assigned to the behavior 2 but it is still undecided whether robot R3 will be assigned to behavior 2 or 4; t_2 represents a changed situation, where R1 after some kicks has lower pressure in its air tank, which leads to changes in the matrix (α_{ij}) and causes a new competition between R1 and R3 to become the attacker; t_3 on the way to a second clear assignment.

In Fig. 2, the α_{ij} change their values dynamically according to equation (2). In the first column the simulator with the robots is shown, in the second and third one the α_{ij} and the resulting ξ_{ij} are presented, whereby the sizes of the circles represent the relative values of the matrix components. In the first situation at t_0 (first row) the assignment is not yet clear. Two robots are possible candidates for becoming the active defender, whereas the remaining robot is likely to take a position on the defense line. In the next situation at t_1, one robot has taken the behavior of the active defender, and the other robots have taken two of the three possible defense positions. At t_2, two robots compete again for becoming active defenders because the already assigned robot has after some kicks lower pressure in its air tank, which results in a change of the parameters α_{ij}. After a short time at t_3, one robot has won the competition and the matrix ξ_{ij} shows distinct assignments again. Here, it is clearly visible that the second target (second column in the ξ-matrix) has only small preferences and therefore this target is not assigned.

6 Conclusion and Outlook

We have shown the applicability of self-organized principles for behavior selection in teams of cooperating robots. The mechanism based on extended coupled selection equations leads to a very flexible system that can be easily parameterized and tuned to the desired overall behavior.

In the near future, we try to extend the system to control more situations in RoboCup and to validate the applicability and performance in real RoboCup competitions. Furthermore, we will use it in other application scenarios for cooperating robots.

7 Acknowledgment

This work is partly supported by the DFG within the priority program SPP 1125 *Kooperierende Teams mobiler Roboter in dynamischen Umgebungen*.

References

1. M. Becht, T. Buchheim, P. Burger, G. Hetzel, G. Kindermann, R. Lafrenz, N. Oswald, M. Schanz, M. Schulße, P. Molnár, J. Starke, and P. Levi. Three-index assignment of robots to targets: An experimental verication. In *IAS-6*, 2000.
2. H. Haken. *Advanced Synergetics*. Springer Series in Synergetics. Springer-Verlag, Heidelberg, Berlin, New York, 1983.
3. H. Haken. *Synergetics, An Introduction*. Springer Series in Synergetics. Springer- Verlag, Heidelberg, Berlin, New York, 1983.
4. M. Lötzsch, J. Bach, H.-D. Burkhard, and M. Jüngel. Designing agent behavior with the extensible agent behavior specication language xabsl. In *7th International Workshop on RoboCup*, 2003.
5. P. Molnár and J. Starke. Control of distributed autonomous robotic systems using principles of pattern formation in nature and pedestrian behaviour. IEEE Transaction on Systems, Men and Cybernetics: Part B, 31(3):433–436, 2001.
6. Daniela Nicklas, Matthias Großmann, Thomas Schwarz, Steen Volz, and Bernhard Mitschang. A model-based, open architecture for mobile, spatially aware applications. Article

in proceedings, Universität Stuttgart : Sonderforschungsbereich SFB 627 (Nexus: Umgebungsmodelle für mobile kontextbezogene Systeme), July 2001.

7. G. Nicolis and I. Prigogine. *Self-Organization in Non-Equilibrium Systems*. Wiley, New York, 1977.
8. Heiko Ottenbacher. Multiagenten planmodellierung. Master's thesis, University of Stuttgart, 2004.
9. RoboCup Ocial Website. http://www.robocup.org.
10. M. Schulèe, M. Schanz, H. Felger, R. Lafrenz, J. Starke, and P. Levi. Control of autonomous robots in the robocup scenario using coupled selection equations. In *Autonome Mobile Systeme*, pages 57–63. Springer, 2001.
11. M. Schulße, M. Schanz, H. Felger, J. Starke, and P. Levi. Dynamic control of autonomous robots using coupled selection equations. In Robotik, 2002.
12. SFB 627: Nexus–Spatial World Models for Mobile Context-Aware Applications. http://www.nexus.uni-stuttgart.de.
13. J. Starke. *Kombinatorische Optimierung auf der Basis gekoppelter Selektionsgleichungen*. PhD thesis, Universität Stuttgart, Verlag Shaker, Aachen, 1997.
14. J. Starke. Dynamical assignments of distributed autonomous robotic systems to manufacturing targets considering environmental feedbacks. In *Proceedings of the 17th IEEE International Symposium on Intelligent Control (ISIC'02)*, pages 678–683, Vancouver, 2002.
15. J. Starke, C. Ellsässer, and Fukuda. Self-organized control in cooperative robots using pattern formation principles. submitted.
16. J. Starke and M. Schanz. Dynamical system approaches to combinatorial optimization. In D.-Z. Du and P. Pardalos, editors, *Handbook of Combinatorial Optimization*, volume 2, pages 471–524. Kluwer Academic Publisher, Dordrecht, Boston, London, 1998.
17. J. Starke, M. Schanz, and H. Haken. Self-organized behaviour of distributed autonomous mobile robotic systems by pattern formation principles. In T. Lueth, R. Dillmann, P. Dario, and H. Wörn, editors, *Distributed Autonomous Robotic Systems 3*, pages 89–100. Springer Verlag, Heidelberg, Berlin, New York, 1998.
18. Oliver Zweigle, Reinhard Lafrenz, Thorsten Buchheim, Hamid Rajaie, Frank Schreiber, and Paul Levi. Cooperative agent behavior based on special interaction nets. Submitted to: Intelligent Autonomous Systems 9, 2006.

Verbesserte Effizienz der Monte-Carlo-Lokalisierung im RoboCup

Patrick Heinemann, Jürgen Haase, Andreas Zell

Wilhelm-Schickard-Institut der Universität Tübingen, Lehrstuhl Rechnerarchitektur, Sand 1, 72074 Tübingen

E-mail: {heinemann, jhaase, zell}@informatik.uni-tuebingen.de

Zusammenfassung: Aktuelle Implementierungen der Monte-Carlo-Lokalisierung benötigen mindestens 100 Samples, was in zeitkritischen Roboter-Systemen, wie z.B. einem RoboCup-Roboter, zu einem Ressourcen-Engpass führen kann. Dieser Artikel beschreibt einen neuen Ansatz für Monte-Carlo-Lokalisierung, bei dem die Anzahl der benötigten Samples adaptiv bis auf ein Minimum von nur einem Sample sinkt, wenn die Positionsschätzung ausreichend exakt ist. Experimente zeigen, dass der vorgestellte Algorithmus sehr schnell in diesen effizienten „Tracking-Modus" übergeht. Durch eine iterative Verbesserung der Positionsschätzung kann sogar eine höhere Genauigkeit der Lokalisierung erreicht werden, als dies mit bisherigen Ansätzen möglich ist.

1 Einleitung

Die Lokalisierung in einem globalen Koordinatensystem ist eine wichtige Aufgabe mobiler Roboter, besonders wenn sie ihr Ziel nur durch Kooperation oder in Konkurrenz zu anderen Robotern erreichen können. Im RoboCup [1] wurden daher in den letzten Jahren viele Ansätze zur Positionsschätzung entwickelt ([2], [3], [4] und [5]). Die aktuelleren Ansätze basieren alle auf der Monte-Carlo-Lokalisierung ([6], [7], [8], [9] und [10]). Diese Ansätze unterscheiden sich hauptsächlich in der Anzahl der benötigten Samples und damit in der Effizienz der Lokalisierung. Der in diesem Artikel beschriebene Ansatz implementiert eine verbesserte Monte-Carlo-Lokalisierung anhand der Spielfeldlinien eines RoboCup-Spielfelds. Dabei wird die Samplezahl abhängig von der Güte der Positionsschätzung adaptiert. Durch eine anschließende iterative Verbesserung dieser Schätzung ist es möglich, den Roboter auch noch mit einem einzigen Sample zu lokalisieren.

2 Verbesserte Monte-Carlo-Lokalisierung

Monte-Carlo-Lokalisierung (MCL) [11] ist eine effiziente, auf Sampling basierende Approximation der kontinuierlichen Wahrscheinlichkeitsverteilung über alle möglichen Positionen eines Roboters, wobei sich die Verteilung der Samples nach der Wahrscheinlichkeit richtet, dass sich der Roboter tatsächlich an der betrachteten Position befindet (Sampling/Importance Resampling).

Der vorgestellte Algorithmus verfügt zu Beginn über kein Vorwissen, obwohl dies möglich wäre, beispielsweise wenn der Roboter immer an der selben Position des Feldes starten würde. Daher wird zunächst die maximale Anzahl von Samples N_{max} generiert und gleichmäßig über den Zustandsraum, hier das befahrbare Feld und alle Ausrichtungen, verteilt.

$$S = \left\{ s_i \mid s_i = ((x_i, y_i, \theta_i), w_i) \right\} \tag{1}$$

repräsentiert die aktuelle Menge der Samples s_i, wobei jedes s_i eine Positionsschätzung $(x_i, y_i, \theta_i) = l_i$ und ein Gewicht w_i beinhaltet, das die Wahrscheinlichkeit ausdrückt, dass sich der Roboter an Position l_i befindet.

In jedem Zyklus bekommt der Algorithmus neue Odometrie-Daten a und ein Bild s der omnidirektionalen Kamera des RoboCup-Roboters. Die Odometrie-Information wird im *Bewegungsmodell* verarbeitet. Dabei werden zunächst alle Samples von S nach der gemessenen Bewegung a verschoben und rotiert. Zusätzlich werden die Samples noch durch ein Gaußsches Rauschen proportional zur Messungenaugkeit der Odometrie verrauscht.

Die Daten der Kamera werden im sogenannten *Sensormodell* verarbeitet. Dieses gibt die bedingte Wahrscheinlichkeit $P(s|l_i)$ an, dass der Roboter das Bild s erhält unter der Voraussetzung, dass er sich an der Position l_i befindet. Zur effizienten Berechnung von $P(s|l_i)$ werden zunächst weiße Pixel, die die Feldlinien des Spielfelds darstellen, aus dem Kamerabild extrahiert und daraus durch inverse perspektivische Transformation die Position der Linienpunkte im Roboter-Koordinatensystem errechnet ([12]). Diese Linienpunkte werden dann an die Positionsschätzung l_i der Samples transformiert. $P(s|l_i)$ kann, wie von Röfer *et al.* [8], [9] und Hundelshausen *et al.* [7] vorgeschlagen, als Summe der quadratischen Distanzen der Linienpunkte zu ihrer nächstgelegenen Linieim Modell der Feldlinien berechnet werden. Da diese Distanzen nur von der Position auf dem Feld abhängig sind, kann man sie vorherberechnen und für schnellen Zugriff in einer zweidimensionalen Tabelle ablegen. Um die Symmetrie, die sich in der Positionsschätzung durch die Linien eines RoboCup-Spielfelds ergibt, aufzulösen, benutzt der Algorithmus zusätzlich noch die Winkel zu den beiden unterschiedlich farbig markierten Toren. Diese Winkel werden mit den erwarteten Torwinkeln an der Position l_i verglichen und ergeben ein weiteres Distanzmaß. Die Gesamtdistanz wird schließlich durch eine Linearkombination der beiden Distanzen errechnet und mit dieser die Gewichte der Samples aktualisiert.

Jetzt kann aus den besten n Samples ein gewichtetes Mittel als Positionsschätzung generiert werden

$$\hat{p} = (x, y, \theta) = n \sum w_n l_n. \tag{2}$$

Diese Positionsschätzung wird nun verbessert, indem iterativ eine Kraft F_k und ein Drehmoment M_k, die das Linienmodell auf die Linienpunkte ausübt, berechnet und auf die Schätzung angewendet werden. Die Kraft ergibt sich aus der Summe der Abstandsvektoren zwischen gemessenen Linienpunkten und der jeweils nächstgelegenen Modelllinie. Das Drehmoment errechnet sich aus dem Kreuzprodukt der Abstandsvektoren mit den Vektoren vom Roboter zum jeweiligen Linienpunkt. Fügt man nun einen Bruchteil der Kraft F_k und des Drehmoments M_k zur Positionsschätzung $\hat{p}$ hinzu, erhält man in jeder Iteration k eine neue Schätzung

$$(x_k, y_k) = (x_{k-1}, y_{k-1}) + \mu F_{k-1} \tag{3}$$

$$\theta_k = \theta_{k-1} + \nu M_{k-1}. \tag{4}$$

Die Iterationen werden fortgesetzt, bis die Verbesserung einen Schwellwert unterschritten hat oder eine maximale Anzahl Iterationen ausgeführt wurde. Diese Idee zur iterativen Positionsverbesserung wurde bereits von Hundelshausen et $al.$ [7] für ein Dead-Reckoning Verfahren zur Lokalisierung benutzt.

Ausgehend von der Gesamtdistanz D der endgültigen Positionsschätzung wird nun noch die Anzahl der Samples für den nächsten Zyklus berechnet.

$$N_{t+1} = \begin{cases} N_{max} & \text{if } \xi D \geq N_{max} \\ \xi D & \text{if } 1 < \xi D < N_{max}, \\ 1 & \text{if } \xi D \leq 1 \end{cases} \tag{5}$$

wobei ξ regelt, wie schnell N_{t+1} auf ein Sample reduziert wird. Zum Abschluss eines Zyklus werden die Samples abhängig von ihrem Gewicht neu gesampled. Hierbei werden schlechte Samples aussterben und gute mehrfach neu erzeugt.

3 Ergebnisse

Die Effizienz des vorgestellten MCL-Algorithmus und die Qualität der errechneten Positionsschätzung wurde in zwei Experimenten untersucht. Die dabei verwendeten Parameter können Tabelle 1 entnommen werden.

Tabelle 1. Die für die Experimente verwendeten Parameter

n	N_{max}	ξ	μ	ν
N	200	2500	0.001	0.0003

Für die Berechnung des gewichteten Mittels aus den besten n Samples wurden alle N Samples herangezogen, um ein aufwändiges Sortieren der Samples zu vermeiden. Die maximale Anzahl von Samples N_{max} wurde so gewählt, dass bei einer schlechten Positionsschätzung gerade so viele Samples generiert werden, wie in einem Standard MCL-Verfahren. Die Wahl von ξ bildet einen Kompromiss zwischen einer schnellen Reduktion auf ein Sample bei guter Schätzung und einer schnellen Relokalisierung beim „Kidnapped-Robot"-Problem durch hohe Sample-Zahlen. Die Regelparameter μ und ν wurden empirisch so ermittelt, dass durch die Iterationen kein Schwingen der Schätzung um das Minimum auftritt, aber gleichzeitig eine möglichst schnelle Konvergenz erzielt werden kann. Im ersten Experiment wurde der Algorithmus mit Bildern eines stehenden Roboters getestet. Die Bilder 1 bis 98 wurden an Position $p_1 = (1.04, 1.07, 2.1)$ aufgenommen, während die Bilder 99 bis 196 an einer anderen Position $p_2 = (1.64, 2.68, 0.0)$ aufgenommen wurden, um das „Kidnapped-Robot"-

Problem zu simulieren. Dieses Experiment wurde sowohl mit dem hier vorgestellten verbesserten MCL-Verfahren, als auch mit Standard MCL-Verfahren ([11]) mit fester Sample-Zahl von $N = 50$, $N = 100$ und $N = 200$ Samples durchgeführt, um die Laufzeit und die Qualität der Schätzung mit bestehenden Verfahren zu vergleichen. Abbildung 1 zeigt die Euklidische Distanz der Positionsschätzungen zur tatsächlichen Position für alle Verfahren. Unabhängig von der verwendeten Sample-Zahl haben alle Verfahren nach spätestens 20 Zyklen die Position auf etwa 10cm genaugeschätzt. Der vorgestellte Algorithmus reduziert seine Sample-Zahl von $N = N_{max} = 200$ auf $N = 1$ in nur 6 Zyklen und erreicht dabei schneller als alle anderen Verfahren einen Positionsfehler von unter 10cm. Bis zur Änderung der Position des Roboters in Zyklus 99 bleibt der Algorithmus bei nur einem Sample und generiert dann sofort wieder $N = N_{max}$ Samples, um auf den großen Schätzfehler zu reagieren. Abgesehen von dem Standard-Verfahren mit 50 Samples, das nach dem Positionssprung nicht mehr auf die neue Position konvergiert, sind alle Algorithmen in der Lage, das „Kidnapped-Robot"-Problem zu lösen. Der gemittelte Positionsfehler über alle Zyklen liegt bei diesen Verfahren bei etwa 20cm, wobei der vorgestellte Algorithmus sogar noch unter diesem Wert bleibt. Zusätzlich reduziert der vorgestellte Algorithmus allerdings in 92.87% aller Zyklen die Samplezahl auf ein einziges Sample, wodurch ein erheblicher Geschwindigkeitsvorteil in der Berechnung der Positionsschätzung erreicht wird. Tabelle 2 fasst diese Ergebnisse zusammen.

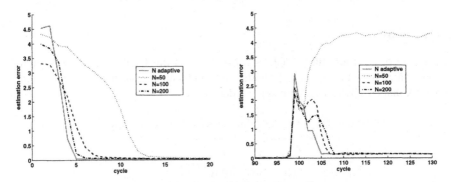

Abb. 1. Comparison of our algorithm to MCL with fixed numbers of samples.

Tabelle 2. Ergebnisse von 196 Zyklen mit verschiedener Samplezahl N.

	Verbesserte MCL	Standard MCL	Standard MCL	Standard MCL
	$N = adaptiv$	$N = 50$	$N = 100$	$N = 200$
Mittlere Rechenzeit	1.7632 ms	3.6426 ms	6.8223 ms	14.2508 ms
Mittlerer Fehler	0.1936 m	2.2515 m	0.2075 m	0.2057 m

Um zu zeigen, dass das vorgestellte Verfahren auch bei einem fahrenden Roboter gute Ergebnisse liefert und dennoch in Echtzeit arbeitet, wurde mit einem Testroboter per Fernsteuerung ein bestimmtes Rechteck abgefahren.

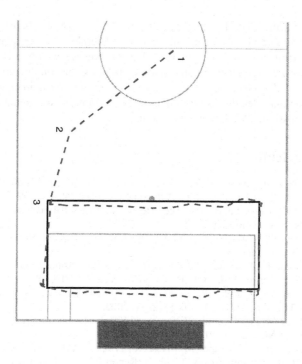

Abb. 2. Die durch den vorgestellten Algorithmus in Echtzeit generierte Positionsschätzung des entlang des schwarzen Rechtecks ferngesteuerten Roboters. Der Roboter steht zu Beginn an der unteren linken Ecke des Rechtecks.

Abbildung 2 zeigt die Ergebnisse dieses Experiments. Im ersten MCL-Zyklus sind die Samples noch gleichmäßig über das Feld verteilt, so dass sich durch das gewichtete Mittel die Positionsschätzung (1) in der Mitte des Feldes ergibt.Bereits ab dem dritten Zyklus befindet sich allerdings die Schätzung schon auf der korrekten Startposition (3) im Rechteck. Über die gesamte Restdauer von 502 Zyklen des Experiments wird die Position des Roboters korrekt geschätzt, wobei in 96.21% allerZyklen nur ein einziges Sample berechnet werden muss. Damit ergab sich eine mittlere Berechnungszeit für einen Zyklus von 0.5413ms auf einem Athlon XP 1800+ System. Leider lässt sich nicht feststellen, ob die gelegentlichen Abweichungen vom perfekten Rechteck durch Ungenauigkeit in der Schätzung oder durch ungenaues Steuern des Roboters hervorgerufen wurden, da ein noch präziseres unabhängiges Lokalisierungssystem nicht zur Verfügung stand.

4 Zusammenfassung

In diesem Artikel wurde eine effiziente Verbesserung der Monte-Carlo Lokalisation vorgestellt. Dieser Algorithmus nutzt einen schnellen Tabellenzugriff um die Fitness der Samples zu bestimmen und eine iterative Verbesserung der Positionsschätzung. Damit kann die Zahl der benötigten Samples bis auf ein einziges Sample reduziert werden. In Experimenten konnte gezeigt werden, dass der Algorithmus dank der variablen Sample-Zahl in der Lage ist, das „Kidnapped-Robot"-Problem zu lösen und die Position eines fahrenden Roboters mit hoher Genauigkeit zu verfolgen. Die dazu benötigte Rechenzeit lässt ausreichend viel Platz für weitere wichtige Berechnungen des Roboters, wie z.B. Objekterkennung und -verfolgung und Pfadplanung. Als weitere Untersuchung der Genauigkeit des Algorithmus sind Experimente mit einem Laser-Scanner geplant, der die exakte Position des Roboters zum quantitativen Vergleich der Ergebnisse des Verfahrens bestimmt.

Literaturverzeichnis

1. Kitano H, Asada M, Kuniyoshi Y, et al.: RoboCup: The Robot World Cup Initiative. AGENTS '97: Proc. of 1st int. conf. on Autonomous agents, 340–347, 1997.
2. Gutmann S, Weigel T, Nebel B: Fast, Accurate, and Robust Self-Localization in Polygonal Environments. Proc. of IROS '99, 1999.
3. Iocchi L, Nardi D: Self-Localization in the RoboCup Environment. RoboCup-99, LNCS, 1856:318–330, Springer, 2000.
4. Marques C, Lima P: A Localization Method for a Soccer Robot Using a Vision-Based Omni-Directional Sensor. Proc. of EuRoboCup Workshop 2000, 2000.
5. De Jong F, Caarls J, Bartelds R, et al.: A Two-Tiered Approach to Self-Localization. RoboCup 2001, LNCS, 2377:405–410, Springer, 2002.
6. Enderle S, Ritter M, Fox D, et al.: Vision-based Localization in RoboCup Environments. RoboCup 2000, LNCS, 2019:291–296, Springer, 2001.
7. Von Hundelshausen F, Schreiber M, Wiesel F, et al.: MATRIX: A force field pattern matching method for mobile robots. Technical Report B-08-03, Free University of Berlin, 2003.
8. Röfer T, Jüngel M: Vision-Based Fast and Reactive Monte-Carlo Localization. Proc. of ICRA 2003, 856–861, 2003.
9. Röfer T, Jüngel M: Fast and Robust Edge-Based Localization in the Sony Four-Legged Robot League. RoboCup 2003, LNCS, 3020:262–273, Springer, 2004.
10. Menegatti E, Pretto A, Pagello E: A New Omnidirectional Vision Sensor for Monte-Carlo Localization. RoboCup 2004, LNCS, 3276:97–109, Springer, 2005.
11. Fox D, Burgard W, Dellaert F, et al.: Monte Carlo Localization: Efficient Position Estimation for Mobile Robots. Proc. of AAAI 1999, 343–349, 1999.
12. Heinemann P, Rückstieß T, Zell A: Fast and Accurate Environment Modelling using Omnidirectional Vision. Dynamic Perception 2004, 9–14, Infix, 2004.

Swarm Embodiment – A New Way for Deriving Emergent Behavior in Artificial Swarms

Sergey Kornienko, Olga Kornienko, Paul Levi

Institute of Parallel and Distributed Systems, University of Stuttgart, Universitätsstr. 38, D-70569 Stuttgart, Germany

Abstract. This paper concerns the emergent properties of collective robotic systems with imposed microscopic and macroscopic constraints. These constraints dramatically impact the emergent behavior of collective systems so that creating desired emergence becomes very challenged. In this paper we present the top-down swarm embodiment methodology that allows obtaining desired collective behavior in systematic way.

1 Introduction

In order to go beyond the current state of the art in the realization of robotic swarms, the European Commission has granted funding to the I-SWARM project. This project is going to produce a large group of micro-robots (about 1000 micro-robots with the proposed size 2×2×1 mm) that are capable to mimic some aspects of social insects. Such a robot swarm is expected to perform a variety of applications, including micro assembling, biological, medical or cleaning tasks.

The micro-robots, due to small size, are very restricted in hardware capabilities, like distance measurement, navigation or communication. These constraints, arising on the microscopic level, hardly limit the emergent behavior. Not only microscopic, but also macroscopic constraints influence the collective behavior. This kind of constraints arises in technically useful behavior because of necessity to emerge collective activities in specific order, with specific parameters.

Both types of constraints change the problem of collective behavior. The question is not only to find the swarm-controlling mechanisms, allowing a solution of typical problems like foraging or division of labors [1]. The question becomes of how to modify and even how to create new mechanisms that generate the desired emergent behavior satisfying all constraints. Trough many scientific domains contribute to solution of this problem, we fail to find a common methodology that consolidate approaches from these domains. Without this, the derivation of swarm mechanisms is often performed in "trial and error" way. The given work suggests the top-down methodology using a swarm embodiment, that enable a systematical derivation of the desired emergent behavior for artificial swarm. The suggested methodology has been tested in real experiments with the group of micro-robots Jasmine (large prototype in the I-Swarm project).

2 Microscopic and macroscopic constraints

As mentioned in the previous section, artificial robotic swarms are constrained from microscopic and macroscopic sides. The microscopic constrains originate mainly from a construction of a robot (see e.g. the micro-robot Jasmine in Fig. 1(r)). The most important constraints are the communication and perception radius, type of sensors, time of autonomous work and so on. These individual capabilities essentially impact a group behavior.

 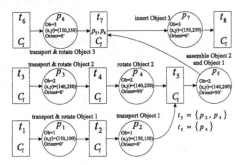

Fig. 1. (r) Micro-robot Jasmine; (l) The assembling plan in the form of Petri-net. P_i are phases, where t_i are transitions with the shown conditions.

Macroscopic constraints arise if the collective systems has to emerge the technically useful behavior. The appearance of these constraints can be demonstrated on a simple assembling example, where robots push three different objects into one defined construction. Assembling of the object has be performed only in the specified order, shown as the Petri-net in Fig. 1(l), otherwise the desired construction will be not obtained. The plan consists of 7 steps, shown as the phases p_1-p_7 with the corresponding positions and rotation angles. The phases p_1, p_3 and p_6 can be started in parallel, other phases have to be proceeded sequentially. The phase p_7 can be started only if p_5, p_6 are finished.

An agent starts transportation or rotation only if its position coincides with the position of an object. This object has not to be processed in this moment by other agents. Moreover the operations defined by the plan have to be applied to the given object only one time (two last problems can be solved by marking). We denote these restrictions as the local restrictions C_l. Each robot looks for objects Ob_i in its own neighborhood and mote to them and reads the mark. If the local and global restrictions are satisfied, the agent executes the required activities. The local rules of an agent have the following form:

```
-   Ob=look for (visible objects); read mark (Ob);
-   if (constraints(Ob)) do (Activity);
```

Agents can start assembling from different initial phases of the plan. Two generated agent-agent cooperation patterns with different initial phases are shown in Figures 2(r) and 2(l). Since these patterns are of different length, we can choose a short assembly by adding rules as e.g.

```
-   at choice → chose phase with smaller number;
```

The cooperation, shown in Fig. 2, emerges without being *preprogrammed*. Emergence arises because of *interactions* between agents that are controlled by *local rules*. The local and global constraints are not only introduced by these rules, but also through parameterization. In the assembly, each operation is *parameterized* by data from the plan. Without knowing these parameters, agents cannot accomplish assembly. Moreover, the emergent cooperation can be of different efficiency and we face the problem of *optimizing the emergent behavior*.

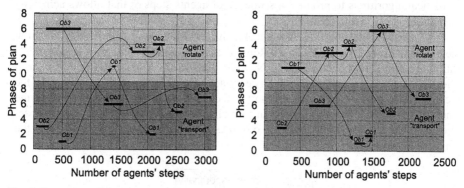

Fig. 2. Examples of emergent "agent-agent" cooperation, generated by the local rules. (r) Initial phases of the plan are $(Ag_1)_{init} = p_6$ and $(Ag_2)_{init} = p_3$; (l) Initial phases are $(Ag_1)_{init} = p_1$ and $(Ag_2)_{init} = p_3$.

The microscopic and macroscopic constraints as well as parameterization and optimization of the emergent behavior appear on the swarm level. All these constraints are closely related with one another and hardly limit the emergent properties of collective systems. In trying to derive the desired emergence, we permanently confront with these constraints so that we identify *the problem of constrained emergent behavior as one of the main problems in artificial swarms* (from the viewpoint of controlling). Without systematic procedure, that allows involving constraints into the collective behavior, the derivation of desired emergence is performed mostly "by trial and error".

3 Top-down methodology

The local rules are in charge of artificial self-organization that appears the desired emergent behavior. There are two strategy to derive such rules. At the bottom-up strategy, the local rules are first programmed into each agent. This rule-based programming [2] originates from the domain of parallel and distributed computing. The general problem of bottom-up approach is that we cannot say in advance, which emergent behavior will be generated by the chosen rules. Especially if this behavior is bounded by constraints. The origin of this problem lies in enormous complexity of nonlinearly interacting system. As pointed out by some authors (e.g. [3]) "*A true emergent phenomenon is one for which the optimal means of prediction is simulation*". It means that *in the worst case* we have to perform really many simulations, gradually changing the local rules, till we receive the desired collective behavior.

Another methodology, consisting in the top-down strategy, shown in Fig. 3.

Using the top-down strategy, the derivation of local rules starts from a definition of the macroscopic pattern Ω and the corresponding constraints. Macroscopic constraints are incorporated directly into Ω, where as microscopic ones influence the so-called "distributing" transformation. The pattern Ω can be defined in statical way, as shown in Fig. 1(1), or in evolutionary way, as e.g. to optimize some value. If the global pattern is determined evolutionary, we can use for "distributing" transformation several heuristical algorithms to produce a sequence of agents steps S_k that allows achieving the pattern Ω [4].

From agent's viewpoint, each agent Ag_k. has a sequence of activities S_k allowing the common group to accomplish Ω. Remark, that all constraints as well as communication are implicitly contained in S_k Now we analyze S_k in order to derive the local rules R_k, that can generate this sequence of activities. More generally, to derive the local rules R_k, we can calculate Kolmogorov complexity of sequence S_k (finding the smallest grammar [5]). In this way can formally derive the set of these rules that defines a cooperation between agents and allows the agents' group jointly to solve the common task Ω.

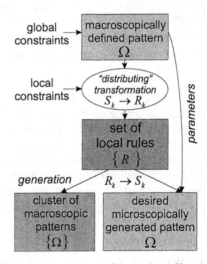

Fig. 3. Top down strategy of derivation of local rules.

3.1 Embodied top-down computational approach

In this section, we show that local rules can be obtained in the mentioned top-down way. Moreover these rules can be integrated (embodied) into specific motion/sensor systems. We consider here a *spatial clusterization* as a basic form of the macroscopic patterns Ω. Particulary, such a cluster can be a n-polygonal shape determined by distances D between corresponding corners. We introduce the local connectivity degree L_{cd} as the number of neighbor agents within the visibility radius R_{vis}. Global connectivity degree $G_{cd} = \sum L_{cd}^i$ is the sum of all local L_{cd}^i and the global compactness Φ is defined as $\Phi = \sum_i^N \sum_j^N D_{ij}$, where D_{ij} is a distance between agent i and agent j, N is the number of agents. In this way, the macroscopic pattern Ω is determined as $min(\Phi)$. The evolutionary algorithms, optimizing Φ, can have the following form

```
-    do {one step in all directions; calculate global compactness;}
-    choose the step which minimum of global compactness;
```

In the case of obtaining n-polygonal spatial formations we introduce additional constrains:

```
-    D=distance(itself-target[pattern]); {do (virtual Activities);
-    D[i]=distance(itself and target[from pattern]);}
-    j=find minimal(delta=D-D[j]); do (Activity j);
```

This algorithm produces a sequence of agents steps S_k allowing building the shapes from Ω. Now we analyze S_k to derive the same behavior, but without using Φ. In Fig. 4 (r), (m) and (l) we plot the global connectivity and compactness, the command of the robot motion controller and, finally, local sensor information. We see that the behavior of whole swarm consists of two phases, that are characterized by different slope of compactness. In the first phase, global compactness rapidly decreases (the global connectivity increases). All neighbor agents during the first phase move in the same direction. In this way, they build small clusters with a homogeneous direction of motion. The size of these local clusters grows whereas the number of them decreases. In the second phase, the rate of building decays and robots no longer move homogeneously. In this phase agents primarily decrease distances only in the cluster. As follows from Fig. 4(l), agents decide to move in the direction with the most high local connectivity degree.

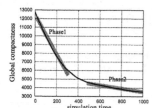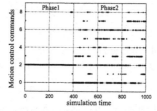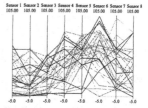

Fig. 4. (r) Global compactness Φ as the function of time; (m) Commands (in the 8-directional DOF motion system) of an agent in the same time range; (l) Sensor information (local connectivity degree L_{cd} in each direction) represented in parallel hierarchical coordinates when an agent decides to move in the direction "6".

In trying to reproduce this behavior (without using Φ), we faced the question of how to replace the gradient (introduced by the global compactness Φ). In the experiments we used two values: the degree of local connectivity L_{cd} and a biologically motivated mechanism based on pAMP-gradient waves, emitted by the fungi *Dictyostelium discoideum* during aggregation phase [6]. In the last case, instead of the pAMP-gradient waves we introduce the following dynamical value k_n based on the L_{cd}: $k_{n+1} = log(\sum_{i=1}^{all\ neighbors} k_n^i)$ with $k_0 = L_{cd}$, where n is the simulation step. The value k_n grows the more rapidly, the larger the cluster is. Based on the values L_{cd} or k_n, robots can decide where the larger cluster is and move in this direction.

The algorithm reproducing the one-cluster-building behavior has the following form (D-direction of motion, nR – neighbor robots with highest L_{cd} or k_n, Dist – distance to nR, CP – adjustment parameter):

```
if (Lcd==0) D=rand; else D=(D of nR); if (D(nR)>CP) D=(D to nR);
```

where (D=(D of nR) and D=(D to nR) are the mentioned two phases of motion. In Fig. 5 we compare the global compactness for the cases of single-phase (only the second phase) motion and two-phases motion based on L_{cd} and k_n.

As followed from Fig. 5, the one-phase motion, that is intuitively the most evident one, builds only small local clusters without bringing them into the bigger one. Both two-phases mechanisms perform building the cluster. However the efficiency of L_{cd} and k_n based mechanisms is different. The biologically motivated mechanism requires less time (and energy) to converge.

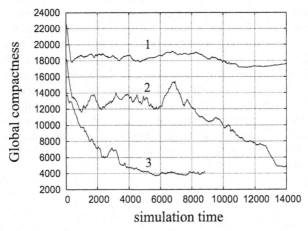

Fig. 5. Comparison of the global compactness for the cases of: (1) single-phase motion, (2) two-phases motion based on L_{cd} and (3) two-phases motion based on k_n. All curves represent typical cases of behavior.

During derivation of local rules R_k we assume some basic functionality F_b, like message transmission or environmental sensing. However the perfectly working simulative sensors essentially differ from real ones. In this way the swarm behavior, generated by R_k, often diverges from our expectations. To get round this problem, we involved the embodiment concept. This says that the same functionality can be implemented in many different ways: Rolf Pfeiffer demonstrated that an "intelligent behavior" can even be implemented when using only some properties of materials [7]. Embodied functionally is also often implemented in some "unusual" way. For example a robot can get a distance to neighbors by sending an IR-impulse and measuring a reflected light. However distances can also be obtaining during communication by measuring a signal intensity. This simple trick saves time and energy: such an unusual functionality is a typical sight of embodiment.

More generally, embodiment means that the system possesses the desired functionality F_b, but this functionality is in a latent form, "it is not appeared". This offers a way of how to get basic functionality for the local rules R_k: the local rules have to influence the hardware development of a robot. The swarm embodiment takes then the following form: definition of the macroscopic pattern Ω and the corresponding micro-scopic/macroscopic constraints; derivation of the local rules R_k; trade off between required functionality and adjustment of hardware; change of the hardware. The local rules have always been considered as a pure software components, however now they are a combination between software and hardware. We can say that in this way *the local rules for the whole swarm behavior are embodied into each individual robot.*

The embodiment in sense *"hardware $\rightarrow$ rules"* has been demonstrated in the work [8]. There we analyzed a dependence between agent's movement and sensor data for the derived S_k. Optimizing local rules to specific motion system, the "top-down"-derived rules can be of 5–20% more efficient than corresponding "bottom-up" rules (see Fig. 6(r)). The embodiment in sense *"rules $\rightarrow$ hardware"* has been demonstrated in the work [9]. In that work we considered context-awareness-related collective ca-pabilities of interacting robotic group and incorporate several local rules into specific sensor system of real micro-robots. The achieved results essentially improve collec-tive robotic behavior and reduced required communication and computational efforts.

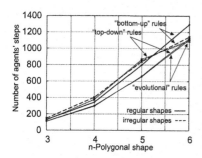

Fig. 6. (r) Comparison between efficiency (the number of steps, needed to reproduce the shapes) of "top-down" and "bottom-up" rules; (l) Preliminary experiments with a small group of micro-robots Jasmine.

We performed several preliminary experiments with a small group of micro-robots Jasmine. The goal was an "embodied top-down" collective perception and spatial information processing (see Fig. 3(l)). The development of collective behavior in-volved a definition of macroscopic patterns, derivation of local rules and redesigning of hardware components. As demonstrated by these experiments, the proposed ap-proach allows creating a specific group's behavioral pattern, whereas robotic behavior still remained flexible (not predetermined). In the further works we will expand this to more generic behavioral types and test in a large robotic swarm.

References

1. Bonabeau E, Dorigo M, Theraulaz G: Swarm intelligence: from natural to artificial sys-tems. New York, Oxford University Press, 1999.
2. Roma G-C, Gamble RF, Ball WE: Formal Derivation of Rule-Based Programs, IEEE Trans. Softw. Eng., 19(3):277–296, 1993.
3. Darley V: Emergent Phenomena and Complexity. In Proc. of Alive IV, 1994.

4. Kornienko S, Kornienko O, Levi P: Multi-agent repairer of damaged process plans in manufacturing environment. In Proc. of IAS-8, Amsterdam, 485–494, 2004.
5. Charikar M, Lehman E, Liu D, et al.: Approximating the smallest grammar. In Proc. of the 34th ACM symposium on Theory of computing, 792–801, 2002.
6. Haken H: Advanced synergetics. Springer, Berlin, 1983.
7. Pfeifer R, Iida F: Embodied artificial intelligence: Trends and challenges. In Iida (ed) et al. Embodied artificial intelligence, 1–26, Springer, 2004.
8. Kornienko S, Kornienko O, Levi, P: Generation of desired emergent behavior in swarm of micro-robots. In Proc. of ECAI 2004, Valencia, Spain, 2004.
9. Kornienko S, Kornienko O, Levi P: Collective AI: context awareness via communication. In: Proc. of IJCAI 2005, Edinburgh, 2005.

Kooperative Multi-Roboter-Wegplanung durch heuristische Prioritätenanpassung

Ralf Regele, Paul Levi

FZI Forschungszentrum Informatik, Heid-und-Neu-Str. 10-14,76131 Karlsruhe

E-mail: {regele,levi}@fzi.de

Zusammenfassung. Im Beitrag wird ein neuer Algorithmus zur verteilten, kooperativen Wegplanung in Multi-Roboter-Systemen vorgestellt. Dabei soll eine große Anzahl von Robotern mit unterschiedlichen Zielpositionen ihre Wege so planen, dass auch in stark verschachtelter Umgebung alle dynamisch entstehenden Konflikte gelöst werden können. Grundidee des Ansatzes ist eine vollständig verteilte Planung ohne zentrale Planungseinheit, bei der die Roboter durch Austausch ihrer Pläne im Raum-Zeit-Konfigurationsraum die Bewegungen der anderen Roboter kooperativ einplanen können. Konflikte werden dabei durch die heuristische Anpassung der Prioritätswerte für die Einzelschritte der Roboter während der Bewegung gelöst. Alle Bewegungsplanungen werden durch fortwährenden Austausch von Nachrichten ständig dynamisch angepasst und erweitert.

1 Einleitung

Die Wegfindung eines einzelnen Roboters wurde in der Vergangenheit bereits intensiv erforscht [8]. Wesentlich komplexer ist das Problem bei einem Multi-Roboter-System, besonders bei Systemen mit einer hohen Anzahl von Robotern bei stark beschränktem Raum. Schon früh entstand die Idee, den Konfigurationsraum des Roboters um eine Zeitdimension zu erweitern [7]. Im weiteren entstanden zwei verschiedene grundsätzliche Ansätze. Zentralistische Ansätze versuchen meist eine vollständige Vorausberechnungen aller Roboterbahnen, sind aber aufgrund der hohen Komplexität des Problems häufig zu zeit- und berechnungsintensiv. Jüngere Ansätze versuchen die Komplexität durch die Verwendung von probabilistischen Roadmaps zu verringern [4]. Verteilte Ansätze auf der anderen Seite reduzieren die Gesamtkomplexität des Problems durch die unabhängige Berechnung der einzelnen Roboterbahnen. Die einzelnen Bahnen können dann unter Zuhilfenahme von Potentialfeldern [6], Geschwindigkeitsplanungen (path coordination)[5] oder der Vergabe von Prioritäten [3] zusammengeführt werden. Die Verfahren sind jedoch meist nicht vollständig, so dass die Gefahr von Verklemmungen und suboptimalen Lösungen besteht. Jüngere Ansätze schlagen deshalb eine stärkere Dynamik bei der Koordination von Bahnen und Prioritäten vor [1,2]. Im vorliegenden Beitrag wird nun ein neuartiges Verfahren zur verteilten kooperativen Wegplanung vorgestellt, welches mit einer hochdynamischen, heuristischen Anpassung der Prioritäten arbeitet. Voraussetzung soll dabei die Fähigkeit zur Kommunikation und Kooperation zwischen den einzelnen Robotern sein. Hauptziel ist das schnelle und verklemmungsfreie Erreichen der individuellen Zielposition. Das System ist speziell für Probleme mit einer hohen Anzahl von Robotern geeignet, da mit einem relativ geringem Aufwand an Kommunikation und Berechnung auch komplexe Probleme mit zahlreichen Einheiten gelöst werden können.

2 Weltmodell

Zur Vereinfachung wird von einem omnidirektionalen, kreisförmigen Roboter ausgegangen. Der Algorithmus verwendet ein sowohl räumlich wie auch zeitlich diskretes Weltmodell. Dabei werden zwei unterschiedliche Karten mit gleichmäßiger Zellaufteilung verwendet. Die erste Karte, die sogenannte Entfernungskarte, besitzt noch keine zeitliche Dimension und berücksichtigt nur statische Hindernisse. Jede Zelle der Karte beinhaltet dabei die Entfernung zur Zielposition des jeweiligen Roboters. Damit kann die Karte sowohl für die statische Wegplanung eines einzelnen Roboters verwendet werden, als auch für die Evaluierung von Wegalternativen. Sofern die statischen Hindernisse bekannt sind, muss die Entfernungskarte nur einmal für jede Zielposition berechnet werden, muss also bei einer Neuplanung des Weges nicht neu berechnet werden. Deshalb ist es möglich, auch mit relativ großen Entfernungskarten zu arbeiten. Die Entfernungskarte wird mit Hilfe einer Breitensuche erstellt, ausgehend von der Zielposition. Für alle dynamischen Planungen wird die sogenannte Umgebungskarte verwendet, welche nun um die zeitliche Dimension erweitert wurde, so dass die korrekte Darstellung von Raum-Zeit-Wegen möglich ist. Die Karte enthält sowohl statische Hindernisse als auch die bereits geplanten Wege der anderen Roboter. Indem diese Wege als Hindernis betrachtet werden, kann im Idealfall bereits ein Weg für den Roboter gefunden werden. Interessanter ist jedoch der Fall bei begrenztem Raum, wenn es zu Konflikten zwischen den Robotern kommt.

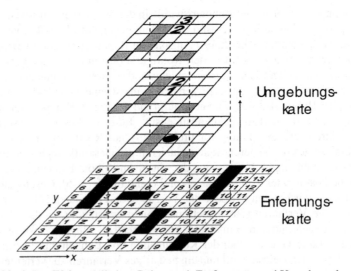

Abb. 1. Das Weltmodell eines Roboter mit Entfernungs- und Umgebungskarte

3 Lokale Wegplanung

An dieser Stelle soll kurz beschrieben werden, wie die Wegplanung eines Roboters im konfliktfreien Fall abläuft. Jeder Roboter besitzt eine lokale Umgebungskarte mit mehr oder weniger vollständigen Informationen über die Wegplanungen der anderen Roboter. Diese Wegplanungen werden nun so ergänzt, dass alle bekannten Wege sich über den vollständigen Zeitraum der Umgebungskarte erstrecken, jeder Roboter führt also die Wegplanungen der anderen Roboter weiter. In einem ersten Durchlauf wird nun eine Breitensuche ausgehend von der momentanen Position des Roboters gestartet, welche sich immer im Zentrum der ersten Zeitebene der Umgebungskarte befindet. Ziel dieses ersten Durchlaufs ist die Ermittlung von allen möglichen Endpositionen im Zeithorizont der Umgebungskarte. Aus allen Zellen der letzten Zeitebene, die erreichbar sind, wird nun die beste Position ausgewählt. Dazu werden die Werte der Entfernungskarte verwendet, der Roboter wählt also die Position mit der geringsten Entfernung zum Ziel aus. In einem zweiten Durchlauf durch die Umgebungskarte werden nun die einzelnen Wegschritte festgelegt, indem ein günstiger Weg aus den zuvor als erreichbar markierten Zellen des Raum-Zeit-Konfigurationsraums ausgewählt wird. Ein günstiger Weg ist dabei ein Weg, der möglichst wenig Bewegungen erfordert und die gewünschte Zielposition möglichst frühzeitig erreicht. Da meistens verschiedene Wege zur Auswahl stehen, können auch andere Faktoren berücksichtigt werden. Zum Beispiel können Wege bevorzugt werden, welche nicht in die Nähe von anderen Robotern führen, oder bestimmte Richtungen werden bei der Schrittplanung bevorzugt (Rechts vor Links).

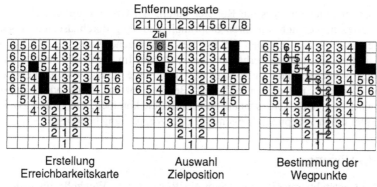

Abb. 2. Ablauf der konfliktfreien Wegplanung in einem Ausschnitt aus der Umgebungskarte entlang der Zeitachse. Die dargestellten Hindernisse könnten zum Beispiel Fragmente der Wege anderer Roboter sein. Die Zahlenwerte bezeichnen die benötigte Schrittanzahl zur jeweiligen Zelle.

4 Kommunikation

Es wird angenommen, dass die Roboter über ein Kommunikationssystem verfügen, um sich gegenseitig Nachrichten verschicken zu können. Der Kommunikationsaufwand soll allerdings nicht zu hoch werden, da ein System mit vielen Robotern und ausgeprägten Verhandlungsabläufen sonst leicht überlastet wird. Für den vorliegenden

Algorithmus genügt ein einfaches Kommunikationssystem, welches in regelmäßigen Abständen Nachrichten an alle Roboter in der Nähe versendet. Hauptbestandteil dieser Nachrichten ist eine Kopie der lokalen Umgebungskarte mit den geplanten Wegen des Roboters. Diese Karten enthalten alle Informationen, die ein fremder Roboter braucht, um die fremde Wegplanung in seiner eigenen Umgebungskarte zu rekonstruieren. Es wird davon ausgegangen, dass die Roboter ihre eigene Position kennen und ihre Nachrichten mit einem Zeitstempel versehen können, welche mit den Zeitstempeln anderer Nachrichten vergleichbar ist. Der allgemeine Ablauf der Verhandlungen findet dann durch den fortwährenden Austausch von Karten statt, in denen die Roboter ihre eigenen Planungen an die Wege der anderen Maschinen anpassen und weiterführen.

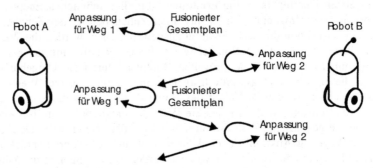

Abb. 3. Verhandlungsschema zwischen den Robotern. Durch fortgesetzten Austausch von veränderten Karten ergibt sich die jeweils aktuelle Wegplanung.

Dabei ist zu beachten, dass die Roboter durchaus Rücksicht aufeinander nehmen und nicht etwa immer nur stur ihren ursprünglichen Weg durchzusetzen versuchen. Als kooperatives System nehmen die Roboter auch eigene Nachteile in Kauf, wenn dadurch die Gesamtsituation verbessert werden kann. Im Idealfall reicht es aus, die gewünschten Wege der anderen Roboter als Hindernisse anzunehmen und einen Weg zu suchen, der nicht mit diesen kollidiert. Bei begrenztem Raum oder zahlreichen Roboter wird es jedoch sehr schnell zu Konflikten kommen, bei denen zwei oder mehr Roboter einen Wegpunkt dringend für ihre Wegplanung benötigen.

5 Konfliktlösung

Ein Konflikt entsteht immer dann, wenn ein Roboter eine Wegzelle für seine Bahn reservieren will, diese aber bereits durch einen anderen Roboter beansprucht wird. In den meisten Ansätzen für Multi-Roboter-Wegplanung wird mit festen Prioritäten gearbeitet, bei denen die Roboter in absteigender Priorität ihren Weg reservieren können [7]. Der vorliegende Ansatz arbeitet mit einer heuristischen Bestimmung der Roboterprioritäten, welche dynamisch für jeden Schritt einzeln festgelegt werden können. Dazu belegt jeder Roboter die einzelnen Schritte seines geplanten Weges mit Prioritätsangaben, bevor er diese an die anderen Roboter versendet. Dabei ist es möglich, unterschiedliche Prioritäten für kritische und weniger wichtige Wegstellen zu verwenden. Jeder Roboter besitzt einen Gesamt-Prioritätswert, der als Basis für die einzelnen Prioritäten der Wegschritte verwendet wird. Diese Gesamt-Priorität wird nun schrittweise an die vorliegende Situation angepasst, bis eine Priorität gefunden wird, die eine erfolgreiche Wegplanung ermöglicht. Alle Roboter gehen dabei ähnlich vor, so dass insgesamt ein hochdynamische System aus sich verändernden Prioritäten entsteht.

Im Einzelnen beginnt jeder Roboter i mit einer Priorität $P(i)=0$. Nach Verarbeitung der übermittelten Karten führt der Roboter eine Wegsuche durch, wobei er alle Wegpunkte als frei betrachtet, die mit einer Priorität $P<P(i)$ belegt worden sind. Alle Wegpunkte, die mit Priorität $P>=P(i)$ belegt wurden, werden als Hindernisse betrachtet. Nach Abschluss der Wegsuche wird der gefundene Weg nun auf Konflikte überprüft. Je nach Ergebnis der Prüfung wird die Priorität des Roboters angeglichen.

– *Fall 1:* Der gewünschte Weg scheint blockiert zu sein. Dies wird ermittelt, indem überprüft wird, ob der ermittelte Weg in den letzten Zeitschritten noch Fortschritte zeigt, oder ob der Roboter zum Warten auf der selben Stelle gezwungen wird. Dies bedeutet, dass ein dynamische Hindernis (ein anderer Roboter) den Weg blockiert. Die Priorität $P(i)$ des Roboters wird um 10 Punkte erhöht, um das verstärkte Interesse des Roboters an der fraglichen Wegstelle zu zeigen.

– *Fall 2:* Der Roboter ist so blockiert, dass er noch nicht einmal eine Warteposition findet. In diesem Fall wird von der Wegsuche kein gültiger Weg gefunden, alle zukünftigen Wegzellen sind bereits belegt. Der Roboter kann also noch nicht einmal seine Ausgangsposition über den Planungszeitraum reservieren. Da der Roboter jedoch physikalisch weiterexistieren muss, ist die vorliegende Planung offenbar nicht sinnvoll. Der Roboter reserviert deshalb sofort einen Notweg mit höchster Priorität, der einfach auf dem Beharren auf der Ausgangsposition beruht. Außerdem wird auch seine allgemeine Priorität $P(i)$ um 19 Punkte erhöht. Die Absicht dabei ist, dass der Roboter sich aus der Blockade befreien kann und den Zuschlag für die kritische Wegstelle bekommt.

– *Fall 3:* Es wird keine Blockade gefunden, der Roboter scheint Fortschritte auf seinem Weg zum Ziel machen zu können. Kleinere Wartephasen während des Planungszeitraums sind dabei akzeptabel. Die Priorität $P(i)$ des Roboters wird um einen Punkt gesenkt, so dass sie langfristig wieder auf 0 sinkt.

Wurde die Priorität des Roboters erhöht, wird nun eine Wegsuche mit der neuen, erhöhten Priorität durchgeführt. Erneute Blockaden bleiben jedoch bestehen, jede Priorität kann nur einmal für jeden Berechnungs- und Kommunikationszyklus erhöht werden. Dies führt zu einer ständigen dynamischen Anpassung der Prioritätswerte an die vorliegende Situation. Ein wichtiger Faktor dabei ist die Akzeptanz von Ausweichbewegungen, solange sie nicht in einem Wartezustand enden. Selbst Bewegungen in die falsche Richtung, die also zu einer Verschlechterung der Zielentfernung führen, werden als blockadefrei akzeptiert. Dadurch wird eine einmal eingeschlagene Ausweichbewegung weitergeführt, der Roboter versucht nicht, sie durch sofortige Erhöhung der Prioritätswerte wieder abzubrechen. Wenn zum Beispiel zwei Roboter in einem engen Gang aufeinandertreffen, dann wird einer der beiden die höhere Priorität erreichen. Welcher Roboter dies ist, wird durch den Ausgangswert der Prioritäten und die genaue Reihenfolge der Nachrichten bestimmt und ist kaum vorhersehbar. Wichtig ist jedoch, dass dieser Roboter nun eine Ausweichbewegung des anderen Roboters erzwingt, die dieser als blockadefrei akzeptieren wird und weiterführt, bis er eine Ausweichstelle findet. Ein Konflikt führt also nur zu einer Blockade, solange noch kein Roboter eine Ausweichbewegung „vorgeschlagen" hat.

Komplexe Verklemmungsmuster, die nicht direkt erkannt werden konnten, führen zu einem immer höheren Ansteigen der Prioritäten. Ein extremer Prioritätswert gilt deshalb als Anzeichen für eine ungelöste Blockade. Zur Lösung wird die Priorität auf einen niedrigen, zufälligen Wert zurückgesetzt, die Prioritäten werden also neu „gemischt".

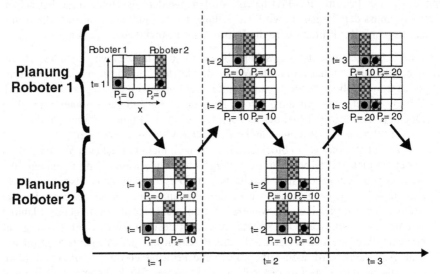

Abb. 4. Ablauf einer Prioritätsverhandlung bei zwei sich entgegenkommenden Robotern. Die Ausschnitte aus dem Raum-Zeit-Konfigurationsraum zeigen, wie der linke Roboter 1 vom rechten Roboter 2 verdrängt wird, sobald ihre Umgebungsbereiche überlappen.

6 Ergebnisse und Ausblick

Der vorgestellte Algorithmus zur kooperativen Wegplanung wurde in einer simulierten Multi-Roboter-Umgebung getestet. Dabei wurde das Verhalten und die Leistungsfähigkeit in unterschiedlichsten Umgebungen und bei verschiedenen Parametrisierungen überprüft. Dabei erwies sich der Algorithmus als leistungsstark und zuverlässig. Auch komplizierte und schwer zu lösende Problemstellungen wie Massenkonflikte, Mehrfachkreuzungen und Blockadedurchfahrten konnten gut gelöst werden. Es wurden Systeme mit bis zu 50 Robotern getestet, durch die vollständig verteilte Struktur des Algorithmus scheint es hier keine Obergrenze für die Anzahl der Roboter zu geben. Durch die heuristische Natur der Wegplanung wird jedoch meist nicht der optimale Weg gefunden. Bei einfachen Szenarien liegt die durchschnittliche Fahrzeit und -strecke dabei nicht wesentlich über dem optimalen Wert, bei komplexeren Situationen mit vielen Robotern bei wenig Ausweichraum muss mit einer um bis zu 50-100% erhöhten Fahrzeit und -strecke gerechnet werden. Nur bei sehr engen Szenarien mit weniger als 50% Freiraum (im Verhältnis zum Raum, der von den Robotern belegt wird) kam es zu Verklemmungen, die erst durch eine Neuordnung der Prioritäten nach Überlauf der Prioritätswerte gelöst werden konnten.

Wichtig ist die geeignete Auswahl der Parameter. Die Größe der Entfernungs- und Umgebungskarten hat direkten Einfluss auf die Qualität der Wegplanung, da bei kleinen Karten Konflikte erst spät erkannt werden und Ausweichmöglichkeiten außerhalb des Sichtbereichs nicht eingeplant werden können. Große Karten erfordern jedoch mehr Zeit und Aufwand bei Kommunikation und Berechnung. Die Parameter sollten deshalb auf die Art der Umgebung angepasst werden. Dabei muss die Umgebung nicht durch eine gleichmäßige Zellverteilung beschrieben werden, der Algorithmus kann für alle Umgebungsmodelle angewendet werden, die durch Graphen beschrieben werden können. Zellbelegungen mit unregelmäßigen Zellen, hierarchische Zellansätze oder Wegnetze können auch weitläufige Umgebungen mit relativ geringem Aufwand beschreiben. Des weiteren kann der Algorithmus durch verfeinerte Strategien bei der Wegplanung in der lokalen Umgebungskarte, bei der Vorhersage von fremden Bewegungsbahnen und bei der Synchronisation der Kommunikation weiter verbessert werden.

Es ist vorgesehen, den Algorithmus an einem realen Multi-Roboter-System zu testen. Dabei werden verstärkt Aspekte der Kollisionsvermeidung, der Kinematik und der Sensorik in das Blickfeld geraten, die bisher eher abstrakt behandelt wurden.

Literaturverzeichnis

1. Clark MC, Rock SM, Latombe JC: Motion Planning for Multiple Mobile Robot Systems using Dynamic Networks. IEEE Int. Conf. on Robotics and Automation, Taipei, Taiwan, 2003.
2. Bennewitz M, Burgard W, Thrun S: Optimizing Schedules for Prioritized Path Planning of Multi-Robot Systems. IEEE Int. Conf. on Robotics and Automation, Seoul, Korea, 2001
3. Azarm K, Schmidt G: Conflict-Free Motion of Multiple Mobile Robots Based on Decentralized Motion Planning and Negotiation. IEEE Int. Conf. on Robotics and Automation, Albuquerque, New Mexico, 1997
4. Svestka P, Overmars MH: Coordinated Path Planning for Multiple Robots. Robotics and Autonomous Systems, 23 (1998).
5. Leroy S, Laumond JP, Siméon T: Multiple path coordination for mobile robots: a geometric algorithm. Int. Joint Conf. on Artificial Intelligence, 1999.
6. Warren CW: Multiple Robot Path Coordination Using Artificial Potential Fields. IEEE Int. Conf. on Robotics and Automation, 1990.
7. Erdmann M, Lozano-Pérez T: On Multiple Moving Objects. IEEE Int. Conf. on Robotics and Automation, 1986.
8. Latombe JC: Robot Motion Planning. Kluwer Academic Publishers, 1991

A Unified Architecture for the Control Software of a Robot Swarm: Design and Investigation Results

Viktor Avrutin, Andreas Koch, Reinhard Lafrenz, Paul Levi, Michael Schanz

Institute of Parallel and Distributed Systems (IPVS), University of Stuttgart,
Universitätstrasse 38, 70569 Stuttgart, Germany

Abstract. In this work a unified approach for modeling of micro-robot swarms and for development of the controlling software architecture for individual robots is presented. The approach leads to swarm models with self-organized behavior and reflects the complexity of the considered swarm scenarios. The application of the presented approach is demonstrated by example scenarios with several complexity levels. Additionally, some techniques for the investigation of the phenomena of self-organization are discussed. These techniques allow for instance the determination of areas in the parameter space, i.e. the parameter settings leading to a specific aimed behavior. Through our simulation experiments we demonstrated, that the hardware requirements for applications using real micro-robots can be determined.

1 Introduction

The kind of control strategies for individual robots represent one of the most significant differences between classical robotics and swarm-robotics. In the classical approach, each robot possesses sufficient capabilities (artificial intelligence), required for completion of tasks. In contrast to this, in swarm-robotics this intelligence is manifested by the robot swarm as a whole, and not by each individual robotic unit. In particular, a swarm is able to form some complex spatial and functional structures, in spite of the fact, that each robot of the swarm follows the same set of rules. Due to the large number of robots in the swarm and their restricted communicational capabilities, it is assumed, that the robots can not be controlled by an external instance. Furthermore, due to their restricted perceptual and computational capabilities, they are not able to perform complex tasks by themselves. Therefore, we suggest, that the capability of the swarm to cooperatively perform tasks has to be achieved by the phenomenon of self-organization [1,2].

According to the theory of self-organization, systems showing phenomena of self-organization, consist typically of a large number of simple subsystems, which interact with each other locally, i.e., within a short-range distance. Obviously, these conditions are fulfilled in the case of robot swarms. Each individual robot can be considered as a simple unit with strongly restricted capabilities, corresponding to the current available hardware. The interactions between robots are restricted by realistic perception and communication capabilities and are therefore local.

Preliminary works [3,4,5,6] show, that simple task assignments can be successfully achieved by self-organization principles in autonomous robots. Furthermore, it is shown in [7,8,9] that several simple tasks, like clustering, can be completed by self-organization within robot swarms.

The main difference between the self-organization in nature and in technical systems is, that the interactions in nature are specified by physical laws. In contrast to this, in technical systems the interactions are based on the perceptual and communication capabilities of the involved subsystems. Therefore the control of self-organization in technical systems is possible by development of proper perceptual strategies and communication protocols.

Based on the possibility to develop the self-organization mechanisms for the robot swarms, we use the challenging and powerful concept on this field, given by hierarchical self-organization. According to this concept, the whole system consists of a hierarchy of subsystems. The most simple subsystems (atomic entities) build formations via a basic self-organization process. Then these formations operate as non-atomic entities and build more complex formations via a self-organization process on the next hierarchy level. Theoretical, the number of self-organization levels in a hierarchical self-organization process is not restricted, however in our approach we restrict it by two levels. This is sufficient in order to demonstrate the applicability of the presented concept and can be extended in the future work.

As an atomic entity in a robot swarm we consider an individual unit (i.e. an individual robot in the swarm). In our approach the first stage of the self-organization is represented by formation of chains with a specified length (see for instance [10,11, 12,13]). Then an already formed chain operates as one entity and represents a part of more complex formations needed in complex scenarios, like cooperative object recognition and object following. Note, that there exists a broad spectrum of complex swarm scenarios, using chains as basic components. Therefore, the first stage for all these scenarios is represented by the chain formation, and the corresponding part of the controlling software can be reused.

2 Two layer architecture

One of the first challenges when dealing with the design of software for swarm units is to find a suitable software architecture. The architectures known from the field of macro-robotics seem to be inapplicable in the case of robot swarms due to the restricted capabilities of the robots with respect to computational power, memory, perception, etc.. This leads to the need of minimalistic solutions for the software architecture. Our suggested solution for the minimalistic software architecture is to structure the software into two layers, namely a *decision layer* and an *operational layer*, like in [14,15].

The responsibility of the decision layer is to switch between several modes of operation. The information which controls this switching behavior is first the internal state of the unit and secondly the environmental information. Hereby, the environmental information is derived from the sensing and communication systems of the swarm units and can be considered within the more abstract concept of environmental events.

For the implementation of the decision layer we use hybrid automata, whereby the transitions of the automaton are controlled by data communicated between the robots and by sensory information. Note that the type of this automaton depends on the complexity of the considered scenario. For simple tests like the random walk of the individual robots, the basic approach based on a degenerated (consisting of one or two states) automaton is sufficient. For scenarios with higher complexity (chain formation, circle formation, straight line formation, etc.) the usual hybrid automata can be used. Finally, for complex scenarios related to the phenomenon of hierarchical self-organ-ization, we use hierarchical hybrid automata [16,17].

The operational layer represents a set of rules, which describe the change of the internal state of the unit corresponding to each mode of operation. Examples for such a change are movements, sensing (which not only provides the robot with information but also reduces its current energy amount), etc. For the implementation of the operational layer we use virtual forces (see Table 1), which are defined for each state of the corresponding hybrid automaton separately as shown in Table 2. Hereby it is possible to use a generic virtual force, which may have, depending on its parameters, attracting and repelling components. It is worth to mention that the desired change of the current state (for instance, the movement) calculated in the operational layer has to be implemented according to the restriction of the assumed target-hardware. In the case of robot swarms it is especially important to distinguish between a more abstract operational layer and the hardware models, because the real hardware of the target robots is currently in the development stage. Related to kinematics of the robots, we use in our experiments two different models, an omnidirectional one and a three-leg resonant locomotion based one.

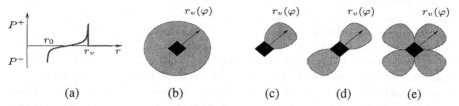

Fig. 1. (a) Generic virtual force $P(r)$. The parameters P and P define the attracting and the repelling components of the force. The parameters r_o and r_v correspond to the robot size and to the visibility range. (b) – (d) Visibility range for the ideal case (uniform 360° visibility, (b)) and for real cases of one, two and four optical sensors.

3 Implemented scenarios

The suggested software architecture is tested for a broad spectrum of scenarios, from simple to complex ones. The most simple scenario is the random walk of individual robots. In order to implement this scenario, only a degenerated hybrid automaton is required, consisting of a single state called hermit. The movement in this state is affected by two virtual forces, which guarantee collision avoidance and obstacle avoidance.

A more complex scenario is the chain formation, which represents a basic sub-task for several more complex scenarios. In this case, the hybrid automaton consists of five states (hermit, single, leader, follower, surrounded) and is shown marked with (I) in Fig. 3. Note, that this automaton controls the formation of chains with a specified nominal length. If this length can be arbitrary, a more simple automaton consisting of three states is sufficient.

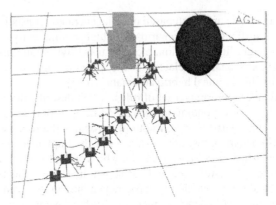

Fig. 2. Object orbiting: in progress.

We implemented the complex scenario of cooperative object orbiting based on the chain formation. This scenario consists of three subsequent tasks: chain formation, collective search for objects and forming a closed chain with a minimal number of robots around the object. This represents an example for a hierarchical self-organization. In this scenario, the chains, build as an example for the scenario with middle complexity, are used as a basic component to achieve a complex behavior. The heads of the chains (units with the mode of operation `leader-2`) search for obstacles in the area using random walk. Note, that it is not the random walk of individual units, but the random walk of complete chains. After having found an obstacle, they will start orbiting it, i.e. they perform a cooperative movement along the obstacle border (see Fig. 2). If the chain is long enough to completely surround the obstacle, the units which are redundant are unlocked in order to fit the chain length to the circumference of the obstacle. In order to achieve the behavior described above, the hierarchical automaton shown in Fig. 3 is used. As one can see, it consists of three meta-modes, which correspond to the above mentioned tasks.

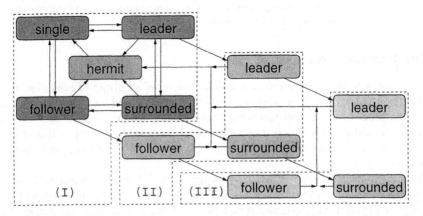

Fig. 3. Hybrid automaton controlling the behavior of the swarm units in the object orbiting scenario and representing the decision layer of the model.

Tab. 1. Design of the virtual forces affecting the behavior

force	description
f_s	Force towards other units: This force is generated by using the generic force function without an attracting component.
f_o	Force towards obstacles: This force function is similar to f_s, in that it just has an repelling component. Nevertheless, it has to be effective on different objects, which is why it is specified separately. It is applied on objects that do not reply to communication.
f_p	Force towards the previous chain member: This force has an attractive and a repelling component. The point of equilibrium is chosen in the middle between robot size and the visibility range.
f_n	Force towards the next chain member: This force is similar to the previous one, but with a greater repelling component. This results in the movement of the chain towards the chain leader.
f_c	Force towards other chain leaders: This force has an attractive and a repelling component. The point of equilibrium is somewhere in between the robot size and the visibility range.

Tab. 2. The virtual forces affecting the behavior in specific modes of operation.

mode of operation	forces affecting the movement
hermit	$f_s + f_o$
single	$f_s + f_o$
leader	$f_n + f_s + f_o$
follower	$f_p + f_s + f_o$
surrounded	$f_n + f_p + f_s + f_o$
leader-2	$f_n + f_s + f_o + f_c$
follower-2	$f_p + f_s + f_o$
surrounded-2	$f_n + f_p + f_s + f_o$
leader-3	$f_n + f_s + \tilde{f}_o + f_c$
follower-3	$f_p + f_s + \tilde{f}_o$
surrounded-3	$f_n + f_p + f_s + \tilde{f}_o$

The first meta-mode of the automaton (marked with (I) in Fig. 3) consists of the modes of operation already implemented for the formation of the chains. If a chain reaches the nominal length, the chain as a whole moves on and the automaton switches to the next meta-mode, i.e., the second task within the scenario. In this second meta-mode (marked with (II) in Fig. 3) the chains start looking for obstacles. If the leader of a formation finds one, it starts orbiting according to the forces which now take effect, and the complete chain follows the leader, which corresponds to the third meta-mode of the automaton (marked with (III) in Fig. 3). Hereby, the virtual force between the leader and the obstacle has not only a repulsive component which is required in order to avoid the collision with the obstacle, but also an attracting component which leads to the fact, that the distance between the leader and the obstacle remains approximatively constant. Because of the pushing forces from behind, the chain leader moves on a line parallel to the obstacle it has detected. If the chain is too long to completely surround the obstacle, members leave the chain, until it fits the obstacle.

We finally remark, that one of the advantages of the suggested approach is, that the components of hierarchical hybrid automata are given by more simple automata developed for the implementation of scenarios with lower complexity. This leads to re-usage of the software components and therefore to reduction of the development costs.

4 Investigation results

In order to determine the efficiency of the swarm, we investigated its ability to perform a given task under variation of several parameters. As a test scenario the chain formation is chosen, because in our approach more complex scenarios are based on the chain formation. The simulation experiments show, that the self-organized behavior is scalable, at least for the investigated swarm sizes up to 150 units no significant dependency of the behavior on the swarm size is observed. It turned out, that the most critical values for the chain formation are the swarm density ρ (that is the number of swarm units per area unit) as well as the the visibility range r_v (that is the range of the proximity sensors).

One of the typical investigation results is shown in Fig. 4. In this experiment, the side length W_f of the quadratic arena is varied up to 500mm. The colors indicate the number N of units that are in complete chains with nominal length of 5. As shown in Fig. 4, there exists an area in the 2D parameter space $W_f \times r$, where the task is completed successfully, i.e. all units are members of chains with the required nominal length. The fact, that outside of this area the task is not completed (at least within a reasonable time) can be explained by taking into account the following: on the one hand, if the swarm density is too low, the units are not able to communicate with each other for a certain time and the chain formation becomes slower. Obviously, this effect is reduced for increasing visibility ranges r_v. On the other hand, if the swarm density is too high, the units constrict each other and the chain formation becomes slower as well.

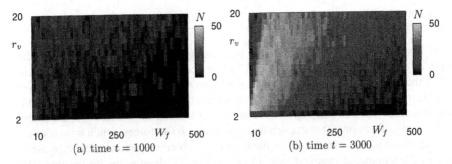

(a) time $t = 1000$ (b) time $t = 3000$

Fig. 4. Progress of the chain formation in a swarm consisting of 50 swarm units under variation of the side length W_f of the arena and visibility range r_v. Results for two different times ($t = 1000$ sec and $t = 3000$ sec).

5 Summary and Outlook

In this work we presented an unified approach for modeling of the micro-robot swarms and for development of the controlling software for individual robots. According to this approach the basic architecture of the controlling software for individual robots consists of the decision- and the operational layers. It is shown, how the operational layer can be implemented using virtual forces, whereas the decision layer can be implemented based on hybrid automata of different types, that depend on the complexity of the considered scenarios. It is also shown, how this approach supports the development of the controlling software for complex scenarios, which are based on the hierarchical self-organization. Hereby the communication and the perception are required within a small range only. Concerning the memory necessary for the storage of variables and parameter of the individual units, the current requirement is about 30 Bytes. The communication volume is about 2 Bytes for each communication activity. Furthermore it is shown, that the developed simulation models can be used in order to obtain the parameter settings for suitable and efficient swarm configurations.

6 Acknowledgment

This work is supported by the EU within the project *Intelligent Small World Autonomous Robots for Micro-manipulation* (I-SWARM, Project N° 507006).

References

1. G. Nicolis and I. Prigogine. *Self-Organization in Non-Equilibrium Systems*. Wiley, New York, 1977.
2. H. Haken. *Advanced Synergetics*. Springer-Verlag, 1983.
3. J. Starke, M. Schanz, and H. Haken. Self-organized behaviour of distributed autonomous mobile robotic systems by pattern formation principles. In T. Lueth, R. Dillmann, P. Dario, and H. Wörn, editors, *Distributed Autonomous Robotic Systems 3*, pages 89–100. Springer Verlag, Heidelberg, Berlin, New York, 1998.
4. M. Becht, T. Buchheim, P. Burger, G. Hetzel, G. Kindermann, R. Lafrenz, N. Oswald, M. Schanz, M. Schulé, P. Molnár, J. Starke, and P. Levi. Three-index assignment of robots to targets: An experimental verification. In *IAS-6*, 2000.
5. M. Schulé, M. Schanz, H. Felger, R. Lafrenz, J. Starke, and P. Levi. Control of autonomous robots in the robocup scenario using coupled selection equations. In P. Levi and M. Schanz, editors, *Autonome Mobile Systeme*, pages 57–63. Springer, 2001.
6. J. Starke. Dynamical assignments of distributed autonomous robotic systems to manufacturing targets considering environmental feedbacks. In *Proceedings of the 17th IEEE International Symposium on Intelligent Control (ISIC'02)*, pages 678–683, Vancouver, 2002.
7. C.W. Reynolds. Steering behaviors for autonomous characters. *Game Developers Conference 1999*, San Francisco, 1999.
8. C.W. Reynolds. Interaction with groups of autonomous characters. *Proceedings of Game Developers Conference 2000*, 2000.
9. C. Saloma, G.J. Perez, G. Tapang, M. Lim, and C. Palmes-Saloma. Self-organized queuing and scale-free behavior in real escape panic. *Proceedings of National Academy of Science*, 100, 2003.

10. H. Yamaguchi and T. Arai. Distributed and autonomous control method for generating shape of multiple mobile robot group. In *Proc. of the IEEE International Conference on Intelligent Robots and Systems*, pages 800–807, 1994.
11. H. Yamaguchi and J.W. Burdick. Asymptotic stabilization of multiple nonholonomic mobile robots forming group formations. In *Proc. of the 1998 IEEE International Conference on Robotics and Automation*, pages 3573–3580, 1998.
12. S. Nouyan and M. Dorigo. Chain formation in a swarm of robots. Technical Report TR/IRIDIA/2004-18, IRIDIA, Université Libre de Bruxelles, March 2004.
13. V. Trianni, S. Nolfi, and M. Dorigo. Cooperative hole avoidance in a *swarm-bot*. *Robotics and Autonomous Systems*, 2005. to appear.
14. R. Volpe, I.A.D. Nesnas, T. Estlin, D. Mutz, R. Petras, and H. Das. "CLARAty: Coupled layer architecture for robotic autonomy". Technical Report D-19975, JPL Technical Report, 2000.
15. I.A. Nesnas, A. Wright, M. Bajracharya, R. Simmons, T. Estlin, and Won Soo Kim. "CLARAty: An architecture for reusable robotic software". In *Proc. of SPIE Aerosense Conference*, Orlando, Florida, 2003.
16. R. Alur, T. Dang, J. Esposito, R. Fierro, Y. Hur, F. Ivanid, V. Kumar, I. Lee, P. Mishra, G. Pappas, and O. Sokolsky. Hierarchical hybrid modeling of embedded systems. In *Embedded Software, First Intern. Workshop*, LNCS 2211. Springer, 2001.
17. Th. Henzinger, M. Minea, and V. Prabhu. Assume-guarantee reasoning for hierarchical hybrid systems. In *Proc. of the Fourth International Workshop on Hybrid Systems: Computation and Control (HSCC)*, LNCS 2034. Springer, 2001.

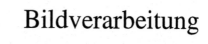

Bildverarbeitung

Kalman Filter based Detection of Obstacles and Lane Boundary in Monocular Image Sequences

Clemens Rabe, Christoph Volmer, Uwe Franke

DaimlerChrysler AG, 71059 Sindelfingen, Germany

E-mail: {clemens.rabe,uwe.franke}@daimlerchrysler.com,
christoph.volmer@stud.tu-ilmenau.de

Abstract. This paper presents a system for monocular obstacle and lane boundary detection running in real-time. A Kalman Filter based depth from motion algorithm is used for the reconstruction of the three-dimensional scene. Using multiple filters in parallel the rate of convergence is significantly higher than in direct methods, especially if the vehicle drives slowly. In addition a pitch correction is introduced which improves the overall estimation in typical road scenarios. Real world examples illustrate the results of the proposed system.

1 Introduction

Intelligent cars of the foreseeable future will be equipped with a camera for tasks such as lane departure protection and Night Vision. However, several applications in robotics and driver assistance such as obstacle detection, obstacle avoidance, and parking require 3D information. Figure 1 displays a lane boundary detection based on the three-dimensional scene reconstruction algorithm presented in this paper as an example for an eligible application in driver assistance.

If only one camera is available, depth must be estimated from the image sequence obtained while driving. Pollefeys [5] gives a comprehensive overview on structure from motion methods. These provide a very precise scene reconstruction but do not meet real-time requirements. Since the mentioned applications require 3D information for as many image points as possible in order not to overlook an important object, fast methods are needed.

Fig. 1. Successfull lane boundary detection using a monocular image sequence.

In systems with very limited computational resources often a direct analysis of the optical flow is propagated. For example, Grünewald et al. calculate in [3] the flow in only one reference line and compare it to the expected flow of a ground plane. Every outlier is interpreted as part of an obstacle. Although this algorithm is fast and works well for the proposed scenarios, it highly depends on the quality of the optical flow algorithm. In case of a slow observer movement direct optical flow methods tend to have difficulties due to small displacement vectors in the image resulting in a small signal-to-noise ratio.

Therefore, an integration of multiple measurements over time promises more robust results with respect to measurement noise. Matthies et al. describe in [4] a Kalman Filter based method to estimate the depth of selected image features for a strict lateral observer movement. Carlsson extends this approach and presents in [1] an algorithm to estimate the observer motion as well as the scene structure using a single Kalman Filter. However, without any additional constraints on the motion orscene structure this can only be solved up to an unknown scale factor.

But if the camera motion is known, in vehicles we know at least speed and yaw rate, the 3D-position of stationary world points can be efficiently estimated using one Kalman Filter per image feature. This reduces the complexity of the calculation in contrast to one global Kalman Filter. In addition, the iterative refinement of the Kalman Filter is optimal with respect to computation time. This raises the hope that a real-time estimation of depth is possible which is robust with respect to measurement noise, especially in case of a slow observer movement.

In this paper we present a system for obstacle and lane boundary detection running in real-time. Image features are tracked using the Kanade-Lucas-Tomasi (KLT) tracker [6]. Kalman Filters estimate the 3d-positions under the assumption of a stationary world. In order to get a high rate of convergence we use a multiple filter approach [2]. Using the Kalman Filter prediction as an initialization of the tracker, its speed and accuracy are improved. To detect obstacles the heights of the 3D-points are analysed. As a pitching error directly induces a wrong height estimation, we added a pitch correction to gain more robust estimates.

The paper is organized as follows. Chapter 2 describes the basic principles of the Kalman Filter based depth reconstruction. The pitch correction is outlined in chapter 3. The obstacle and lane boundary detection is described in chapter 4. Results for real sequences are finally presented in chapter 5, followed by the conclusion in chapter 6.

2 Kalman Filter based 3D from Motion

Assuming that a tracked image feature corresponds to a stationary world point a Kalman Filter is used to estimate its 3d-position. In the following we use a right handed coordinate system with the origin at the road. The lateral x-axis points to the left, the height axis y points upwards and the z-axis representing the distance of a point is straight ahead. This coordinate system is fixed to the car, so that all estimated positions are given in the coordinate system of the moving observer. The camera is at $(x, y, z) = (0, height, 0)$.

2.1 System Model

The movement of a vehicle with constant velocity v and yaw rate $\dot{\psi}$ over the time interval Δt, measured by inertial sensors, can be described in this car coordinate system by the translation vector $\Delta \underline{x}_c$ and the rotational matrix R_y around the y-axis. The position of a static world point $\underline{x} = (X, Y, Z)^T$ after the time Δt can be described in the car coordinate system at time step k as

$$\underline{x}_k = R_y \left(\underline{x}_{k-1} - \Delta \underline{x}_c \right). \tag{1}$$

This yields to the discrete system model equation

$$\underline{x}_k = A_k \underline{x}_{k-1} + B_k v + \underline{w}_{k-1} \tag{2}$$

with the state transition matrix $A_k = R_y$ and the control matrix

$$B_k = \frac{1}{\dot{\psi}} \begin{pmatrix} 1 - \cos(\dot{\psi}\Delta t) \\ 0 \\ -\sin(\dot{\psi}\Delta t) \end{pmatrix}. \tag{3}$$

The noise term $\underline{w}$ is assumed to be a gaussian white noise with covariance Q.

2.2 Measurement Model

Image coordinates u and v of a feature are measured using an appropriate point tracker. In our current implementation we use a Kanade-Lucas-Tomasi tracker [6]. Assuming a pin hole camera the nonlinear measurement equation for a point given in the camera coordinate system is

$$\underline{z} = \begin{pmatrix} u \\ v \end{pmatrix} = \frac{1}{Z} \begin{pmatrix} X f_u \\ Y f_v \end{pmatrix} + \underline{v} \tag{4}$$

with the focal lengths f_v. The measurement noise term $\underline{v}$ is also assumed to be a gaussian noise with covariance R.

2.3 Initialization

The Kalman Filter given by the equations 2 and 4 estimates the world position of a static point in relation to the moving car. Before the filter can begin its work, it has to be initialized. Using the first two measurements an initial position can be calculated by triangulation. However, this is highly susceptible to measurement noise, especially if the measured displacement vector is small. Therefore, we get the initial position as the intersection of the focal ray with a pre-defined plane, e.g. the ground plane.

This initial guess will be refined by the filter over time. The more the first guess deviates from the correct value, the longer it takes until the estimate error is below a given threshold. To achieve a higher rate of convergence, multiple filters are used in parallel initialized using different pre-defined planes. Each filter represents a hypothesis in the state space of one world point. All hypothesis are then combined in a weighted average using the innovation errors as a goodness-of-fit criterium [2].

In a running system new features are continuously added. Under the asumption that nearby image features are laos closely related in the world an additional hypothesis can be formulated using the already existing estimation results of the neighbours. Therefore, a fixed area around the new feature is searched and an average world position is calculated. For each found feature its estimated position is weighted with the covariance and the distance to the new feature in the image.

3 Pitch Correction

The Kalman Filter provides an estimate which is robust against gaussian noise in the measurement. However, a camera orientation change, such as pitch, yaw or roll, which is not measured by additional sensors, has a dramatic influence on the whole estimation process. Therefore, the correction of such errors is a fundamental step to improve the 3D-reconstruction.

In our scenarios the main component is pitching. Therefore we developed a pitch correction which takes place between the Kalman Filter prediction and measurement update step. Assuming a correct estimated 3D-point the predicted and the measured image position should be identical. However, any remaining innovation in vertical direction is interpreted as a pitch movement. To be robust against outliers we average the pitch deviations of all features. As the state covariance gives the uncertainty of the 3D-estimation, we use it as a weighting factor. The resulting pitch angle is used in the Kalman Filter measurement update step for all tracked features.

The benefit of the pitch correction can be seen in the right diagram of Fig. 2. It shows the estimated 3D-points in a top view. Without correction the pitching in this sequence causes wrong estimates, shown as green boxes. Especially points at a large distance tend to have serious problems. Here, the expected innovation caused by the forward motion is very small in contrast to the innovation due to pitching. On the other hand, the 3D-reconstruction using the pitch correction leads to stable estimation results, indicated by red crosses. This significantly improves obstacle and lane boundary detection.

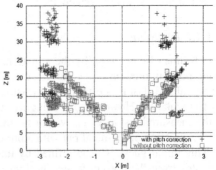

Fig. 2. 3d-estimation stabilized by pitch correction (left). The warmth of the color indicates the estimated height of the tracked image features. The improvement of the pitch correction is illustrated by the top view (right).

4 Obstacle and Lane Boundary Detection

Assuming a planar road all estimated world points with a significant height above the ground are interpreted as part of obstacles. For a robust object detection all these points are mapped into a two dimensional histogram according to their x-z-coordinates and a connected component analysis is performed to identify the obstacles.

The two dimensional histogram is also used to determine the lateral boundary of the free corridor in front of the vehicle. We start on the vehicles lateral position and search in each line for the left and right boundary respectively. Since we are interested to locate the first significant slope the derivative of this one dimensional distribution is calculated. The first maximas to each side in the derivative function now correspond to the lateral position of the boundary.

5 Real World Results

In this final section we discuss three real-live examples. First, a sequence of 60 frames containing a static obstacle is investigated. The sequence was taken from a truck driving at about $1\frac{m}{s}$ towards a standing person. Due to this very slow speed, the lengths of the optical flow vectors are small and the integration performed by the Kalman Filter is of paticular importance.

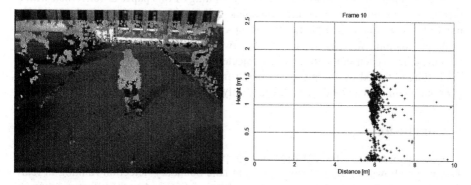

Fig. 3. Estimation result for the person at frame 10 (driven distance 40 cm). The warmth of the color encodes the estimated height.

Figure 3 shows on the left the estimated heights after 10 frames (i.e. 40 cm driving distance) qualitatively. We illustrate the height with the warmth of the color. The estimation of about 5000 image features was performed by three Kalman Filters per feature, initialized on the heights 0, 1 and 2 m. On the right the corresponding 3D-reconstruction result is given in a lateral view.

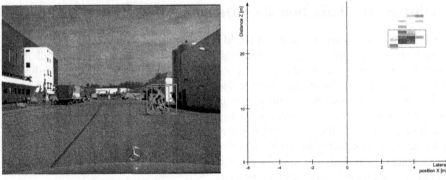

Fig. 4. Result of the obstacle detection. The two dimensional histogram on the right clearly shows the standing bicyclist as an obstacle.

The result of the obstacle detection method is illustrated by Fig. 4. Here we drive with a passenger car at about $3{,}5\mathrm{m}\frac{m}{s}$ towards a standing bicyclist. The purpose of this experiment is to stop the vehicle in case of an obstacle. The right figure shows the two dimensional histogram used for the obstacle detection with a quantisation of 0,5 m. The result of the closed component analysis is given by the red rectangle and is reprojected as a green box in the camera image.

The last example shown in Fig. 1 at the beginning of this paper is concerned with a typical construction site. Here, the actual lane is limited by beacons and other vehicles in a traffic jam. In contrast to the first examples, ourcar is driving at about $10\frac{m}{s}$.

After mapping the 3D-points into the two dimensional histogram, the 3D-lane boundary have been computed and reprojected into the image. The red fences show the results obtained by the presented depth-from-motion algorithm. This proves that 3D road boundary can be detected even if only one camera is installed in the vehicle.

6 Conclusion

A robust way to solve the 3D-from-motion problem is the usage of Kalman Filters. This paper shows a system for obstacle and lane boundary detection running in real-time. The proper combination of multiple filters initialized with different states as well as the proposed initialization based on the neighbourhood speed up the rate of convergence. The presented pitch correction stabilises the 3d-reconstruction on typical road scenes significantly. In our current implementation, the 3D-positions of 1100 points are calculated on a 3 GHz Pentium 4 at 15 Hz. This includes the image acquiring process, the tracking of the features on images with QVGA resolution, the Kalman Filter estimation using three filters per feature and the obstacle and lane boundary detection.

References

1. Carlsson, S.: Recursive Estimation of Ego-Motion and Scene Structure from a Moving Platform. Proceedings of the 7th Scandinavian Conference on Image Analysis, Aalborg, pp. 958–965, 1991
2. Franke, U. and Rabe, C.: Kalman Filter based Depth from Motion with Fast Convergence. Proceedings of the IEEE Intelligent Vehicles Symposium 2005, Las Vegas, NV, June 2005
3. Grünewald, M. and Sitte, J.: A Resource-Efficient Approach to Obstacle Avoidance via Optical Flow. Proceedings of the 5th International Heinz Nixdorf Symposium: Autonomous Minirobots for Research and Edutainment, Paderborn, pp. 205–214, October 2001
4. Matthies, L. and Kanade, T. and Szeliski, R.: Kalman Filter-based Algorithms for Estimating Depth from Image Sequences. International Journal of Computer Vision, Nr. 3, pp. 209–236, 1989
5. Pollefeys, M.: Self-Calibration and metric 3D reconstruction from uncalibrated image sequences. Leuven, 1999
6. Tomasi, C. and Kanade, T.: Detection and Tracking of Point Features. Technical Report CMU-CS-91-132, School of Computer Science, Carnegie Mellon University, Pittsburgh, PA, April 1991

Komponentenbasierte Bildanalyse zur Identifikation von Objektkategorien

Florian Bley, Karl-Friedrich Kraiss

Lehrstuhl für Technische Informatik,
Rheinisch-Westfälische Technische Hochschule Aachen (RWTH),
Ahornstr. 55, 52074 Aachen

E-mail: {bley,kraiss}@techinfo.rwth-aachen.de

Zusammenfassung. Autonome mobile Greifarme benötigen zur sichtgeführten Manipulation komplexer Gegenstände Vorwissen. Dies kann beispielsweise die exakte geometrische Form der zu manipulierenden Objekte oder die genaue Lage markanter Punkte auf deren Oberfläche sein. Die Handhabung von unbekannten Objekten erfordert einen vorgeschalteten Analyseschritt, bei dem die Form des Objektes erkannt und ein optimaler Griff ermittelt wird. Wissenschaftliche Untersuchungen in den 80er Jahren propagierten, dass das menschliche Gehirn komplexe Objekte erkennt, indem es sie in elementare Teilkomponenten zerlegt. Der vorliegende Beitrag nutzt diese Erkenntnisse zur automatischen Bildanalyse, mit deren Hilfe ein mobiler Manipulator in die Lage versetzt wird, unbekannte Objekte nach einem Rekonstruktionsschritt zu klassifizieren und zu greifen.

1 Einleitung

Die Anwendungsbereiche der Robotik reichen von der industriellen Fertigung bis zu mobilen Assistenzsystemen im häuslichen Umfeld und verheißen hohes Potenzial für zukünftige Absatzmärkte. Die autonome Interaktion mit der Umgebung stellt eine wichtige Funktionsklasse für diese Systeme dar. Für einen erfolgreichen Einsatz in dynamischen Umgebungen ist die korrekte Wahrnehmung der Umwelt von entscheidender Bedeutung. Autonome Manipulatoren werden oft mit einer so genannten *eye in hand* Konfiguration ausgestattet, bei der ein kompakter optischer Sensor auf den Greifer des Manipulators montiert wird. Dieser ermöglicht die sichtgeführte Manipulation von Gegenständen. Eine aktuelle Entwicklungen dieser Art, die im häuslichen Umfeld eingesetzt werden kann, wird in [1] beschrieben. Der Manipulator greift jedoch nur solche Objekte, die aus einfarbigen Quadern zusammengesetzt sind. Der in [2] beschriebene mobile Manipulator erfasst Objekte rein farbbasiert, setzt aber eine rotationssymmetrische Form voraus, die einen Griff aus jeder Richtung ermöglicht. Zudem wird die Position der Objekte zusätzlich durch einen drucksensitiven Untergrund erfasst. Um die Alltagstauglichkeit und damit die Vermarktungsfähigkeit der referenzierten Systeme zu erhöhen, müssen diese auch für komplexe Geometrien in unbekannter Lage einen stabilen Griff planen können.

Im folgenden Abschnitt werden zunächst gängige Methoden zur Ermittlung der Objektgeometrie aufgeführt. Der Abschnitt 3 verdeutlicht, wie komplexe Objekte in einfache Grundkörper zerlegt werden können und wie mit Hilfe eines semantischen

Netzes Objektkategorien beschrieben werden. Im Abschnitt 4 wird die Implementation und der Versuchsaufbau beschrieben. Die Zusammenfassung zieht ein kurzes Fazit und gibt zusätzlich einen Ausblick auf weitere Entwicklungsschritte.

2 Ermittlung der Objektgeometrie

Eine Möglichkeit zur Ermittlung der genauen Objektform ist deren Erfassung und Speicherung in einem Vorverarbeitungsschritt. Diese Methode kommt bei der *ansichtenbasierten* ([3], [4]) und der *modellbasierten Erkennung* [5] zur Anwendung. Das Manipulationssystem wird hierbei allerdings in seiner Flexibilität eingeschränkt, da

- die Daten für jedes einzelne Objekt mit hohem Aufwand in eine Datenbank eingepflegt werden müssen und
- das System zur Laufzeit nur mit den zuvor eingepflegten Objekten interagieren kann.

Auch bei der Beschränkung der Interaktion mit einer bestimmten Objektkategorie (z.B. „Tassen") müssen alle Instanzen dieser Kategorie in die Datenbank aufgenommen werden. Bei der hohen Diversifikation in Farbe und Form (siehe Abb. 1) ist dies ein unvertretbar hoher Aufwand.

Abb. 1. Mehrere Objekte einer Kategorie. Trotz unterschiedlicher Größen und Seitenverhältnissen können alle Formen durch einen Zylinder und einen seitlichen halben Rundbogen angenähert werden.

Eine zweite Möglichkeit besteht in der Erfassung der Geometrie zur Laufzeit. Hier wird der Gegenstand von allen Seiten mit Sensoren abgetastet und seine Form rekonstruiert. Mit dem in [6] beschriebenen System können einfache geometrische Grundformen wie Würfel erkannt und in ein Gittermodell überführt werden. Der beschriebene Versuchsaufbau verwendet jedoch fest platzierte Kameras zur Bildakquise. Obwohl das System für einen humanoiden Roboter konzipiert wurde, werden keine Angaben darüber gemacht, ob die beschriebene Objektrekonstruktion erfolgreich zur Interaktion mit Objekten eingesetzt wurde.

Auch wenn das Rekonstruktionsverfahren prinzipiell eine hohe Flexibilität verspricht, so ist es aufgrund der Hardwareeinschränkungen in der mobilen Robotik nur schwer einsetzbar. Die Sensorik muss auf der beweglichen Plattform angebracht werden. In den meisten Fällen wird eine Kamera auf dem Greifer des Manipulators montiert, die per *Active Vision* zur Szenenanalyse eingesetzt wird. Die relativ hohe Positionierungenauigkeit von kommerziell verfügbaren und preislich attraktiven Manipulatoren erschwert die Rekonstruktion jedoch erheblich.

3 Beschreibung von Objektkategorien

Der hier verfolgte Ansatz besteht aus einem Mittelweg zwischen der Erkennung bereits bekannter Objekte und der Rekonstruktion unbekannter Geometrien. Dabei wird die Tatsache ausgenutzt, dass die für die Interaktion interessanten Objekte oft in begrenzten funktionalen Kategorien zu finden sind (z.B. „greifbare Gegenstände aus dem häuslichen Umfeld"). Statt die geometrische Form oder das Aussehen von einzelnen Objekten in einer Datenbank abzulegen, wird Konzeptwissen über eine Objektkategorie mit Hilfe eines semantischen Netzes gespeichert. Semantische Netzen wurden bereits zur reinen Funktionsbeschreibung von Objekten verwendet [7], häufiger jedoch um komplexe Szenen zu beschreiben (z.B. [8],[9]). Wie in Abb. 2 gezeigt, umfasst das hier verwendete Netz Wissen über Geometrie (die Form der einzelnen Grundkörper, erwartete Verhältnisse zwischen Höhe und Breite sowie topologische Beziehungen zwischen den Grundkörpern), Regeln für die Handhabung und mechanische Eigenschaften wie Gewicht und Schwerpunktlage.

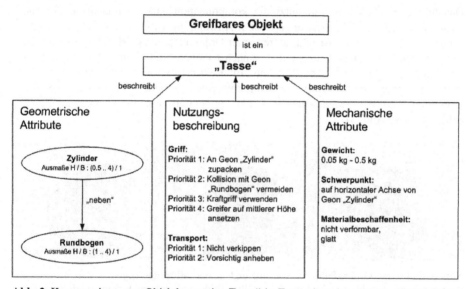

Abb. 2. Konzeptwissen zur Objektkategorie „Tasse" in Form eines semantischen Netzes. Die geometrische Beschreibung setzt sich aus elementaren Komponenten, so genannten Geons, nach [10] zusammen. Anhand der Beschreibungen kann für die meisten Instanzen der Kategorie ein stabiler Griff geplant werden.

Es ist daher möglich, Vorwissen über die Geometrie der zu erwartenden Objekte zum Ausgleichen von Rekonstruktionsfehlern zur verwenden. Die bislang verwendete Objekterkennung wird somit zu einer Objektkategorisierung, bei der ein erkanntes Objekt einer bestimmten Kategorie zugeordnet wird.

Um die Erweiterung der Datenbank einfach zu gestalten, werden keine Punkt- oder Gittermodelle gespeichert, sondern nur die Grundkörper aus denen die Objektklasse besteht. Die diesem Ansatz zugrunde liegende Theorie wurde Ende der 80er Jahre þvon Biederman [10] entwickelt. Seine Untersuchungen legen nahe, dass Menschen Objekte nicht durch mentalen Vergleich mit blickrichtungsabhängigen 2D-Ansichten wieder erkennen, sondern diese mit einer dreidimensionalen Form abgleichen. Weiterhin zeigen seine Untersuchungsergebnisse, dass Menschen komplexe Objekte auf gleiche Art und Weise in Untereinheiten zerlegen. Für diese volumetrischen Grundprimitive prägt er den Namen Geons. Sie zeichnen sich dadurch aus, dass sie sich von den meisten Blickwinkeln aus leicht voneinander unterscheiden lassen und ihre Identifikation unempfindlich gegenüber Rauschen ist. Biedermans Theorie wurde mehrfach in reale Objekterkennungssysteme umgesetzt ([11],[12],[13]). Dabei hielten sich die jeweiligen Arbeitsgruppen streng an die von Biederman postulierte Kantendetektion zur Erkennung und Klassifizierung der Geons. Die Robustheit der Algorithmen hing damit stark von einfach zu detektierenden Kanten zwischen den einzelnen Geons ab. Diese Voraussetzung ist unter realen Bedingungen allerdings nicht immer gegeben und erschwerte den Einsatz in der Praxis.

4 Bildanalyse

Das hier entwickelte System nutzt die komponentenbasierte Bildanalyse in abgewandelter Form. Das Ziel der angestrebten Anwendung ist es, zuvor unbekannte Instanzen einer begrenzten Anzahl interessanter Objektklassen zu erkennen. In dem hier verwendeten Versuchsaufbau werden zunächst Gegenstände betrachtet, die nur aus einer geringen Anzahl Geons bestehen. Im Rekonstruktionsschritt wird zuerst ein grober dreidimensionaler Umriss des Objektvolumens gebildet. Dazu wird das Kamerabild mit dem *Watershed* Verfahren [14] segmentiert und danach eine Hüllkurve gebildet (siehe Abb. 3).

Abb. 3. Links das Originalbild der Kamera und in der Mitte das Ergebnis der Watershed-Segmentierung. Die im rechten Bild dargestellte Hüllkurve wird zur Rekonstruktion des Objektvolumens verwendet.

Aus den Hüllkurven verschiedener Ansichten wird in der Rekonstruktionsphase mit Hilfe von *Volume Intersection* [15] ein dreidimensionales Modell des Objektes erzeugt. Diese Methode ist relativ robust gegenüber Fehlern bei der Bestimmung des Blickwinkels und somit für den Einsatz in der mobilen Robotik geeignet. Es können auch Kameras verwendet werden, deren intrinsische Parameter unbekannt sind. Nach Durchführung dieser Schritte liegt ein ungenaues Volumenmodell des Objektes vor, dessen Oberfläche durch Ebenengleichungen und Schnittgeraden zwischen diesen Ebenen beschrieben werden. Zur Vereinfachung der späteren Berechnungen wird dieses Volumenmodell nun in eine Punktrepräsentation überführt (siehe Abb. 4).

In der darauf folgenden Identifikationsphase wird die Punktrepräsentation mit den in der Datenbank vorhandenen Referenzmodellen verglichen. Dazu werden im ersten

Schritt die größten Geons der Referenzen über die rekonstruierte Punktwolke gelegt und versucht, beide in Deckung zu bringen.

In der ersten Entwicklungsphase des Projektes ist das semantische Netz auf vier Modellkategorien beschränkt.

Tasse: Auf Kreisfläche stehendes Geon „Zylinder" neben Geon „Rundbogen"
Becher: Auf Kreisfläche stehendes Geon „Zylinder"
Eckige Verpackung: Geon „Quader"
Korb mit Henkel: Geon „Zylinder" unter Geon „Rundbogen"

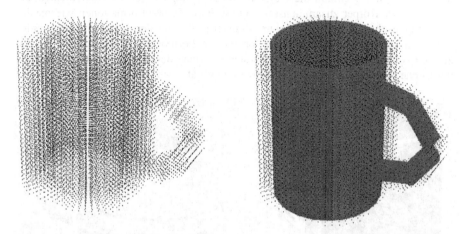

Abb. 4. Links die Punktrepräsentation des Objektes. In der Identifikationsphase wird versucht, einzelne Teilkomponenten möglicher Modellkategorien mit den Objektpunkten zur Deckung zu bringen (rechts).

Daher werden zuerst Zylinder und Quader in Deck gebracht, deren Kantenlängen solange vergrößert werden, bis sie über die Punktrepräsentation hinaus wachsen. Aus unterschiedlichen Faktoren wie dem Verhältnis zwischen überdeckten Punkten und der Gesamtzahl der Objektpunkte oder der Lage noch nicht überdeckter Punkte wird die Wahrscheinlichkeit für das hypothesierte Geon berechnet. Dieser wahrscheinlichkeitsbasierte Ansatz erlaubt auch die Erkennung von Komponenten, die durch Linsenverzerrungen oder Fehlern bei der Rekonstruktion deformiert wurden.

Unter Berücksichtigung der zuvor ermittelten Wahrscheinlichkeiten wird nun die Existenz von benachbarten Teilkörpern untersucht. Dazu ermittelt der Algorithmus zunächst wo sich die größte Anhäufung nicht überdeckter Punkte befindet. Diese Punktcluster müssen möglichst vollständig mit den möglichen Nachbargeons in Deckung gebracht werden. Zur Vereinfachung wird der Rundbogen hierbei durch vier Zylinder angenähert, bei denen sowohl Größe als auch Ausrichtung im Raum variieren kann. (siehe Abb. 4).

Eine Modellkategorie gilt als identifiziert, wenn alle im semantischen Netz enthaltenen Geons mit einer Mindestwahrscheinlichkeit gefunden wurden. Nach erfolgter Identifizierung werden aus der Datenbank Informationen über mögliche Angriffsflächen auf dem Gegenstand ausgelesen. Danach berechnet das System Trajektorien von der aktuellen Greiferposition zu möglichen Zielorten und überprüft, ob der Manipulator wenigstens eine dieser Trajektorien abgefahren kann. Sollte dies nicht der Fall sein, bleibt dem System nur die Option durch Rangieren der mobilen Plattform eine geeignetere Greifposition einzunehmen.

5 Zusammenfassung

Das erläuterte Verfahren zeigt einen Weg auf, mit dem sich Objekte durch allgemeine Beschreibung auf Zugehörigkeit zu einer Objektkategorie prüfen lassen. Diese allgemeine Beschreibung enthält die Objektform durch Angabe von Grundelementen, aus denen sich der Körper zusammensetzt, und weitere Zusatzinformationen wie mechanische Eigenschaften und Handhabungshinweise, die bei der späteren Manipulation berücksichtigt werden. Die unscharf formulierte Formbeschreibung erlaubt die Erkennung verschiedener Objekte einer Kategorie ohne detaillierte Merkmalsbeschreibung (siehe Abb. 5) und erhöht damit die Flexibilität des Systems.

Abb. 5. Die erfolgreiche Deckung der Topologie „Zylinder neben Rundbogen" führt zu der Annahme, dass die beiden betrachteten Objekte zur Kategorie Tasse gehören.

Literaturverzeichnis

1. Matsikis A.: Bildgestütztes Teach-In eines mobilen Manipulators in einer virtuellen Umgebung. Disseration, RWTH Aachen, 2005.
2. Volosyak I, Kouzmitcheva O, Ristic D, Gräser A: Improvement of Visual Perceptual Capabilities by Feedback Structures for Robotic System FRIEND. IEEE Transactions on Systems, Man and Cybernetics, 2005.
3. Kragic D, Christensen H: Robust Visual Servoing. The International Journal of Robotics Research, Vol. 22, 2003.
4. Leibe B, Schiele B: Analyzing Appearance and Contour Based Methods for Object Categorization. International Conference on Computer Vision and Pattern Recognition, 2003.
5. Borges D: 3D Recognition by Parts: A Complete Solution using Parameterized Volumetric Models. IX Simpósio Brasileiro de Comp. Gráfica e Proc. de Imagens (SIBGRAPI), 1996.
6. Beyer U, Smieja F: A model-based approach to recognition and measurement of partially hidden objects in complex scenes. GMD technical Report, 1996.

7. Stark L, Bowyer K: Achieving Generalized Object Recognition through Reasoning about Association of Function to Structure. IEEE Transactions on Pattern Analysis and Machine Intelligence, 1991.
8. Growe S, Liedtke C, Pakzad K: A Knowledge Based Approach to Sensor Fusion Applied to Multisensory and Multitemporal Imagery. Fourth Int. Airborne Remote Sensing Conference, 1999.
9. Kasprzak W: Adaptive Computation Methods in Digital Image Sequence Analysis. Elektronika series, vol. 127/2000, Warsaw University of Technology Press, 2000.
10. Biederman I: Recognition-by-components: a theory of human image understanding. Psychological Review, 1987.
11. Dickinson S, Pentland A, Rosenfeld A: 3-D Shape Recovery Using Distributed Aspect Matching.IEEE Transactions on Pattern Analysis and Machine Intelligence, vol. 14, no. 2, pp. 174-198, 1992.
12. Bergevin D, Levine M: Generic Object Recognition: Building and Matching Coarse Descriptions from Line Drawings. IEEE Transactions on Pattern Analysis and Machine Intelligence, vol. 15, no. 1, pp. 19-36, 1993.
13. Pilu M: Part-based Grouping and Recognition: A Model-Guided Approach. PhD Thesis, University of Edinburgh, 1996.
14. Moya J: Segmentation of color images for interactive 3d object retrieval. Ph.D. dissertation, RWTH Aachen University, 2004.
15. Martin W, Aggarwal J: Volumetric descriptions of objects from multiple views. IEEE Transactions on Pattern Analysis and Machine Intelligence, vol. 5, no. 2, pp. 150-158, 1983.

Gesichtsanalyse für die intuitive
Mensch-Roboter-Interaktion

Hans-Joachim Böhme, Torsten Wilhelm, Horst-Michael Groß

Technische Universität Ilmenau, FG Neuroinformatik und Kognitive Robotik
PF 00565, 98684 Ilmenau

E-mail: {hans-joachim.boehme,torsten.wilhelm,horst-michael.gross}@tu-ilmenau.de

Zusammenfassung. Der Beitrag beschreibt Verfahren zur visuellen Gesichtsanalyse im Kontext der Mensch-Roboter-Interaktion. Dabei besteht das mittel-
fristige Ziel darin, neben der Identifizierung des aktuell angemeldeten Nutzers
auch dessen Geschlecht, das Alter sowie den Gesichtsausdruck zu ermitteln und
damit über möglichst reichhaltige Statusinformationen über den menschlichen
Interaktionspartner zu verfügen, um seitens des Roboters die Dialogführung
entsprechend anpassen zu können. Um diese Informationen aus dem Gesichts-
bild zu extrahieren, werden die Verfahren Elastic Graph Matching, Independent
Component Analysis sowie Active Appearance Models in Verbindung mit verschiedenen Klassifikatoren vergleichend untersucht, wobei hier zunächst der
Schwerpunkt auf der Personenidentifikation und der Geschlechtsschätzung liegt.
Anschließend wird die Nutzung dieser Informationen anhand der Dialogführung eines interaktiven Shopping-Assistenten diskutiert.

1 Einleitung

Die hier vorgestellten Forschungsergebnisse ordnen sich ein in die Entwicklung eines
interaktiven mobilen Shopping-Assistenten für den Einsatz in einem Baumarkt [3].
Neben der Detektion und dem kontinuierlichen Verfolgen von Personen in der
Einsatzumgebung eines mobilen Assistenzsystems ist es auch in diesem Szenario
wünschenswert, detailliertere Informationen über den aktuellen Interaktionspartner zu
erhalten, die den Roboter befähigen, im Interaktionsprozess diese Informationen zu
nutzen und bspw. den Dialog entsprechend zu adaptieren. So konnte nachgewiesen
werden, dass sich beispielsweise Frauen beim Einkauf völlig anders verhalten als
Männer. Auch das Wissen darüber, ob es sich beim aktuellen Interaktionspartner nach
wie vor um die gleiche Person handelt, ist insbesondere bei kurzzeitigen Unterbre-
chungen der Interaktion sehr hilfreich. Im konkreten Einsatzszenario besteht die besondere Herausforderung darin, dass die eingesetzten Verfahren ihre Robustheit unter
allen denkbaren Widrigkeiten hinsichtlich Beleuchtung, Szenenhintergrund, unterschiedlichen Kopfposen etc. beweisen müssen.

2 Datenmaterial

Für die zu erstellende Datenbasis bestand die Anforderung, ein möglichst breites
Spektrum an Informationen (Geschlecht, Altersgruppen, verschiedene Mimiken) zu

beinhalten. Der *Neutral-Datensatz* umfasst 70 Personen (Gleichverteilung hinsichtlich Alter und Geschlecht, Gesichter mit frontaler und um 5 Grad in horizontaler Richtung gedrehter Ansicht, neutralem Gesichtsausdruck und verschiedenen Beleuchtungen). Für die Erstellung des Mimik-Datensatzes wurden Probanden gebeten, sieben Basisemotionen (Neutral, Freude, Überraschung, Wut, Angst, Ekel, Trauer) darzustellen. Um Bilder mit unglaubhaft dargestellten Gesichtsausdrücken auszuschließen, erfolgte nach der Aufnahme eine Befundung der Bilder durch 10 Personen. Der Mimik-Datensatz umfasst 30 Personen (Gleichverteilung hinsichtlich Alter und Geschlecht).

3 Verfahren zur Gesichtsanalyse

Für die Gesichtsanalyse haben sich insbesondere das Elastic Graph Matching, die Independent Component Analysis sowie die Active Appearance Models als favorisierte Methoden herauskristallisiert.

3.1 Elastic Graph Matching (EGM)

Anwendungen des EGM sind hauptsächlich aus dem Bereich Personenidentifikation [6], jedoch zunehmend auch für die Problemkreise Geschlechtsschätzung und Mimikanalyse, bekannt. Für die Erstellung des Modells werden Beispielbilder mit Gabor-Wavelets unterschiedlicher Frequenz und Orientierung gefaltet und anschließend an markanten Punkten die Filterantworten zu Jets zusammengefasst. Mehrere Jets zusammen bilden einen Face-Graph, der eine Beschreibung für jeweils ein Gesicht darstellt. Eine genügend große und vielfältige Auswahl an so erstellten Graphen bildet die sogenannte General-Face-Knowledge (GFK). Der Average-Graph, eine Art Durchschnittsgesicht, wird erstellt, indem über alle Beispielgraphen der GFK gemittelt wird. Für die Analyse eines unbekannten Eingangsbildes (Erzeugung eines Bildgraphen) wird nach erfolgter Gaborfilterung der Average-Graph im Bild positioniert und so lange lokal angepasst, bis die Ähnlichkeit der Jets des Average-Graphen und der Jets des Bildgraphen maximal wird. Im Falle der Identifikation erfolgt anschließend der Vergleich mit einer Galerie, die die Graphen aller zu prüfenden Personen enthält. Im Falle der Geschlechtsschätzung werden für jeden Knoten des Bildgraphen die n ähnlichsten Jets aus der GFK ermittelt. Anschließend wird überprüft, ob diese ähnlichsten Jets mehrheitlich aus Graphen von Männern oder Frauen innerhalb der GFK stammen. Daraus wird dann die Entscheidung abgeleitet, ob es sich bei der aktuell vorliegenden Person um einen Mann oder um eine Frau handelt.

3.2 Independent Component Analysis (ICA)

Werden Bilder, also zweidimensionale Signale betrachtet, müssen diese für die Anwendung der ICA zunächst in eine eindimensionale Darstellung überführt werden. Die Bilder werden als Zeilen in einer Beobachtungsmatrix X angeordnet, wobei jede Spalte von X als Beobachtungsvektor, also als Resultat eines Zufallsexperiments, aufgefasst wird [1]. Ausgehend von m Bildern der Größe $x \times y$ werden jeweils die i-ten Pixel aller Bilder zu einem Beobachtungsvektor zusammengefasst (Bildraumdarstellung). Anschließend erfolgt für diese Matrix mittels des FastICA-Algorithmus die Berechnung der unabhängigen Basisbilder. Die Projektion des aktuellen Eingangsbildes auf die Achsen dieses neuen Basissystems liefert einen Merkmalsvektor, der anschließend klassifiziert wird. Für diesen letzten Verarbeitungsschritt wurden folgende Klassifikatoren untersucht: Nearest Neighbor (NN), Multilayer Perceptron (MLP), Radial Basis Function Network (RBF) sowie verschiedene Varianten der Learning Vector Quantization (LVQ).

3.3 Active Appearance Model (AAM)

Ein AAM besteht aus einem Form- und einem Grauwertmodell [2]. 116 Punkte entlang markanter Gesichtsbereiche bilden die Markierungspunkte für das AAM. Für die Modellerstellung wird zunächst anhand dieser Punkte ein mittleres Formmodell berechnet und die Variation der Markierungspunkte mittels PCA bestimmt. In nächsten Schritt erfolgt ein Warping der Trainingsbilder auf das mittlere Formmodell. Analog zum mittleren Formodell wir nun ein mittleres Grauwertmodell berechnet, woran sich die Bestimmung der Grauwertvariation wiederum via PCA anschließt. Abschließend wird eine Prädiktormatrix ermittelt, die den Effekt einer Parametervariation auf das Differenzbild zwischen formnormiertem Eingangsbild und synthetisiertem Grauwertbild beschreibt. Bei der Anwendung des Modells wird dann versucht, die Parameter des mittleren Modelles so zu verändern, dass das unbekannte Eingangsgesicht mit minimalem Fehler synthetisiert werden kann. Der dabei entstehende Parametervektor wird mit den o.g. Klassifikatoren abschließend klassifiziert. Alle genannten Modelle werden in Abb. 1 veranschaulicht.

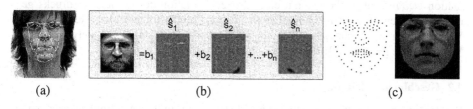

(a)   (b)   (c)

Abb. 1. Veranschaulichung der verwendeten Modelle zur Gesichtsdanalyse: (a) Gesicht mit eingezeichnetem Face Graph (EGM), (b) Darstellung des Eingangsbildes als Kombination unabhängiger Basisbilder (ICA), (c) Form- und Grauwertmodell des AAM.

4 Ergebnisse

4.1 Identifikation

Die vorgestellten Verfahren wurden anhand der beschriebenen Datenbasis vergleichend untersucht. Für die möglichst exakte Positionierung und Skalierung der Gesichter wurden die Augenpositionen mit einem eigens dafür trainierten Detektor (Adaboost [4]) automatisch ermittelt.

Beim EGM erfolgt ein Vergleich des Bildgraphen mit allen Galeriegraphen, während für ICA und AAM das normierte Skalarprodukt zwischen dem Modell des Eingangsgesichtes und den Modellen der Galerie als Ähnlichkeitsmaß herangezogen wurde.

Zur Beurteilung der Güte der verschiedenen Verfahren dienen die Equal Error Rates (EER), die sich aus dem Schnittpunkt der False-Acceptance-Rate(FAR)-Kurve und der False-Rejection-Rate(FRR)-Kurve ergeben, siehe Abb. 2.

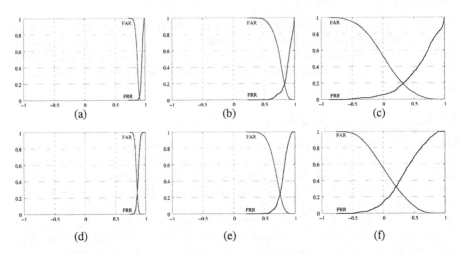

Abb. 2. FRR/FAR-Kurven bei der Personenidentifikation: (a) EGM, (b) ICA und (c) AAM auf dem Datensatz mit neutralen Gesichtsausdrücken; (d) EGM, (e) ICA und (f) AAM auf der Vereinigung von Neutral- und Mimik-Datensatz. Die Verwendung von Neutral- und Mimikdaten führt bei allen drei Verfahren zu einem signifikanten Anstieg der Equal-Error-Rates im Vergleich zur alleinigen Verwendung der Neutraldaten.

Die Verwendung von Neutral- und Mimikdaten kommt der realen Anwendungssituation weitaus näher, da hier mit einer großen Variation des Gesichtsausdrucks gerechnet werden muss. Aus Abb. 2 lässt sich somit schlussfolgern, dass sich das ICA-Verfahren für den praktischen Einsatz am besten eignet, da es im Vergleich aller Verfahren die geringsten EER-Werte erreicht.

4.2 Geschlechtsschätzung

Bei der Geschlechtsschätzung wurden neben den Tests auf dem Neutraldatensatz ebenfalls Tests auf der Vereinigung von Neutral- und Mimikdatensatz durchgeführt, um die Abhängigkeit der Erkennungsraten vom Gesichtsausdruck zu bestimmen, siehe Abb. 3.

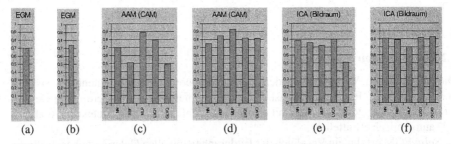

Abb. 3. Erkennungsraten bei der Geschlechtsschätzung (a) EGM (b) ICA und (c) AAM auf dem Datensatz mit neutralen Gesichtsausdrücken. Bei den AAMs wird die beste Erkennungsrate 93% mit dem MLP erreicht, während bei der ICA die NN- und LVQ-Klassifikatoren am besten abschneiden. Das EGM erreicht mit 74% nur eine relativ niedrige Erkennungsrate. (d) EGM (e) ICA und (f) AAM auf dem Gesamtdatensatz mit Neutral- und Mimikdaten.

Aus Abb. 3 wird ersichtlich, dass sich hier das AAM in Kombination mit einem MLP als Klassifikator für den praktischen Einsatz empfiehlt. Angesichts der im Vergleich zur ICA deutlich höheren Rechenzeit (durchschnittlich 2200 ms, ICA 21 ms, EGM 900 ms) wurde jedoch auch hier auf das ICA-Verfahren in Kombination mit einem LVQ1-Klassifikator zurückgegriffen.

Die bislang dargestellten Ergebnisse hinsichtlich Identität und Geschlecht beziehen sich ausschließlich auf die Verwendung einer abgeschlossenen Datenbasis. Damit sind weitergehende Aussagen zur Robustheit unter Umgebungsbedingungen, deren Variabilität über die in den Daten enthaltene hinausgeht, nicht möglich.

5 Praktische Anwendung im Dialogprozess

Die Trainings- und Testdaten enthalten frontale Ansichten mit out-of-plane Rotationen (horizontal) bis 5 Grad. Bei der in diesem Abschnitt diskutierten praktischen Anwendung werden in-plane-Rotationen durch eine Augendetektion und eine anschließende affine Transformation behandelt. Durch ein vorgeschaltetes Tracking-System kann sich die Person frei vor dem Roboter (Abb. 4) bewegen, wobei die Blickrichtung (wie in den Trainingsdaten) etwa frontal sein sollte.

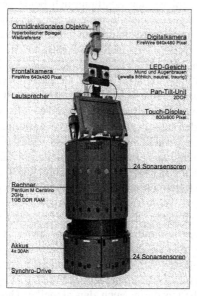

Abb. 4. Der Roboter PERSES ist mit einer omnidirektionalen Kamera, zwei frontal orientierten Kopf-Kameras, zwei Ebenen von Sonarsensoren, einem Touchdisplay und einem Gesicht, bestehend aus mehreren Arrays von Leuchtdioden, ausgestattet. Das Gesicht kann Emotionen wie Fröhlichkeit, Traurigkeit und Wut darstellen. Der Kopf, der neben dem Gesicht die beiden frontal ausgerichteten Kameras umfasst, wird mit Hilfe einer separaten Pan-Tilt-Unit bewegt.

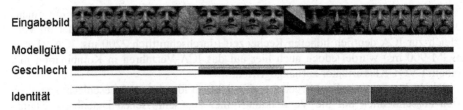

Abb. 5. Die obere Reihe zeigt die durch die Kopfkamera aufgenommenen und normalisierten Bilder. Die Modellgüte beschreibt die Ähnlichkeit zu dem mittleren ICA-Gesichtsmodell. Die Grauwerte kodieren die Ausgabe des Klassifikators: weiß steht für 0 und schwarz für 1. Für die Geschlechtsschätzung kodiert der obere Balken männlich, der untere Balken weiblich. Als aktueller Nutzer ist die erste dargestellte Person beim System angemeldet. Zum Beginn der Interaktion wird ein Modell mit drei Bildern für den aktuellen Nutzer erstellt, das im weiteren Verlauf zur Personenidentifikation herangezogen wird. Bei der Identitätsschätzung kodieren die Grauwerte die Hypothese, dass es sich um den aktuellen Nutzer handelt (schwarz: 1, weiß: 0).

Identifikation und Geschlechtsschätzung wurden bereits in den Dialogprozess integriert. Mittels eines peripheren Sehsystems (Omnikamera) werden potentielle Nutzer detektiert und durch das foveale Sehsystem (Kopfkamera) verifiziert und verfolgt [5]. Im Bild der Kopfkamera werden kontinuierlich beide genannten Informationen gewonnen. Abbildung 5 zeigt Ausschnitte aus einem entsprechenden Experiment. Die dargestellte Sequenz beinhaltet Ausschnitte aus einem längeren Experiment, in dem der aktuelle Nutzer sowie weitere Personen versuchten, mit dem Roboter zu interagieren. Um der Dynamik einer realen Mensch-Roboter-Interaktion Rechnung zu tragen, wird sowohl für die Identitäts- als auch für die Geschlechtsschätzung eine gleitende gewichtete Mittelwertbildung eingesetzt, wobei sich das Gewicht eines Klassifikationsergebnisses aus der Güte des dabei verwendeten Modells, das während der Interaktion online erstellt werden muss, ergibt. Dabei beschreibt die Modellgüte die Ähnlichkeit zu dem mittleren Gesichtsmodell. D. h. die Klassifikationsergebnisse von Modellen mit niedriger Güte haben entsprechend kleineren Einfluss auf das Gesamtergebnis. Weiterhin wird die Klassifikation des aktuellen Gesichtsbildes nur dann durchgeführt, wenn die Modellgüte eine vorgegebene Schwelle (Mindestgüte) überschreitet.

Erste Experimente in der realen Einsatzumgebung bestätigten, dass sich sowohl für die Identitäts- als auch für die Geschlechtsschätzung das ICA-basierte Verfahren als robust und praktisch einsetzbar erweist.

Quantitative Aussagen lassen sich aufgrund der noch zu geringen Zahl an Interaktionen jedoch noch nicht treffen. Neben den hier behandelten Fragestellungen besteht ein weiteres Ziel unserer Arbeiten darin, auch den aktuellen Gesichtsausdruck und das Alter des Interaktionspartners, ebenfalls mit den hier genannten Verfahren, zu schätzen.

Literaturverzeichnis

1. M.S. Bartlett. *Face image analysis by unsupervised learning*. Kluwer Academic Publishers, 2001.
2. Cootes, T.F., Edwards, G.J., and Taylor, C.J. Active appearance models. *Lecture Notes in Computer Science*, 1407:484–??, 1998.

3. Gross, H.-M., Boehme, H.-J., Key, J, and Wilhelm, T. The perses project – a visionbased interactive mobile shopping assistant. *Künstliche Intelligenz*, 4:34–36, 2000.
4. Viola, P. and Jones, M. Robust real-time object detection. In *International Journal of Computer Vision*, 2001.
5. Wilhelm, T., Böhme, H.-J., and Gross, H.-M. Looking closer. In *Proc. of the 1st Eur. Conf. on Mobile Robots (ECMR)*, pages 65–70. ZTUREK Research-Scientific Institute, 2003.
6. Wiskott, L. and Fellous, J.-M. Face recognition by elastic bunch graph matching. *Online Publikation, Research Project*, 1994.

Klassifizierungsaspekte bei der 3D-Szenenexploration mit einer neuen 2D/3D-Multichip-Kamera

Klaus Hartmann, Otmar Loffeld, Seyed Eghbal Ghobadi, Valerij Peters,
T.D. Arun Prasad, Arnd Sluiter, Wolfgang Weihs, Tobias Lerch, Oliver Lottner

Zentrum für Sensorsysteme (ZESS), Universität Siegen, Paul-Bonatz-Straße 9–11,
57076 Siegen

Zusammenfassung. Die hier präsentierten Arbeiten stehen im Zusammenhang mit der Entwicklung einer neuen 2D/3D-Kameratechnologie. Unter Verwendung eines neuen phasenmessenden Bildsensors und verschiedener Bildsensoren zur Erzeugung von 2D-Bildern entstehen verschiedene Kamerakonfigurationen. Für unsere Arbeiten verwenden wir derzeit die PMD-Bildsensoren. Mit Hilfe der System on Programmable Chip (SoPC) Technologie können alle Aspekte der verschiedenen Bildsensoren für eine Echtzeitanwendung berücksichtigt werden. Die in Entwicklung befindliche Plattform bietet neben der Echtzeitfähigkeit eine transparente Nutzung aller Daten und Parameter aus einer Anwendung heraus. Für den Bereich der autonomen Systeme erwarten wir wesentliche Fortschritte hinsichtlich der Bereitstellung von visuellen Informationen aus der Umgebung und der Möglichkeit bestehende Auswerteverfahren deutlich zu verbessern und mit neuen Algorithmen weitergehende Leistungsmerkmale zu erreichen. Als Basis für eine neue leistungsfähige Bilddatenauswertung verwenden wir eine modularisierte Partitionierung von hardwarebasierter Vorverarbeitung und der Fusionierung verschiedener Bildinformationen auf der nächsten Ebene. Dadurch sind insbesondere auch Pixelsynchronisierte Auswerteverfahren für das dynamische 3D-Sehen möglich.

1 Einleitung

So wie die Bedeutung des visuellen Systems durch den Vergleich zur menschlichen Wahrnehmung hergeleitet werden kann, ist es nahe liegend, ein entsprechendes bioanaloges Sichtsystem für künstliche Systeme vorzusehen. Ein wichtiger Aspekt bei einem technischen Auge ist z.B. die Bildstabilisierung in Verbindung mit einer Bewegungssteuerung. Für eine Fahrzeugkamera können über Inertialsensoren die Fahrzeugbewegungen detektiert und für die Bewegungssteuerung (Ausgleichsbewegungen) bereitgestellt werden. Dies ermöglicht dann die Blickrichtungsstabilisierung deren Vorgaben aus der entsprechenden Bildverarbeitung kommen. Primäre Zielsetzung ist die Erkennung und Verfolgung von besonders wichtigen Szeneninhalten oder die Abstandsregelung beziehungsweise die Kollisionsvermeidung. Beim Menschen werden verschiedene Signale des Auges und des Gleichgewichtsorgans für die Bildstabilisierung im Gehirn fusioniert um entsprechende Stellsignale an die Augenmuskulatur abzugeben (vestibulookulärer Reflex). Dabei ist festzustellen, dass hohe Geschwindigkeiten und Beschleunigungen auftreten. Bei Blicksprüngen und der Verfolgung von bewegten Objekten ergeben sich unterschiedlichen Bewegungsgeschwindigkeiten für das Auge. Dadurch ergeben sich zum Teil unterschiedliche Anforderungen an ein entsprechendes mechatronisches System und an die Bildsensorik. Bei Blicksprüngen können die Bewegungszeiten ausgeblendet werden, da die nächste Information an der neuen Blickposition erwartet wird. Für die Verfolgung von bewegten Objekten in einer Szene ist eine kurze Integrationszeit, eine daran angepasste Beleuchtungsstärke sowie eine möglichst große Tiefenschärfe zu fordern.

Abweichend von einem bioanalogen technischen Auge ist das Bemühen, zusätzlich 3D-Informationen aus den visuellen Daten als diskrete einzelne Werte zu ermitteln. Die vorgenannten beispielhaften Aspekte gelten prinzipiell für alle technischen Bilderfassungssysteme und sind hinsichtlich ihrer Anwendbarkeit als technisches Auge zu beachten. Für die 3D-Erfassung der Umgebung wäre somit eine möglichst kompakte Lösung n anzustreben. Wir befinden uns auf dem Weg, unter Berücksichtigung neuer phasenmessender Bildsensoren eine 2D/3D Kameratechnologie zu entwickeln, die neue Leistungsgrenzen definiert. Die Zielsetzung, die Technologie grundsätzlich auch als technisches Auge in kompakten autonom agierenden Systemen einsetzen zu können, ist für die Zukunft ein wichtiger Aspekt unserer Forschungsarbeiten. Die nachfolgenden Ausführungen können aufgrund des begrenzten Rahmens nur einen kleinen Überblick hinsichtlich einiger Zusammenhänge geben.

2 2D/3D-Kameratechnologie

Die 3D-Bildsensoren ermöglichen die direkte Erfassung von räumlich geordneten Informationen. unter Berücksichtigung einer speziellen Lichtquelle [3]. Da die 3D-Bildsensoren mit eine hinreichenden Qualität noch nicht solange am Markt verfügbar sind, stecken die Forschungen und Entwicklungen bezüglich einer 2D/3D-Kameratechnologie noch in den Anfängen.

2.1 3D-Kamera

Alle verfügbaren „reinrassigen" 3D-Bildsensortechniken erlauben prinzipiell die gleichzeitige Erfassung von Intensitäts-Bildern und der Erfassung von Laufzeitinformationen mit jedem einzelnen Pixel. Die heutigen Bildsensoren haben relativ große Pixel, um eine Mindestempfindlichkeit für praktische Anwendungen zu erreichen. Dies begrenzt massiv die laterale Auflösung solcher Bildsensoren. Die maximale Pixelanzahl liegt derzeit bei 160x120 Pixel. Neben der Qualität der 3D-Pixel, die im Vergleich zu moderner CMOS/CCD-Pixeln hinsichtlich der Intensitätsinformationen schlechter sind, ist die Begrenzung der 2D-Auflösung ein wesentlicher Nachteil bezüglich der Nutzung existierenden Bildverarbeitungsverfahren bei der Fusionierung von 2D- und 3D-Informationen.

Das Prinzip der verwendeten optischen Entfernungsmesssysteme beruht auf der Messung der Signallaufzeit des gesendeten Signals zum Empfängerpixel. Unter der Voraussetzung einer kontinuierlichen sinus- oder rechteckförmigen Modulation (cw - Modulation) ergibt sich mit der Modulationsfrequenz f_{mod} für den Abstand R unter der Annahme, dass sich die Lichtquelle am selben Ort wie der Empfänger befindet und c für die Lichtgeschwindigkeit steht, folgender Zusammenhang:

$$R = \frac{c \cdot \Delta\varphi}{4 \cdot \pi \cdot f_{mod}} \tag{1}$$

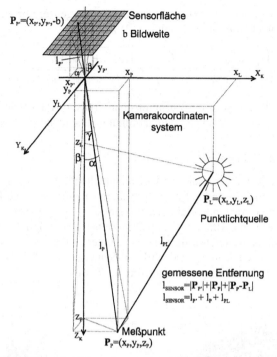

Abb. 1. Bestimmung des Abstands zwischen Lichtquelle und Bildsensorpixel

Der Phasenunterschied $\Delta\varphi$ ergibt sich aus der Laufzeit des Lichtsignals zum Objekt und zurück zum Empfänger. Der Bildsensor ermittelt einen Wert, mit dem $\Delta\varphi$ bestimmt werden kann.

Es ist zu berücksichtigen, dass der Abstand zu einem 3D-Raumpunkt nur durch eine Kalibrierung in Verbindung mit einer Beleuchtungsquelle, die zum Bildsensor eine feste definierte Position hat, ermittelt wird. Bei einer ausgedehnten, flächigen Beleuchtungsquelle und einer strukturierten Objektoberfläche ergeben sich unterschiedliche Abstandsdaten, da keine genaue Zuordnung von Punktlichtquelle und dem jeweiligen phasenmessenden Pixel existiert. Die nachfolgende Darstellung gibt einen Zusammenhang für die Ermittlung der Entfernungsbestimmung einer Punktlichtquelle und einem phasenmessendem Pixel des Bildsensors, wobei diese Zusammenhänge auch sehr starke Parallelen zu bistatischen Aufnahmesituationen beim Synthetic Aperture Radar (SAR) aufweisen [1], [2], [4].

$$l_{SENSOR'} = l_{SENSOR} - \sqrt{x^2{}_{P'} + y^2{}_{P'} + b^2} \qquad (2)$$

$$l_{PL} = l_{SENSOR'} - l_P = |P_L - P_P| \qquad (3)$$

Mit verschiedenen Vereinfachungen und Umformungen ergibt sich die gesuchte Entfernung zum Objektpunkt mit Gleichung (3) zu:

$$l_P = \frac{\left(x_L{}^2 + y_L{}^2 + z_L{}^2 - l_{SENSOR'}{}^2\right)\sqrt{x^2{}_{P'} + y^2{}_{P'} + b^2}}{2\left(x_L x_{P'} + y_L x_{P'} + z_L b - l_{SENSOR'}\sqrt{x^2{}_{P'} + y^2{}_{P'} + b^2}\right)} \quad P_L \neq P_{P'} \tag{4}$$

Insbesondere bei Kanten ist der Zusammenhang für die Klassifizierung der Entfernungsdaten je nach Situation und Systemanordnung entsprechend zu berücksichtigen.

2.2 Multichip 2D/3D-Kamera

Die 2D/3D-Kameratechnologie ist eine viel versprechende Lösung für die gleichzeitige Nutzung von hochauflösenden Intensitätsbildern (monochrom und Farbe) und zum 2D-Bild registrierten Abstandsbildinformationen. Eine interessante Weiterentwicklung solcher Ansätze sind Bildsensorchips, die neben der Möglichkeit, 3D-Informationen zu erfassen, auch gleichzeitig eine hohe Lateralauflösung im 2D-Bild liefern. Dabei kann man von sogenannten Mixed Pixel Sensoren sprechen, d. h. der Bildsensor besteht aus einer Anordnung unterschiedlicher Pixel.

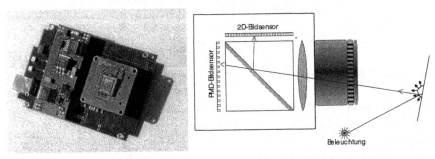

Abb. 2. 2D/3D-Kameraelektronik und Prinzipdarstellung der Bildsensoranordnung

3 2D-Registrierte 3D-Messung

Basierend auf den Bilddaten eines hochauflösenden 2D-Bildsensors können die Bildinhalte mittels der bekannten Verfahren der 2D-Bildverarbeitung extrahiert werden. Für einen monokularen Aufbau erhält man mit Hilfe entsprechender Kalibrierungsroutinen eine pixelbasierte Koregistrierung. Es besteht somit ein direkter Bezug zu den Oberflächenbereichen der Objekte und einem Pixel des PMD-Bildsensors und einer Menge von Pixeln auf dem 2D-Bildsensor.

In Abb. 3 sind einzelne Bildinformationen einer 2D/3D-Kamera zu sehen. Durch die Kalibrierung sind die verschiedenen Größen und Bereiche auf dem Testobjekt gut zu erkennen. Aufgrund der Größe des PMD-Chip (Version 64x16 Pixel) ist das PMD-Bild größer und wurde bei der Überlagerung beschnitten. Für die Auslegung eines Systems sind die unterschiedlichen Empfindlichkeiten der Sensoren zu berücksichtigen. Für die Auswertung von einem IR-Intensitätsbild des PMD-Bildsensors wäre die Blende möglichst weit zu schließen, um eine möglichst gute Tiefenschärfe zu erhalten. Für die 3D-Messung wird anderseits möglichst viel Licht benötigt. Bei geöffneter

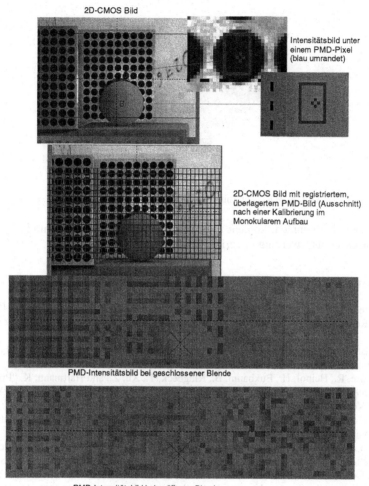

Abb. 3. Aufnahmen mit einer 2D/3D-Kamera (3D-Kalibrieraufbau mit einem Ball)

Blende hat das Intensitätsbild des 3D-Bildsensors allerdings eine schlechtere Qualität. Die Analyse der Objektregionen, auf denen Tiefendatenpunkte liegen (die Gitterstruktur auf dem mittleren Bildteil entsteht durch die überlagerten PMD-Pixel), können für eine weitergehende Analyse und Klassifikationen herangezogen werden. Der 2D-Sensor ist prinzipiell für die modulierte Beleuchtung des PMD-Sensor empfindlich.

Die Datenfusion auf Pixelebene basiert auf der direkten Fusion der einzelnen Datenquellen anhand spezieller Operatoren, welche die Verknüpfung dieser beiden Datenquellen beschreibt. Die Originalbilder werden dazu pixelweise miteinander fusioniert, d.h. sie müssen registriert sein. Diese Aussage bedeutet nicht gleichzeitig, dass beide Bilder in derselben lateralen Auflösung vorhanden sein müssen. Die Fusion kann auch auf vereinzelte Punkte angewendet werden. Dies ist auch aus Sicht der Echtzeitfähigkeit interessant. Durch die Bewertung von Tiefendaten, die auf Kanten liegen, ergeben sich Verbesserungen. Kanten, die im Grauwertbild nicht eindeutig

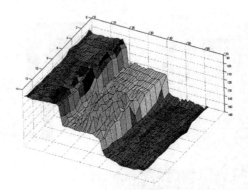

Abb. 4. 3D-Daten mit einigen Messpunkten auf dem Ball

detektiert werden konnten, können z.T. im korrespondierenden Tiefenbild extrahiert und in die Gesamtauswertung integriert werden.

Literaturverzeichnis

1. D.Justen, K.Hartmann; "3D-Image Processing through Grayscale Images and Distance Data Related to Individual Pixels"; SPIE; Visual Information Processing VIII; Vol. 3716; 1999
2. D. Justen, „Untersuchung eines neuartigen 2D-gestützten 3D-PMD-Bildverarbeitungssystems" Dissertation Universität Siegen, 2001.
3. Schwarte, R., Heinol, H., Buxbaum, B., Ringbeck, T., Xu, Z., and Hartmann, K. "Principles of Three-Dimensional Imaging Techniques", Handbook of Computer Vision and Applications, Volume 1, Sensors and Imaging, Jähne, B., Haußecker, H., Geißler, P. (Hrsg.), 1999
4. O. Loffeld, H. Nies, V. Peters, S. Knedlik, Models and useful relations for bistatic SAR processing, IEEE Transaction on Geoscience and Remote Sensing, 42(10), 2004
5. Sluiter, A., Hartmann, K., Hasenmaier, B., Weihs, W., Stieler, D., Rost, M.; 'Multilevel test and validation of algorithms implemented in a PSoC-vision-node"; Proceedings of SPIE Bd. 5267; Conference Photonics East 2003 Providence; Rhode Island; USA 27.-29. Oct.; 2003
6. 3D-Bildsensoren: www.pmdtec.de, www.swissranger.ch, www.canesta.com.

Segmentation of Independently Moving Objects Using a Maximum-Likelihood Principle

Martin Clauss, Pierre Bayerl, Heiko Neumann

Dept. of Neural Information Processing, University of Ulm, 89081 Ulm, Germany

E-mail: {martin.clauss,pierre.bayerl,heiko.neumann}@uni-ulm.de

Abstract. Detection of independently moving objects (IMOs) is a demanding task especially in situations, where the observer is moving himself. In such situations detection of IMOs as well as estimation of egomotion depend on each other and thus have to be handled simultaneously. We present an algorithm based on the Expectation/Maximization algorithm, which is capable of sharply separating background and independently moving objects, whilst the observer itself is moving. Furthermore it incorporates temporal integration of extracted information to improve estimation.

1 Introduction

When an observer (camera) is moving freely through a dominantly rigid scene, the detection of independently moving objects (IMOs) is a difficult task. In this situation, optical flow may result from either self-motion or from IMOs. On the one hand, detection of IMOs requires knowledge of the observer's egomotion to eliminate those flow components induced by the observer's own motion. On the other hand, estimation of the observer's motion is based on the global optical flow pattern, which is disturbed by the presence of IMOs in the current scene. Since both problems depend on each other, they have to be dealt with simultaneously in order to achieve a robust solution.

Unlike other previous proposals, we propose an approach that employs a single camera to segment global flow patterns (due to self-motion) from motion that is induced by other moving objects in the scene. The proposed approach may be utilized to feed further applications like collision warning, autonomous robot navigation, guidance, etc.

We will outline some previously proposed approaches for the problem scenario in section 2. Subsequently, we present our new method to simultaneously solve scene segmentation and egomotion estimation in section 3, together with a brief outline of the expectation/maximization algorithm and a short description of the underlying data representation. Some results of our algorithm will be presented in section 4, followed by a brief conclusion in section 5.

2 Related work

Approaches have been proposed to detect independent motion utilizing multiple cameras or object shape constants (e.g. [11]). Algorithms that utilize a single camera to detect IMOs during observer motion can be basically categorized into two groups: The first group are algorithms based on motion similarity. Smith and Brady [12] presented an approach that groups similar flow vectors returned by a feature tracking algorithm. The flow magnitude needs to contrast the background estimates in order to be considered as candidate. Their approach uses information extracted over time to learn the contour and the motion parameters of detected IMOs. The method does not employ knowledge of the observer's egomotion.

The other group are algorithms which detect IMOs based on knowledge of egomotion. Pauwels and van Hulle [10] iteratively estimate the observer's motion and remove data points not matching the observer's current motion-estimate. The method does not use any knowledge obtained from earlier image frames, but only uses flow information of the current frame. In order to circumvent an estimation of egomotion, several approaches (e.g. [1]) employ additional sensors to recover egomotion.

Woelk and Koch [14] utilize a particle filter [4] for sampling of optical flow estimation. The focus of expansion (FOE) and the translational component of the observer's motion are estimated, while relying on an inertial sensor to determine the rotational component. After removal of the rotational component, the flow vectors are classified depending on their deviation from the radial flow direction, pointing outwards from the FOE.

MacLean et al. [8] apply a subspace method to segregate rotation from translation. The constraints on the translation vectors, which are obtained from the subspace method, are then associated to a dynamical number of processes using the EM-algorithm. The number of processes is estimated according to the total fit of the constraint vectors to the estimated translation vectors.

Our approach combines previous proposals ([8][10][12]) to develop a framework for robust segmentation of background and IMOs during egomotion.

3 Detecting IMOs using the EM-Algorithm

Optical flow calculated from an image stream that is captured by a moving camera represents a superposition of four components: $F = T + R + I + \xi$. Translational motion T and rotational motion R of the observer, motion I induced by independently moving objects and noise ξ resulting from either camera measurement or the optical flow algorithm. Being able to only observe F by means of a single camera, these four components cannot be easily split.

The EM algorithm provides an iterative framework for finding the corresponding unobservable data association. After introducing our underlying data representation, we briefly introduce the EM algorithm and how its principle can be utilized in our scenario, leading to our new approach.

3.1 Data representation

We utilize a spatially sparse multimodal token-based data representation. This data representation is based on the biologically motivated work of Krüger et al. [6], which elaborates on the primal sketch introduced by Marr [9]. Different local feature attributes are stored in a common symbolic token T:

- position $\mathbf{x}$ of the token within the input image,
- optical flow $\mathbf{u}$ (Lucas-Kanade, [7]),
- greyscale structure λ_1, λ_2, ω (corner-/edgeness and orientation, [13]),
- status information (initial position x_0, age t).

In sum, a token T can be stored as vector

$$\mathbf{T} = (position(x), flow(u), greyscale(\lambda_1, \lambda_2, \omega), status(x_0, t))$$

This token-based representation reduces the memory requirements, whilst preserving relevant information.

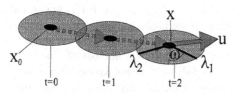

Fig. 1. Token-based symbolic data representation. Shown is one token at three successive time steps. Each token holds information on its position $\mathbf{x}$, its movement $\mathbf{u}$ (from optical flow), underlying greyscale information (from structure tensor) given by λ_1, λ_2, ω and token status information. All values are coupled with confidence values.

3.2 EM-Algorithm

When observing data samples y, with each sample originating from exactly one out of n models M_n, the maximum-likelihood (ml) parameters are those parameters for the generating models M_n, that maximize the probability of observing the data samples y. In general, however, the association of the data samples y to the models M_n cannot be observed, that is, the complete data x is unavailable. The Expectation/Maximization (EM) algorithm [3] seeks to iteratively find an optimized solution for model estimation from such incomplete observed data in an ml-fashion. The algorithm exploits, that the ml-parameters as well as the association of data samples to the models can be computed depending on each other. Thus, the EM algorithm iteratively solves the problem as follows: Be y the observation and x the corresponding complete data. The probability density function is $f(x|\theta)$ with θ denoting the parameters of the density. In the beginning, theparameters θ are initialized randomly.

In the **Expectation** step, the algorithm estimates the probability that the data sample originates from a model, for each model and every data sample, given the current model parameter estimates:

$$Q(\theta|\theta_k) = E[\log f(x|\theta)|y, \theta_k]$$

Where θ_k is the parameter set after the k-th iteration of the algorithm.

In the **Maximization** step, the model parameters are optimized in a maximum-likelihood fashion given the current data association estimate from the Expectation step:

$$\theta_{k+1} = \arg \max_{\theta} Q(\theta|\theta_k)$$

Expectation and Maximization step are alternately executed until precision is regarded sufficient, e.g. if $\|\theta_k - \theta_{k-1}\| < \epsilon$ for suitable ϵ and $\| \cdot \|$. The EM algorithm is guaranteed to converge [3] with respect to a local maximum of the likelihood function. One drawback of the EM algorithm is, that the number of underlying models n has to be known in advance. In our scenario, the number of IMOs is not known a priori and, therefore, the number of models is unknown as well, which is one of the major issues solved by our new approach.

3.3 Method

The EM-principle can be utilized for detecting IMOs via an indirect approach based on the following reasoning: If we could estimate the translational and rotational flow field that is induced by the observer's self-motion, then any deviating flow must arise from independent motions of an unknown number of objects. As the number of models (that is, the number of IMOs+1) is not known in advance for our scenario, we propose a simplified version of the EM algorithm in which only one model needs to be estimated. This model is the one associated with the observer's motion.

Expectation Step: During the expectation step, the fit of data samples is determined only for the model corresponding to the observer. First, the rotational component of the flow field is removed. As the rotational component only depends on the actual motion parameters and not on the scene geometry, the rotational parameters are sufficient to achieve this. These parameters are calculated by the subspace method employed during the maximization step. Under ideal conditions in a rigid environment without presence of IMOs, a flow field radially expanding from the focus of expansion (FOE) is obtained. The location f_0 of the FOE depends on the observer's translational parameters. The angular deviation (Fig. 2) from the expected radial flow field is then calculated for each flow vector (compare [14][10]):

$$\alpha = \arccos \frac{u \cdot x}{||u|| \cdot ||x||}$$

where x denotes the token's position relative to the focus of expansion f_0 and u is the optical flow component stored in the corresponding token. The deviation represents the grade of fit to the observer's motion model for a given flow vector.

Maximization Step: In the maximization step, the motion parameters are estimated using the subspace method presented by Heeger and Jepson [5]. By transforming to a subspace the rotational component vanishes and constraints for translation are obtained. Translational parameters are then computed from these constraints in a least-squares sense. As this subspace method does not support a continuous weighting of the input data samples per se, the model association of data points is first transformed to a binary membership using a fixed threshold.

Instead of initializing the model parameters randomly, we determine an initial guess by starting with the maximization step. For the first input frame, all data samples are assumed originating from the observer's motion model. For subsequent frames, modelassociations obtained from previously analysed video frames are used. This temporal feedback results in a better initial guess as well as a reduction in computational cost. In order to ensure that the correct model is estimated, we need to assume that more than 50% of the tokens represent the scene background that yield to estimates corresponding with the observer's egomotion. After convergence of the proposed algorithm, the segmentation of the image into background and IMOs is given by the association of data points to the observer's motion model obtained in the final expectation step.

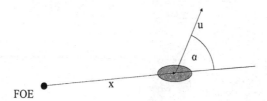

Fig. 2. Deviation of flow from to the expected motion pattern. Vector **x** points from the focus of expansion to the position of the token. Vector **u** is the movement of the token, i.e. the optical flow corresponding to the token. The angular error α employed in the segmentation algorithm is calculated as the angle between vector **x** and vector **u**

4 Results

We show results of our proposed algorithm from three scenarios. The input data originates from a moving car on a freeway and in an inner city environment, as well as from an autonomous mobile soccer robot (Fig. 3 left). The model association of the flow vectors obtained during the final expectation step are plotted. These associations represent the probabilities that a flow vector originates from an IMO. The sparse data representation is transformed into pixel-maps, which are shown on the right hand side (Fig. 3, right). In the top row a van is overtaking the observer's vehicle. The middle row is taken from an observer approaching an intersection. In this sequence, the observer moves while another vehicle is moving to the right. All other vehicles are waiting at traffic signs and were therefore not detected as IMOs. In the bottom row a soccer ball crosses the robot's path. The plots on the right hand side show that the IMOs (cars, ball) are detected correctly by the algorithm. There are some false positives, especially in the freeway and the robot sequence. These yield from incorrectly determined optical flow, which partly occurs because tracked features moved out of the image (e.g. freeway). Partly this is also due to the aperture problem (e.g. robot), which can be solved by advanced motion algorithms [2].

Fig. 3. Results of the proposed algorithm. Left: Input images taken from moving observers (car/robot). Right: Probability of IMO presence. We transform the sparse data by adding gaussians in a pixel-map at the corresponding locations. The height of thesegaussians is proportional to the probability of the optical flow of originating from an IMO. The width is proportional to the flow vector's distance to its nearest neighbour.

5 Conclusion

We presented an algorithmic approach for separating IMOs from background using techniques based on the expectation/maximization principle. The algorithm can handle arbitrary motion including rotation of the observer by iterative estimation of egomotion from optical flow. It features a spatially sparse data representation together with usage of feedback processing, which yields in a reduction of the amount of data points to be processed over time as well as an enhancement regarding the estimation of the observer's egomotion. The approach only assumes that the majority of estimated flow vectors is induced by the static background.

6 Acknowledgements

This work has been supported by a grant from the Ministry of Science, Research and the Arts of Baden-Württemberg (Az: 23-7532.24-12-19) to Martin Clauss and Heiko Neumann and a scholarship funded by the University of Ulm granted to Martin Clauss.

References

1. D. Baehring, S. Simon, W. Niehsen, and C. Stiller. Detection of close cut-in and overtaking vehicles for driver assistance based on planar parallax. *Proc. Intelligent Vehicles*, pages 289–294, 2005.
2. P. Bayerl and H. Neumann. Disambiguating visual motion through contextual feedback modulation. *Neural Computation*, 16:2041–2066, 2004.
3. A.P. Dempster, N.M. Laird, and D.B. Rubin. Maximum likelihood from incomplete data via the em algorithm. *Journal of the Royal Statistical Society*, Ser. B 39:1–38, 1977.
4. Michael Isard and Andrew Blake. Condensation – conditional density propagation for visual tracking. *Intl. Journal of Computer Vision*, 29:5–28, 1998.
5. Allan D. Jepson and David J. Heeger. Linear subspace methods for recovering translational direction. In *Proceedings of the 1991 York conference on spatial vision in humans and robots*, pages 39–62, New York, NY, USA, 1993. Cambridge University Press.
6. N. Krüger, M. Lappe, and F. Wörgötter. Biologically motivated multi-modal rocessing of visual primitives. *Journal of Artificial Intelligence and Simulation of Behaviour*, 1:417–428, 2004.
7. B.D. Lucas and T. Kanade. An iterative image registration technique with an application to stereo vision. In *Int. Joint Conf. on Artificial Intelligence*, pages 674–679, 1981.
8. J. MacLean, A. Jepson, and R. Frecker. Recovery of egomotion and segmentation of indepedent object motion using the em-algorithm. In *British Machine Vision Conference*, pages 175–184, 1994.
9. D. Marr. *Vision: A Computational Investigation into the Human Representation and Processing of Visual Information*. W. H. Freeman, New York, 1982.
10. K. Pauwels and M. van Hulle. Segmenting independently moving objects from egomotion flow fields. *Early Cognitive Vision Workshop*, 2004.
11. Harpreet S. Sawhney, Yanlin Guo, and Rakesh Kumar. Independent motion detection in 3d scenes. *IEEE Trans. Pattern Anal. Mach. Intell.*, 22(10):1191–1199, 2000.
12. S. M. Smith and J. M. Brady. Asset-2: Real-time motion segmentation and shape tracking. *IEEE Trans. Pattern Anal. Mach. Intell.*, 17:814–820, 1995.
13. Emanuele Trucco and Alessandro Verri. *Introductory Techniques for 3-D Computer Vision*. Prentice Hall PTR, Upper Saddle River, NJ, USA, 1998.
14. Felix Woelk and Reinhard Koch. Fast monocular bayesian detection of independently moving objects by a moving observer. In *DAGM-Symposium*, volume 3175 of *Lecture Notes in Computer Science*, pages 27–35. Springer, 2004.

Region-based Depth Feature Map for Visual Attention in Autonomous Mobile Systems

Muhammed Zaheer Aziz, Ralf Stemmer, Bärbel Mertsching

GET Lab, Faculty of Computer Science, Electrical Engineering and Mathematics,
University of Paderborn, Pohlweg 47-49, 33098 Paderborn

E-mail: {aziz,stemmer,mertsching}@get.upb.de

Abstract. In this contribution we present a fast region based approach for distance estimation within a bottom-up visual attention model. The feature "depth" is utilized to focus the attention on objects that are closer to the mobile robot in order to accelerate the overall process of visual attention. Due to the fact that in the stereo algorithm we only search for corresponding segments determined by color segmentation, we provide a fast method for usage in mobile systems. Although we do not achieve very dense depth maps, the accuracy is sufficient for collision avoidance and the integration of the obtained depth map into the attention model.

1 Introduction

Visual input is a major way of sensing the environment with mobile robots. With the availability of improved quality camera devices, it is possible to obtain full color and high resolution images for machine vision. But processing time can exponentially increase when working on images of big size. This raises the need of fast algorithms that are able to handle information of complex scenes.

A promising approach towards a decrease of computation time is to reduce the amount of data to be processed using visual attention models before actually performing complex vision oriented tasks such as object recognition. The visual attention system locates only those areas in a scene that have a certain significance in some respect [12]. Driven by the study of biological vision systems the nature of significance is decided. Ultimately, small windows are extracted as focus of attention for detailed processing instead of performing vision procedure on all of the input. This leads to lower computational cost and so, this approach can be exploited for real-time processing of high resolution input [11].

One of the key problems for mobile systems is the 3D reconstruction of the environment. It is important for local navigation tasks like collision avoidance as well as pose estimation for objects which have to be grasped autonomously by a robotic hand eDaMa. Lots of mobile robots use stereo vision for depth reconstruction. But the computation of dense and accurate disparity maps conflicts with the requirement of real time control. Depth maps from the stereo visual input have been utilized as an ingredient for computing focus of attention in some attention models such as [1] and [2]. In literature, stereo algorithms can be found which compute highly accurate depth maps with the help of massive computational costs. On the other hand, one can find real time algorithms for low resolution depth maps.

This paper proposes a fast method to tackle the problem of computing depth saliency maps in the bottom-up process of artificial visual attention. A region based approach is adopted to compute the depth map from the input provided by color segmentation. A comparison of a small set of features is performed to establish fast correspondence between regions in the left and right frame. This leads to region based disparity and finally a region based depth map is constructed. The resulting disparity map is, as expected, not dense but this method leads to significant acceleration of the computation time for the visual attention process such that detailed tasks of computer vision can be performed only on the relevant regions in a given scene.

2 Visual Attention for Mobile Robots

Models of artificial visual attention are constructed keeping human or biological vision systems as a role model. It is established that visual attention plays a major role in the human information processing because it influences the distribution of resources, the selection of relevant information, and the prioritization of tasks. Therefore, visual attention acts as an important control unit for realizing action-perception-cycles [4]. The distance between an object and the viewer plays a major roleas closer objects draw more attention than farther ones.

Most of the recently developed models for artificial visual attention are based on the approach of Koch and Ullman [10]. The base of these models is the analysis of different features like color contrast, eccentricity, symmetry, and depth etc teBaMeBo. The feature maps are then combined to form a master map which represents the degree of saliency at every image location. Figure 1 shows results of this process for a sample image [1]. In a Winner Takes All (WTA) process the focus of attention (FoA) is determined and an action, like directing the camera head towards the FoA, is executed.

Fig. 1. (Left to right) Input image and feature maps for eccentricity, color contrast, symmetry, depth, and master map by our former attention model. Brighter gray values represent higher saliency.

One significant drawback in existing techniques is that some of the feature maps such as symmetry and depth are constructed on pixel based functions in the frequency domain. This results in a master feature map with cloudy region shapes. It is difficult to perform shape based recognition with such a master map. The system has to go through a segmentation process again for the salient portions of a given scene. Our proposed approach to attention constructs all saliency maps in such a way that they highlight prominent regions in shape of actual objects so that the same regions could be directly conveyed to the higher level of computer vision procedures. Hence, we propose to perform all saliency computations on regions. Our approach starts with a robust region extraction from colored input images with cancellation of light effects such as shades and shadows. Then, fast methods to compute saliency maps, for features recommended in literature, on these regions are explored. The ultimate target is to produce fast results in bottom-up attention in order to support real-time top-down behaviors.

3 Existing Methods for Depth Computation

According to the classical definition in [15], as the y coordinates of corresponding pixels are exactly the same, the correspondence between a pixel (x, y) in a reference image and a pixel (x', y') in the matching image is computed as

$$x' = x + sd(x, y), y' = y, \tag{1}$$

where the value of s is chosen as +1 or -1 to keep disparities always positive. According to a recent taxonomy of pixel-based stereo correspondence algorithms [18], these algorithms generally perform some or all of four basic steps namely, computation of matching cost, aggregation of cost, computation of disparity, and refinement of disparity. Various methods have been proposed for computing the matching cost such as squared intensity differences [9], absolute intensity differences e Ka, or truncated quadratics [17] etc. Aggregation of matching cost is performed by summing or averaging over a prescribed window in the disparity space image. Then, the disparity associated with the minimum cost is chosen for each pixel. Some algorithms such as those described in [16] and [19] go for a sub-pixel refinement to obtain continuity in solid objects in the scene.

There are a couple of methods that work on spatially interconnected regions, named as layers or surfaces. The work in [3] presents recovery of layers by techniques used for motion estimation, image registration, and mosaicing and then refine the layer estimates using a resynthesis step. The method proposed in [14] estimates scene structure as a set of smooth surface patches. Disparities within each patch are modeled by a spline and pixels of patch extents are labelled. Segmentation is done by graph cuts, aided by image gradients.

Some of the work done in our lab on stereo algorithms includes [5] and [6]. In this research, the general similarity measurement is computed by applying a normalized cross-correlation on vertical contour features and then ambiguities are eliminated via a special relaxation procedure formulated in terms of a cost function. Once the minimum of the cost function is found, the valid disparity of a pixel is retrieved by a maximum search across the parameters belonging to that pixel.

The pixel-based methods discussed above do produce a dense and accurate depth map but the computation time required for them is a big overhead for the visual attention process. On the other hand, the surface based approaches are mostly concerned in constructing the layers from stereo information rather than producing depth map from regions. For our application we need a method to produce a fast and rough depth map using the regions obtained from the segmention procedure so that all the feature maps for the visual attention are constructed on the same processing pipeline. Hence, we propose a region based approach for this purpose as explained in the following section.

4 Region Based Depth Map

The basic concept of the proposed approach is to establish the depth map by finding groups of contiguous points in left and right frames that belong to the same location in the image. One object may be composed of many of such regions. An object with the same color having different faces can have different depths at its various locations. These faces usually have diverse shades of the object color resulting in multiple regions such that points belonging to one region will have approximately the same depth. The depth of points constituting these small contiguous regions may be estimated using disparity between the two corresponding regions containing these points. Hence, if point clustering is carried out based on adjacency and similarity of color value, then a fairly correct depth map can be achieved.

The proposed method for construction of depth maps for visual attention purposes begins with a color segmentation of the stereo input images. The segmentation is performed in RGB color space in order to apply strict criteria for region growing. Other color spaces such as normalized RGB and HSI have a high tolerance to shades of colors [7] and hence are not suitable for this application. Information regarding regions including their average color value, size, and coordinates of location are loaded in two separate lists. Elements of the left list are picked one by one and corresponding regions in the right list are found as follows. For an element in the left list, only those elements are accessed from the right list for which no strong correspondence have been established yet and those which have almost the same horizontal coordinate of their centres. If one of the accessed right regions matches closely in color as well as size then a strong correspondence is recorded for this pair. On the other hand, if any one of the size or color features differ largely, but under a certain threshold, then a weak correspondence is recorded. Right regions with a weak correspondence are not exempted from comparison with next coming left ones as they may find a strong correspondence later. When the end of the left list is reached, disparity is computed for all regions for which a strong correspondence was found. In the case of absence of a strong correspondence, weak correspondence is considered. Figure 2 demonstrates the architecture of this method diagrammatically.

Finally, a depth map is constructed that will be useful for the attention system in order to attend those areas from camera input that are more close to the robot. This feature map will be combined together with other saliency maps such as those for

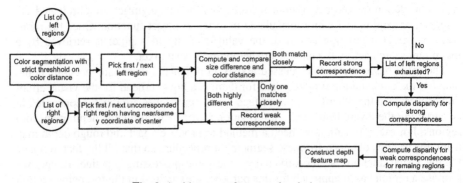

Fig. 2. Architecture of proposed technique

eccentricity, symmetry, color contrast, etc. The master map will lead to some salient region of the whole image that need specific attention. Now, detailed processing with accurate algorithms of depth detection or object recognition may be performed only on the selected attention windows.

5 Experimental Results

The proposed method of depth feature map construction was implemented in a C++ program and a number of test stereo images were used as input data. The input image pairs included some benchmark pictures that are normally used in experiments on stereo vision. Some other real life and artificial images, which are used in our experiments on visual attention, were also tested upon. Figure 3 shows the output and the intermediate steps in which the program highlights a pair of regions in left and right frames for which correspondence has been found by a rectangle. The rightmost image is the depth map in which brighter gray values represent regions closer to the observer. Theoretically, it was feared that a region based technique may lead to very coarse results of depth estimation but practically, it can be observed, the results are very encouraging with all of the input set. Obviously, some errors can also be spotted but other specialized stereo techniques also suffer from a small percentage of erroneous results.

Fig. 3. Input frames, regions correspondence and depth map for a benchmark input

6 Discussion

The new technique in depth estimation has provided the advantage of short computation time compared to an existing method used for the same purpose [2]. Average processing time, for an image pair, with the new approach is around 90 m/s while the previous technique took 580\,ms averagely. The processing time of the proposed method largely depends upon the number of segments obtained. Further speedup can be achieved by applying more efficient data storage for maintaing the region lists and quicker search technique for finding correspondences between regions. The second advantage is that the depth map is gained according to shapes of regions. Hence, further processing of detailed vision can directly be applied on these regions.

On the other hand, pixel based techniques result in cloudy and distorted feature maps that are not directly useful for the next steps of machine vision. Figure 4 shows a comparison of the output for an artificial test image with the proposed method and one of the existing methods. Another advantage is that our approach is independant on any model information in contrast to layered approaches available in literature [3]. It is also free of dependancy on texture [14], hence leading to emergence of those objects in a given scene which are of more interest for the robot in order to avoid collisions.

Fig. 4. Left input frame, depth map by an existing method, and by proposed method

One disadvantage found so far in the proposed technique is that its performance mainly depends upon the segmentation method utilized. Hypothetically, a pair of stereo images should result in very similar regions after segmentation because the viewing points differ only in a linear translation. But practically this does not occur due to object occlusions and shadow shifts because of the disparity in camera locations. This leads to some regions with false or no correspondence. If false correspondence occurs, it applies to the whole region hence causing inaccuracy for many points. Customization in segmentation technique may be explored in future to obtain more accurately similar regions for both images.

References

1. G. Backer: Modellierung visueller Aufmerksamkeit im Computer-Sehen: Ein zweistufiges Selektionsmodell für ein Aktives Sehsystem. Ph.D Thesis, University of Hamburg, Oct. 2003
2. G. Backer, B. Mertsching, M. Bollmann: Data and Model-Driven Gaze Control for am Active-Vision System. IEEE Trans. on PAMI, 23 (12), Dec. 2001, pp. 1415–1429
3. S. Backer, R. Szeliski, P. Anandan: A layered approach to stereo reconstruction. Conference on Computer Vision and Pattern Recognition, Santa Barbara, CA, June 1998, pp. 434–441
4. R. Bajcsy : Active perception. Proceedings of the IEEE Vol. 76, Issue 8, Aug. 1988 pp. 966–1005
5. R. Brockers, M. Hund, B. Mertsching: Stereo matching with occlusion detection using cost relaxation. Accepted for: IEEE International Conference on Emerging Technologies, 2005
6. R. Brockers, R. Stemmer, S. Drüe, G. Hartmann: A Stereo Camera System at a Robot's Hand for Exhausting Engines. Proceedings of CISST 2000, Las Vegas, June 2000, vol. 1, pp. 123–130
7. H. D. Cheng, X. H. Jiang, Y. Sun, J. Wang: Color segmentation: advances and prospects. Pattern Recognition, vol. 34, Elsevier-Pergamon, 2001, pp. 2259-2281
8. E.P. Dadios, O. A. Maravillas: Cooperative mobile robots with obstacle and collision avoidance using fuzzy logic. Proceedings of the 2002 IEEE International Symposium on Intelligent Control, Oct. 2002, pp. 75–80
9. M. J. Hannah: Computer matching of areas in stereo images. PhD thesis, Stanford University, 1974
10. L. Itti, G. Rees, J. K. Tsotsos: Neurobiology of attention. Elsevier, San Diego, CA, Jan 2005, pp. 576–582

11. L. Itti, C. Koch, E. Niebur: A model of saliency-based visual attention for rapid scene analysis. IEEE Trans. on PAMI, Vol. 20, No. 11, Nov. 1998, pp. 1254–1259
12. L. Itti, C. Koch: A saliency-based search mechanism for overt and covert shifts of visual attention. Vision Research, vol. 40, 2000, pp. 1489–1506
13. T. Kanande: Development of a video-rate stereo machine. In Image Understanding Workshop, Monterey, CA, 1994, pp. 549–557
14. M. H. Lin, C. Tomasi: Surfaces with occlusions from layered stereo. IEEE Transactions on PAMI, Vol. 26, No. 8, August 2004, pp. 1073–1078
15. M. Okutomi, T. Kanade: A locally adaptive window for signal matching. International Journal for Computer Vision, 7(2), 1993, pp. 143–162
16. T. W. Ryan, R. T. Gray, B. R. Hunt: Prediction of correlation errors in stereo-pair images. Optical Engineering, 19(3), 1980, pp. 312–322
17. D. Scharstein, R. Szeliski: Stereo matching with nonlinear difussion. IJCV, 28(2), 1998, pp. 155–174
18. D. Scharstein, R. Szeliski: A Taxonomy and Evaluation of Dense Two-Frame Stereo Correspondence Algorithms. IJCV, 47(1–3), 2002, pp. 7–42
19. Q. Tian, M. N. Huhns: Algorithms for subpixel registration. CVGIP, No. 35, 1986, pp. 220–233
20. Y. Yang, A. Yuille, J. Lu: Local, global, and multilevel stereo matching. CVPR, 1993, pp. 274–279

Lokalisierung und Kartographierung

Bearing-Only SLAM with an Omnicam
An Experimental Evaluation for Service Robotics Applications

Christian Schlegel, Siegfried Hochdorfer

University of Applied Sciences Informatics Faculty, Prittwitzstr. 10, D-89075 Ulm

E-mail: {schlegel,hochdorf}@fh-ulm.de

Abstract. SLAM mechanisms are a key component towards advanced service robotics applications since knowing the pose allows goal directed movements or can ensure full coverage of cleaning areas. In particular, SLAM reduces the deployment efforts of a service robot. Currently, a major hurdle are the high costs of suitable range measuring devices which do not fit into the budget of many proposed applications. A solution are bearing-only SLAM approaches since these can be used with cheap sensors like omnicams.

This paper applies a bearing-only SLAM approach to features extracted from images of an uncalibrated omnicam. Experiments with an omnicam on-board a Pioneer-3DX robot prove the suitability of EKF-based bearing-only SLAM for service robotics applications.

The contribution of this work is to evaluate bearing-only SLAM with a cheap sensor on a real robot in scenarios that cover characteristic requirements of advanced service robotics applications. The results show that bearing-only SLAM can be succesfully used with cheap sensors to overcome a major hurdle towards advanced service robots.

1 Introduction

Many advanced service robotics applications require goal directed motions or systematic coverage of free space. Thus, a key component towards widespread use of service robots is localization, either pose tracking, relocalization or SLAM. Although a large body of work proved that the problem is solvable, most approaches require distance measuring sensors. These are either too expensive for most service robots (e.g. laser range finders) or do not show the required performance (e.g. ultrasonic sensors in large open space like gyms or lobbies).

An omnicam provides feature rich information on the surrounding of the robot with high update rates. As long as no calibrated system is needed, they are cheap and small and thus perfect for service robotics applications. As drawback, one does not get range information to landmarks as needed with many algorithms. However, bearing-only SLAM mechanisms can take advantage out of an omnicam image since these only require angles to landmarks. To determine angles to vertical landmarks, even no methods to correct image distortion or to correct perspective have to be applied.

Fig. 1. Test environment   **Fig. 2.** Robotics testbed   **Fig. 3.** Omnicam setup

2 Related Work

Bearing-only SLAM is gaining more and more attention. This paper is based on the method described in [1] that is also used in [3]. The focus of this work is on adapting and evaluating that approach in a real-world scenario with an omnicam. Pros and cons of the approach of delayed landmark initialization are discussed in [6]. An algorithm for undelayed landmark initialization is proposed in [8]. A forward looking camera is used for SLAM in [2]. However, the focus is on colour-based feature tracking. The approach in [5] uses visual features in 3D for bearing-only SLAM. The work presented in [4] uses a Gaussian Sum Filter to represent the initial feature state.

3 Bearing-Only SLAM

According to [1], the state vector is given by $\mathbf{x} = [\mathbf{x}_v^T, \mathbf{x}_{v_m}^T, ..., \mathbf{x}_{v_1}^T, \mathbf{x}_{f_1}^T, ..., \mathbf{x}_{f_n}^T]^T$ with $\mathbf{x}_v = [x_v, y_v, \phi_v]^T$ (vehicle pose), $\mathbf{x}_{v_i} = [x_{v_i}, y_{v_i}, \phi_{v_i}]^T$ and $\mathbf{x}_{f_i} = [x_{f_i}, y_{f_i}]^T$ (pose of landmark i). Each pose $\mathbf{x}_{v_i}$ corresponds to the time and location where a set of measurements $\{\theta_{v_i}^1, ..., \theta_{v_i}^k\}$ was obtained. Thus, measurements of a possible landmark can be accumulated over time until sufficient information is available for a reliable landmark initialization. The deferred landmark initialization is necessary since more than one observation is needed for bearing-only landmark initialization. Multiple observations also reduce the risk of collapsing the EKF. The processing order per time-step adapted from [1] is as follows:

1. get image from omnicam, get robot pose from odometry, extract features and determine angle to features
2. measurements of existing map features are processed first in a batch update
3. if there exists a well-conditioned pair of measurements for a non-initialized landmark, the initial landmark estimate is added to the SLAM state vector
4. for each newly initialized feature, the remaining accumulated measurements are applied in a batch update
5. robot poses $\mathbf{x}_{v_i}$ are removed from the SLAM state vector as soon as they no longer have assigned observations that are not yet processed

6. in case the current robot pose provides a measurement for a not yet initialized landmark, the observation is stored and the current robot pose is added to the SLAM state vector by *stochastic cloning* [7], $[\hat{\mathbf{x}}_v^T, \hat{\mathbf{x}}_m^T]^T \rightarrow [\hat{\mathbf{x}}_v^T, \hat{\mathbf{x}}_v^T, \hat{\mathbf{x}}_m^T]^T$ and

$$\begin{bmatrix} \mathbf{P}_{vv} & \mathbf{P}_{vm} \\ \mathbf{P}_{vm}^T & \mathbf{P}_{mm} \end{bmatrix} \rightarrow \begin{bmatrix} \mathbf{P}_{vv} & \mathbf{P}_{vv} & \mathbf{P}_{vm} \\ \mathbf{P}_{vv} & \mathbf{P}_{vv} & \mathbf{P}_{vm} \\ \mathbf{P}_{vm}^T & \mathbf{P}_{vm}^T & \mathbf{P}_{mm} \end{bmatrix}$$

7. perform odometry pose update according to current SLAM state

Motion Model

The motion of the vehicle between two poses as reported by the odometry is approximated by a linear motion according to $[x_v, y_v, \phi_v]_{new}^T = \mathbf{m}(x_v, y_v, \phi_v, d, \Delta\alpha_1, \Delta\alpha_2, \Delta\beta) = [x_v + d\cos(\phi_v + \Delta\alpha_1), y_v + d\sin(\phi_v + \Delta\alpha_1), \phi_v + \Delta\alpha_1 + \Delta\alpha_2 + \Delta\beta]^T$. The robot first turns towards $\mathbf{x}_{v,new}$ by $\Delta\alpha_1$, moves then distance d to arrive in $\mathbf{x}_{v,new}$ and then turns by $\Delta\alpha_2$ to achieve the final heading. This motion model always considers the total amount of rotation even for S-shaped motions. Given $\mathbf{x}_v$ and $\mathbf{x}_{v,new}$ from odometry, the parameters of the motion model are $d = \sqrt{(y_{v,new} - y_v)^2 + (x_{v,new} - x_v)^2}$ (traveled distance), $\delta = \arctan(\frac{y_{v,new} - y_v}{x_{v,new} - x_v})$ (direction of d), $\Delta\alpha_1 = \delta - \phi_v$ (first turn in $\mathbf{x}_v$), $\Delta\alpha_2 = \phi_{v,new} - \delta$ (second turn in $\mathbf{x}_{v,new}$), and $\Delta\beta = s(d)$ (distance dependent drift). The Jacobian of the motion model $\frac{\partial \mathbf{m}(\ldots)}{\partial(x_v, y_v, \phi_v, d, \Delta\alpha_1, \Delta\alpha_2, \Delta\beta)}$ to calculate the covariance matrix of $\mathbf{x}_{v,new}$ is

$$\begin{bmatrix} 1 & 0 & -d\sin(\phi_v + \Delta\alpha_1) & \cos(\phi_v + \Delta\alpha_1) & -d\sin(\phi_v + \Delta\alpha_1) & 0 & 0(1) \\ 0 & 1 & d\cos(\phi_v + \Delta\alpha_1) & \sin(\phi_v + \Delta\alpha_1) & d\cos(\phi_v + \Delta\alpha_1) & 0 & 0(2) \\ 0 & 0 & 1 & 0 & 1 & 1 & 1 \end{bmatrix}$$

Observation Model

The observation model is used to integrate another observation of an already initialized landmark and can be expressed by $z_i = h(\mathbf{x}_v, \mathbf{x}_{f_i}) = \arctan(\frac{y_{f_i} - y_v}{x_{f_i} - x_v}) - \phi_v + n_i$ with n_i the measurement error modeled as zero-mean white gaussian noise with variance σ_θ^2. The EKF update equations require the Jacobian of the observation model with d_{vf_i} the euclidean distance between vehicle and landmark. The position of $\mathbf{H}_{f_i}$ in $\mathbf{H}$ corresponds to the position of the landmark in the state vector.

$$\mathbf{H} = \nabla_{\mathbf{x}} h(\hat{\mathbf{x}}) = \begin{bmatrix} \mathbf{H}_v & 0 & \ldots & 0 & \mathbf{H}_{f_i} & 0 & \ldots & 0 \end{bmatrix}$$

$$\mathbf{H}_v = \begin{bmatrix} \frac{\partial h}{\partial \hat{x}_v} & \frac{\partial h}{\partial \hat{y}_v} & \frac{\partial h}{\partial \hat{\phi}_v} \end{bmatrix} = \begin{bmatrix} \frac{\hat{y}_{f_i} - \hat{y}_v}{\hat{d}_{vf_i}^2} & -\frac{\hat{x}_{f_i} - \hat{x}_v}{\hat{d}_{vf_i}^2} & -1 \end{bmatrix}$$

$$\mathbf{H}_{f_i} = \begin{bmatrix} \frac{\partial h}{\partial \hat{x}_{f_i}} & \frac{\partial h}{\partial \hat{y}_{f_i}} \end{bmatrix} = \begin{bmatrix} -\frac{\hat{y}_{f_i} - \hat{y}_v}{\hat{d}_{vf_i}^2} & \frac{\hat{x}_{f_i} - \hat{x}_v}{\hat{d}_{vf_i}^2} \end{bmatrix}$$

3.1.1 Feature Initialization

For proper initialization of a landmark in SLAM with bearing-only information, at least two bearing measurements z_i and z_j from two different vehicle poses $\mathbf{x}_{v_i}$ and $\mathbf{x}_{v_j}$ are needed. In case of no errors, the true location of the landmark would then be given by intersecting two lines given in point-slope form as $(y_f - y_{v_i}) = \tan(\theta_i + \phi_{v_i})(x_f - x_{v_i})$ and $(y_f - y_{v_j}) = \tan(\theta_j + \phi_{v_j})(x_f - x_{v_j})$ (see [1]):

$$\mathbf{x}_f = \begin{bmatrix} x_f(3) \\ y_f \end{bmatrix} = \mathbf{g}(\mathbf{x}_{v_i}, \mathbf{x}_{v_j}, \theta_i, \theta_j) = \begin{bmatrix} g_x(..)(4) \\ g_y(..) \end{bmatrix} = \begin{bmatrix} \dfrac{x_{v_i} s_i c_j - x_{v_j} s_j c_i + (y_{v_j} - y_{v_i}) c_i c_j}{s_i c_j - s_j c_i} \\ \dfrac{y_{v_j} s_i c_j - y_{v_i} s_j c_i + (x_{v_i} - x_{v_j}) s_i s_j}{s_i c_j - s_j c_i} \end{bmatrix}$$

where we abbreviate $s_i = \sin(\phi_{v_i} + \theta_i)$ and $c_i = \cos(\phi_{v_i} + \theta_i)$. In case of noise-corrupted vehicle poses and measurements, the landmark estimate is given by $[\hat{x}_f, \hat{y}_f]^T = \mathbf{g}(\hat{\mathbf{x}}_{v_i}, \hat{\mathbf{x}}_{v_j}, z_i, z_j)$ with $z_i = \theta_i + n_i$ and $z_j = \theta_j + n_j$ where n_i and n_j denote zero mean white gaussian noise. Using first-order Taylor series expansion for $\mathbf{g}(...)$ and taking into account that the measurement noise is independent of the vehicle position, one can approximate the 2×2 covariance matrix $\mathbf{P}_{LL}$ of the landmark position estimate by $\mathbf{P}_{LL} = \mathbf{G}\mathbf{P}\mathbf{G}^T + \mathbf{W}\begin{pmatrix} \sigma_\theta^2 & 0 \\ 0 & \sigma_\theta^2 \end{pmatrix}\mathbf{W}^T$ and the correlation with the other entries in the SLAM state vector, $\mathbf{P}_{LX}$ and $\mathbf{P}_{XL}$, by $\mathbf{P}_{LX} = \mathbf{P}_{XL}^T = \mathbf{H}\mathbf{P}$ assuming uncorrelatedness between system and measurement error [3] where $t = (s_i c_j - s_j c_i)^2$ and

$$\mathbf{G} = \nabla_{\mathbf{x}}\mathbf{g}(\hat{\mathbf{x}}) = \begin{bmatrix} \mathbf{G}_{v_i} & 0 & ... & 0 & \mathbf{G}_{v_j} & 0 & ... & 0 \end{bmatrix}$$

$$\mathbf{G}_{v_i} = \nabla_{\mathbf{x}_{v_i}}\mathbf{g}(\hat{\mathbf{x}}) = \begin{pmatrix} \dfrac{\partial g_x}{\partial x_{v_i}} & \dfrac{\partial g_x}{\partial y_{v_i}} & \dfrac{\partial g_x}{\partial \phi_{v_i}}(6) \\ \dfrac{\partial g_y}{\partial x_{v_i}} & \dfrac{\partial g_y}{\partial y_{v_i}} & \dfrac{\partial g_y}{\partial \phi_{v_i}} \end{pmatrix}$$

$$= \frac{1}{t}\begin{pmatrix} s_i c_j(s_i c_j - s_j c_i), & -c_i c_j(s_i c_j - s_j c_i), & (\hat{x}_{v_j} - \hat{x}_{v_i})s_j c_j - (\hat{y}_{v_j} - \hat{y}_{v_i})c_j^2 (7) \\ s_i s_j(s_i c_j - s_j c_i), & -s_j c_i(s_i c_j - s_j c_i), & (\hat{x}_{v_j} - \hat{x}_{v_i})s_j^2 - (\hat{y}_{v_j} - \hat{y}_{v_i})s_j c_j \end{pmatrix}$$

$$\mathbf{G}_{v_j} = \nabla_{\mathbf{x}_{v_j}}\mathbf{g}(\hat{\mathbf{x}}) =$$

$$\frac{1}{t}\begin{pmatrix} -s_i c_j(s_i c_j - s_j c_i), & -c_i c_j(s_i c_j - s_j c_i), & (\hat{x}_{v_i} - \hat{x}_{v_j})s_i c_i + (\hat{y}_{v_j} - \hat{y}_{v_i})c_i^2 (8) \\ -s_i s_j(s_i c_j - s_j c_i), & -s_j c_i(s_i c_j - s_j c_i), & (\hat{x}_{v_i} - \hat{x}_{v_j})s_i^2 + (\hat{y}_{v_j} - \hat{y}_{v_i})s_i c_i \end{pmatrix}$$

$$\mathbf{W} = \nabla_\mathbf{n}\mathbf{g}(\hat{\mathbf{x}}) =$$

$$\frac{1}{t}\begin{pmatrix} (\hat{x}_{v_j} - \hat{x}_{v_i})s_j c_j - (\hat{y}_{v_j} - \hat{y}_{v_i})c_j^2, & (\hat{x}_{v_i} - \hat{x}_{v_j})s_i c_i + (\hat{y}_{v_j} - \hat{y}_{v_i})c_i^2 (9) \\ (\hat{x}_{v_j} - \hat{x}_{v_i})s_j^2 - (\hat{y}_{v_j} - \hat{y}_{v_i})s_j c_j, & (\hat{x}_{v_i} - \hat{x}_{v_j})s_i^2 + (\hat{y}_{v_j} - \hat{y}_{v_i})s_i c_i \end{pmatrix}$$

The decision on when a pair of measurements is well-formed is based on the *Kull-back-Leibler* distance exactly as described in [1].

3.1.2 Feature Extraction

Figure 4 shows the result of the image processing step. Currently, simple features like color and size ratios are used to detect artificial landmarks and to extract the angles to the landmarks as shown in Fig. 5. The angle is defined by the polar line starting in the center of the image and going through the color blob such that the lateral error of the color blob boundaries with respect to the line is minimal. Of course, as soon as more elaborate feature extraction and classification mechanisms are used, one can get rid of artificial landmarks. However, one then has to face the extremely hard data association problem.

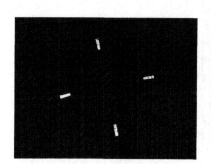

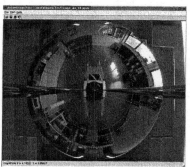

Fig. 4. Extracted landmarks   **Fig. 5.** Angles to landmarks

4 Experiments and Results

To perform the experiments on a real world platform, we implemented all necessary steps such that a closed loop system including a Pioneer-3DX robot, an omnicam and the bearing-only SLAM algorithm is available. The omnicam is a Sony DFW-X710 camera (1024x768, 1/3 inch, progressive scan, firewire, YUV color, 15 images/second) with a hyperbolic glass mirror (H3G, Neovision). The robot is equipped with an on-board Pentium M class small sized PC. The software is Linux based and is a combination of C++ code with calls to a Matlab server for the SLAM algorithm. The image processing is based on the Intel Computer Vision Lib. The parameters of the motion model are $(0.05m)^2/1m$ (distance error), $(5\deg)^2/360\deg$ (rotational error) and no drift error. The sensor model uses $\sigma_\theta^2 = (1.5\deg)^2$ as angular error of the landmark detection. The threshold of the distance metric to decide on the initial integration of a landmark is set to 12. The robot typically moved about $1m$ between measurements.

Figure 6 shows the evolution of the landmark covariance matrices as they are tracked by the robot via its omnicam. The images are taken out of a run performed in the lab consisting of 36 sensing steps. The run contains a loop that has to be closed by the SLAM mechanism and covers an area of $11\times6m^2$. Not all landmarks are always

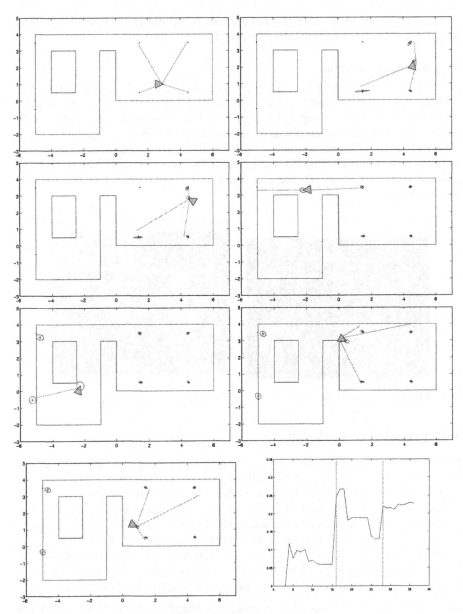

Fig. 6. The sensing steps 1, 4, 5, 14, 23, 28 and 31 of a 36 step run with closing a loop. The map rotates since there is no absolute initial reference. The lines indicate which landmark was been by the robot in that step. The last figure shows the standard error over all relative landmark distances.

in the field of view. The last figure shows the standard error over all relative landmark distances extracted from the estimated landmark poses. Ground truth is available for relative landmark distances. Of course, the overall error increases with each newly added landmark and then shrinks again as further observations are made.

5 Conclusion and Future Work

The conducted experiments on a real platform prove that EKF-based bearing-only SLAM methods can be applied to features extracted from an omnicam image. However, the covariances of the estimates still seem to shrink too fast, mostly due to the linear approximation of the EKF.

Currently, the experiments are conducted in a prepared indoor environment. However, we now have all the prerequisites to gather further experience with typical service robotics scenarios. Then, one can replace sub-components like the feature extraction mechanisms by algorithms tailored to these scenarios to end up with a balanced solution in terms of costs, performance and robustness.

Another approach to supply the landmark viewing angles is to use a laser system instead of an omnicam. A rotating laser pointer and a photo diode can be used to detect retroreflectors that are placed as artificial landmarks. These could even encode an absolute identifier similar to barcodes. The angle is detected by a shaft encoder of the rotating laser unit. The important point is that we only need the angle and *not* the distance.

To summarize, bearing-only SLAM can successfully be exploited with cheap sensors for service robots to overcome a major hurdle, namely the high costs of range measuring sensors. The next steps are to perform tests in much larger environments. Future work is mainly focusing on the image processing part where SIFT features might be an interesting approach.

6 Acknowledgement

We would like to thank Tim Bailey for providing an initial implementation of the approach described in [1] that formed the starting point of our work.

References

1. T. Bailey. Constrained Initialisation for Bearing-Only SLAM. In: *Proc. IEEE Int. Conf. on Robotics and Automation (ICRA)*, 2003.
2. T. Fitzgibbons and E. Nebot. Bearing-Only SLAM Using Colour-based Feature Tracking. In: *Proc. of the Australian Conf. on Robotics and Automation*, Auckland, 2002.
3. J. Hesch and N. Trawny. Simultaneous Localization and Mapping using an Omni-Directional Camera. Available: www-users.cs.umn.edu/ ~ joel/_files/BoS.pdf.
4. N. M. Kwok, G. Dissanayaka and Q. P. Ha. Bearing-only SLAM Using a SPRT Based Gaussian Sum Filter. In: *Proc. IEEE Int. Conf. on Robotics and Automation (ICRA)*, 2005.
5. T. Lemaire et. al. A practical 3D Bearing-Only SLAM algorithm. In: *IEEE Int. Conf. on Intelligent Robots and Systems (IROS)*, Edmonton, 2005.

6. J. Ortega et al. Delayed vs Undelayed Landmark Initialization for Bearing Only SLAM. In: *ICRA 2005 Workshop W-M08: Simultaneous Localization and Mapping*. Available: web-diis.unizar.es/%7Ejdtardos/ICRA2005/SolaAbstract.pdf
7. S. Roumeliotis and J. Burdick. Stochastic cloning: A generalized framework for processing relative state measurements. In: *Proc. IEEE Int. Conf. on Robotics and Automation (ICRA)*, 2002.
8. J. Sola et al. Undelayed Initialization in Bearing Only SLAM. In: *IEEE Int. Conf. on Intelligent Robots and Systems (IROS)*, Edmonton, 2005.

Automatic Generation of Indoor VR-Models by a Mobile Robot with a Laser Range Finder and a Color Camera

Christian Weiss, Andreas Zell

Universität Tübingen, Wilhelm-Schickard-Institut für Informatik, Lehrstuhl Rechnerarchitektur,
D-72076 Tübingen,

E-mail: cweiss,zell@informatik.uni-tuebingen.de

Abstract. We present an intuitive method to generate visually convincing 3D models of indoor environments from data collected by a single 2D laser range finder and a color camera. The 2D map created from the laser data forms the basis of a line model which serves as floor plan for the 3D model. The walls of this model are textured using the color images. In contrast to panoramic images where the resolution of the textures is substantially decreased by the panoramic mirror, our standard camera can provide high quality textures at moderate image resolutions.

1 Introduction

Traditional 2D maps built using range sensors such as laser range finders or ultrasonic sensors provide enough data for navigation tasks, but lack visual information. This visual information can be very useful, especially when humans have to interpret the data collected by the robot. Textured 3D models of the environment give a faster and more detailed overview of a scene than a 2D map, and therefore can help to simplify tasks such as surveillance or rescue and also provide a way to interchange models of building interiors over far distances.

In this paper, we present a method to create 3D models based on 2D maps and color images, both obtained using a mobile robot. First, we extract a set of line segments from the 2D map to determine walls. This set serves as a floor plan for the 3D model. Then the 3D model is enriched by textures created from the color images. Note that our approach does not intend to include small objects like plants or chairs in the 3D model. On the one hand it is not possible to determine the height of objects with the horizontal laser range finder; objects located entirely below or above the scan plane are not detected anyway. On the other hand, the objects are visible in the textures which suffices for a fast overview.

A method that also uses a single horizontal laser range finder to build 3D models was presented by Biber et al. [1]. They use a panoramic camera to acquire the images from which the textures are created. In combination with a method by Fleck [2], which can create 3D point clouds from pairs of panoramic images, some more 3D information is added to the model [3]. A different approach was presented by Thrun et al. [4] who use a second vertical laser range finder to collect 3D range data. Planar surfaces extracted from this data are textured with color images recorded by a panoramic camera. Due to the wide field of view of the panoramic cameras used in these approaches, the resolutions of the textures are low. Our method uses a standard camera, so we can create detailed textures although the image resolution is moderate (e.g. 640×480 pixels).

The rest of this paper is organized as follows: Section 2 presents the acquisition of the data. Section 3 describes how the 2D map is transformed into a line model. The creation of the 3D model is described in Section 4 and Section 5 presents some experimental results. Finally, Section 6 gives a short conclusion.

2 Data Acquisition

In this work, we use an RWI (Real World Interface) ATRV-Jr mobile robot to collect the laser and image data. The robot is equipped with a SICK LMS 200 laser range finder and a stereo camera system, of which we use only one camera. The camera system is mounted on a pan tilt unit (PTU).

The robot is remotely driven around to build a 2D map using a SLAM (Simultaneous Localization And Mapping) technique proposed by Biber [5]. The final 2D map consists of a set of laser scan points in a global coordinate frame. Fig. 1(a) shows a 2D map generated at our institute. Note that the horizontal left wing is not at a right angle with the vertical main axis corridor in the real environment.

During mapping, the robot stops approximately every meter and takes three color images; one to the front of the robot, one 90° to the left and one 90° to the right. The different orientations of the camera are achieved by turning the PTU. Each image is saved together with the current pose estimate of the robot and the orientation of the PTU relative to the robot's orientation.

3 Line Model

The next step is to transform the 2D map, which is a set of unrelated 2D points, into a set of line segments. Each line segments represents one wall of the 3D model, i.e. each line segment is the projection of a wall onto the floor.

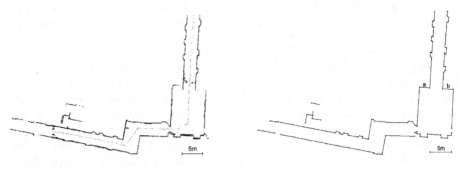

(a) 2D laser map consisting of laser scan points. The path of the robot is indicated by a dotted line.

(b) Final line model of the same environment.

Fig. 1. 2D laser map and corresponding line model.

To extract lines from the 2D map, we use the Hough Transform [6,7]. Additionally, each line is associated with the set of 2D scan points which contribute to its creation. These point are orthogonally projected onto the line. The lines are then split into line segments at every location where two neighbouring points on the line have a distance larger than 10 cm.

After that, line segments which seem to belong to the same linear structure of the environment are merged. We use adaptions of methods proposed by Sack [8] and Schröter [9]. As Sack uses operations that are geometrically more exact and involve non-linear regression, his method is five to 12 times slower than Schröter's technique, but creates results that are more suitable for our purposes. More suitable means that the environment can be represented by fewer but longer line segments.

Next, the line model is further improved to meet our requirements. The steps to improve the model are the following, where steps 1 and 4 are adaptions of techniques proposed in [8]:

1. We adjust the length of segments whose endpoints are near each other. This ensures that walls forming the corner of a room meet in a common point.
2. Line segments shorter than 20 cm are deleted; for line segment that are not connected to another one, the minimal length is 80 cm. This step deletes many line segments that represent small structures like plants or chairs.
3. The orientation of each wall is determined, i.e. we decide which side of the wall faces the corridor. We use the scan points associated with the line segment and the directions towards the global poses of the corresponding scans to solve this task. This step is very important as the orientation of the texture on the wall is determined by the orientation of the wall.
4. Short line segments are inserted to close gaps. Only gaps smaller than 60 cm are closed.
5. Segments that share a common endpoint with more than one other segment are deleted. This step implements the constraint of the real environment that no more than two walls meet in the corner of a room.
6. Certain sequences of connected line segments are deleted. The overall length of these sequences must be below a threshold (1.5 m) and the individual segments must be short.

The improvement steps are fully automatic and provide a good representation of the walls in the environment. As it is possible that some walls are fully occluded during the acquisition process and therefore are not present in the laser map, the user is given the possibility to manually insert missing walls or to remove undesired ones. Figure 1(b) shows the final line model for the map shown in Fig. 1(a). The line segments *a* and *b* were manually inserted as the corresonding walls never were inside the range of the laser range finder during data acquisition.

4 3D Model

The line model serves as a floor plan for the 3D model, where each line segment represents one wall. As the horizontal laser range finder cannot determine the height of the walls automatically, the height must be given as parameter. The images taken by the color camera are used to texture the walls.

The outline of the texture creation process is as follows:

1. Long walls are split into shorter subwalls.
2. For each subwall, a suitable image is determined from which the texture can be created.
3. The texture for the wall is determined as the section of the image which shows nothing but the wall.
4. The intensities of neighbouring textures are adjusted to hide seams.

In contrast to Biber et al. [1], who merge several panoramic images by multiresolution blending to create textures for long walls, we split long walls into shorter subwalls of fixed length l. If l is chosen properly, the texture for a subwall can be created from a single image which shows the entire subwall. In corridors, l is around 1.5 meters, in wider rooms it can be larger.

To be suitable as a texture for a wall, an image must meet at least three conditions. First, the photograph position p of the image must be located such that p is on the corridor or room side of the wall. This fact can be verified using the orientation of the wall. Second, the wall must not be occluded by other walls when looking from p towards the wall. Third, the wall must be seen entirely on the image. This can be verified by projecting the corner points of the wall onto the image and checking if

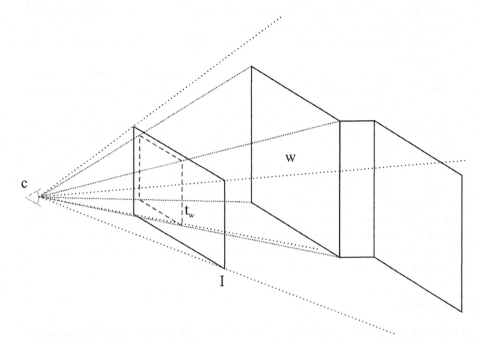

Fig. 2. Generation of the texture for a wall w. The image I, shot from camera position c, was identified as the most suitable image for w. The section t_w of I is cropped out and serves as texture for w. Note that in general, t_w is not a rectangle but a trapezoid.

they are inside the image. Finally, it is beneficial if the wall is displayed nearly frontally on the image; the angle between the viewing direction of the image and the orientation of the wall provides a measure for this condition. For such images, the perspective does not have to be corrected much. A perspective correction can drastically decrease the quality of the image in some image regions. If no frontal image is available, this condition is ignored. If no image is suitable at all, i.e. does not meet at least the first three conditions, the wall is left untextured. In situations where there is more than one suitable image, e.g. when the robot frontally approaches a wall, the image with the closest distance to the wall is selected.

After a suitable image for a wall has been found, we calculate the texture as the section of the image which shows nothing but the wall. This calculation is done by projecting the 3D corner points of the wall onto the image and cropping out the resulting image section. Fig. 2 illustrates the procedure. Note that in general, the image section is a trapezoid. A rectification of this trapezoid corrects the perspective.

To build the projection matrix, we need the intrinsic and extrinsic parameters of the camera. The intrinsic ones are obtained by camera calibration; the extrinsic ones are given by the pose of the robot, the height of the camera above the floor and the orientation of the PTU at the time the image was taken.

5 Results

Figure 3(a) partly shows the 3D model of a single room, the robots laboratory; the 36 color images on which the textures are based have a resolution of 1280×960 pixels. The creation of the line and 3D models took a total of about 10.4 s on a 3Ghz Pentium 4 PC (512MB of RAM). Once the 3D model is created, real-time walkthroughs are possible. The same holds for the next data set.

This data set is based on the map shown in Fig. 1. It contains a total of 195 images (640×480 pixels), which were taken at 65 different locations. The resulting 3D model was created from the laser map in about 43.2 s using the adaption of [8] for line merging. Figure 3(b) shows a bird's eye view of the 3D model and Fig. 3(c) shows a detailed view. As our camera is not capable of adjusting its exposure automatically, the intensities of the textures differ throughout the 3D model due to different lighting conditions in the real environment. In Fig. 3(c) you can also see that in reality, indentations for doors are only as high as the doors themselves. Our horizontal laser range finder cannot detect such situations, so the geometry of the virtual walls is wrong, but the 3D model still looks good.

However, in more complex environments with many objects in the middle of a room or occluding entire walls, our approach would fail to detect the walls. Here, additional 3D information from another source like a second laser range finder or a stereo camera is needed.

Figure 3(d) shows a detailed view of a small part of the 3D model presented by Biber et al. in [1]. In the picture, you can see a desk and a monitor standing on it, but the rest is not recognisible. In our model, small objects like ring binders or the checkerboard used for camera calibration are identifiable (Fig. 3(a) on the right). In Fig. 3(c), posters are visible in detail. Note that in Biber's 3D model, the textures of most other walls are worse than the ones shown in Fig. 3(d).

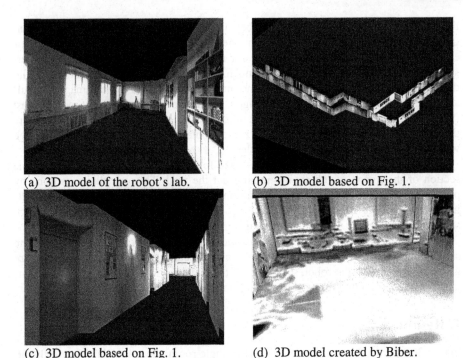

(a) 3D model of the robot's lab.   (b) 3D model based on Fig. 1.

(c) 3D model based on Fig. 1.   (d) 3D model created by Biber.

Fig. 3. Different views of automatically generated 3D models. (a)-(c): generated using our method. (d): taken from a 3D model presented by Biber et al. in [1].Thanks to Peter Biber for the permission to use the image of his 3D model.

6 Conclusion

We presented a method to create visually convincing 3D models of indoor environments. The 3D model is based on a 2D map, which is built using a single laser range finder. We then transform the point-based 2D map into a set of line segments, where each segment represents a rectangular wall. Textures for the walls are created from images obtained by a standard color camera. The relatively high-quality textures provide more visual detail than the textures used in other work. Furthermore, the creation of the 3D model is fast and needs no or few manual intervention.

As the depth information gathered by the horizontal laser range finder is bound to a single plane, our method is best suited to corridor-like environments and rooms with few objects occluding the walls. Objects different from walls are not geometrically represented in the 3D model but visible in the textures. To overcome this limitation, further 3D information could be used to improve the model. This additional information could be obtained e.g. by a stereo camera system or a second laser range finder.

References

1. Biber, P., Andreasson, H., Duckett, T. and Schilling, A.: 3D Modeling of Indoor Environments by a Mobile Robot with a Laser Scanner and Panoramic Camera. Proc. IEEE/RSJ Int. Conf. on Intelligent Robots and Systems (IROS), 2004
2. Fleck, S., Busch, F., Biber, P., Andreasson, H. and Straßer, W.: Omnidirectional 3D Modeling on a Mobile Robot using Graph Cuts. Proc. IEEE Int. Conf. on Robotics and Automation (ICRA), 2005
3. Biber, P., Fleck, S., Busch, F., Wand, M., Duckett, T. and Straßer, W.: 3D Modeling of Indoor Environments by a mobile Platform With a Laser Scanner And Panoramic Camera. Proc. 13th European Signal Processing Conference (EUSIPCO), 2005
4. Thrun, S., Martin, C., Liu, Y., Hähnel, D., Emery-Montemerlo, R., Chakrabarti, D. and Burgard, W.: A Real-Time Expectation Maximization Algorithm for Acquiring Multi-Planar Maps of Indoor Environments with Mobile Robots. IEEE Trans. on Robotics and Automation, 20(3), pp. 433-442, 2003
5. Biber, P. and Straßer, W.: The Normal Distributions Transform: A New Approach to Laser Scan Matching. Proc. IEEE/RSJ Int. Conf. on Intelligent Robots and Systems (IROS), 2003
6. Hough, P.C.V.: Methods and Means for Recognising Complex Patterns. US Patent 3069654, 1962
7. Ballard, D. H. and Brown C. M.: Computer Vision. Prentice Hall, Englewood Cliffs, NJ, USA, 1982
8. Sack, D.: Lernen von Linienmodellen aus Laserscannerdaten für mobile Roboter. Diplomarbeit, Albert-Ludwigs-Universität Freiburg, Institut für Informatik, Arbeitsgruppe Autonome Intelligente Systeme, 2003 (german)
9. Schröter, D., Beetz, M. and Gutmann, J.-S.: RG Mapping: Learning Compact and Structured 2D Line Maps of Indoor Environments. Proc. Int. Workshop on Robot and Human Interaction Communication (ROMAN), 2002

References

Elastic View Graphs:
A new Framework for Sequential 3D-SLAM

Peter Kohlhepp, Marcus Walther

Forschungszentrum Karlsruhe in der Helmholtz-Gemeinschaft,
Institut für Angewandte Informatik (IAI), Postfach 3640, 76021 Karlsruhe,
Universität Karlsruhe, Institut für Technische Informatik (ITEC),
Haid-und-Neu-Straße 7, 76131 Karlsruhe,

E-mail: kohlhepp@iai.fzk.de, mwalther@ira.uka.de

Abstract. This paper presents a new approach for incremental 3D SLAM from segmented range images with unknown feature association. For feature and motion tracking, an any-time interpretation tree is compared to a new algorithm called Orthogonal Surface Assignment. Both algorithms may utilize uncertain pose estimates from vehicle sensors, and yield several hypotheses. An elastic graph of surface submaps handles global pose correction. By stochastic error propagation and spatial indexing, loops are detected within bounded time. Loop closing with hypothesis exchange is priorized on a measure of node ambiguity. Indoor experiments performed with a rotating laser scanner demonstrate the effectiveness of our approach.

1 Introduction

Creating 3D models of complex scenes from range images has experienced a boom with the advent of commercially affordable laser scanners and the advances in computing power. Usually, the goal is to produce some kind of a CAD model, a single well-defined output that is useful to *humans* in facility management, telepresence, or entertainment. Achieving the goal will require human interaction. Mobile platforms and robots can aid in such imaging and mapping tasks. This paper deals with the case where the main purpose of the map is to serve the robot itself in its decision making *while* being created. Exploring work spaces inaccessible or dangerous to humans such as abandoned mines [10,14] or disaster areas are important applications. A stream of visual inputs must be processed *autonomously* and *sequentially* to produce a map that should *at any stage* be up-to-date, effective, and efficient to *answer questions*: where is the robot, the closest obstacle, an open passage towards the goal? Does this part need to be inspected more closely? Simultaneous localization and mapping (SLAM) means to build a global map from many local geometric measurements where the observer path is, at best, imprecisely known. With map entities bearing on estimated view points, and view points bearing on recognized features in the map, the *uncertainty* of both tends to grow. Being aware of its magnitude and acting accordingly is seen as a key feature discerning SLAM from mere 3D reconstruction [15]. Just aligning a view optimally to evenly distribute an error [10] may eventually lead to an inconsistent map when closing loops.

2 State of 3D SLAM and new approach

In general, a probability density function (pdf) of the state, comprising the view path and map features, is estimated under which the observed and noisy data become most likely. The Extended Kalman filter (EKF) and related methods, dealing with Gaussian pdf's and linearized system models, are heavily used in SLAM problems [2,3] because they provide simple running updates for states and covariances. Resolving global pose ambiguities however requires multi-modal methods such as Monte Carlo localization (MCL [13]) or particle filters (PF [7]). FastSLAM [15] combines particle filters with Kalman filters: each particle models its own version of the view path, enabling the n map feature positions to be estimated by n decoupled EKF's, one for each feature.

Today's SLAM algorithms mostly reside in two dimensional work spaces, but there have been recent attempts in 3D, e.g. Früh's city models built from ground and aerial views [4], Liu's expectation maximization (EM) algorithm [9], and the mine mapping systemGroundhog by Thrun et.al. [14]. Motion tracking, as the backbone of the 3D process, relies on a fixed horizontal laser scanner and on 2D *scan matching*. A sequence of *planar* pose transformations (x_i, y_i, Θ_i) thus results.

Any inconsistencies, e.g. when closing loops, are resolved by a *global pose correction* algorithm, for example, MCL [4] or the Extended Information Filter (EIF) [14]. Here, the entire motion sequence is processed again in *batch mode*. After transforming all scanner data, a dense point model or a triangular mesh results. Higher-level modelling such as approximating planes or mapping textures comes last, after pose correction. For decision making and planning, a geometric-topological model is needed, but it could hardly be re-generated every time step, from a growing point model. We conclude that today's 3D SLAM algorithms do not yet work fully three-dimensionally (3D view path, 3D geometry) and at the same time sequentially and globally consistent.

This paper presents a new approach towards incremental 3D mapping, working entirely on surface models which seem more effective for decision making. A single rotating scanner scans the entire work space. Its stream of scan lines is packaged into images and reduced to partial surface models (*range views*) right away. Vehicle motion between view pairs is recovered in six degrees of freedom, applying 3D *range image understanding* methods [5,6]. Compared to scan matching and ICP, highly discriminative constraints of *visibility* and *pose invariant* relations are explored and multiple hypotheses offered.

At the global level, the range views are arranged in an *elastic view graph* (EVG). By error propagation and spatial indexing, loops are detected within bounded time. Loop closing with hypothesis exchange is priorized on a measure of node ambiguity. Different local pose tracking algorithms can be plugged into the scheme, and additional prior pose estimates from vehicle sensors, e.g. wheel encoders and inclinometers, [3,6] may be used or not. In this sense, our approach is a framework similar to theATLAS framework [1] for 2D SLAM.

3 Sequential capturing and pose tracking of range views

For range data capturing, a new 3D laser scanner, the *Rotating SICK RoSi* [12], has been developed at Karlsruhe University. A commercial 2D laser scanner (SICK LMS) freely rotates around its optical axis (fig. 1) and captures a spherical cap with an opening angle of 100° or 180° in front of the vehicle. Owing to the scan line rotation, no orientations are preferred, but objects in the center receive the highest resolution and at the border of the view field the lowest (*foveal* vision). Also, the continuous rotation allows fast scanning with low mechanical stress. Each half resolution of RoSi yields a closed range image. The ordering within and between scan lines (*parameter space*) must be exploited to extract surface features in real time from maximally connected, geometrically homogeneous pixel regions. But the parameter space topology is a twisted cylinder mantle (*Moebius band*) and not simply a rectangle. For details of the segmentation algorithm see [6]. A partial surface model of planar and conical patches is generated with several features like 3D boundary polygon, center of gravity, direction of normal vector, extent, weight, shape etc. The relations between patches, based on relative distances and orientations, form the second main component of the feature model. From this information, the free, the occupied, and the unknown space is derived, and special surfaces such as the floor can be tracked and classified. One short example will be given in section 5. Pose tracking means to estimate a rigid transformations between pairs of consecutive range views. Two different methods have been implemented, compared and tested. The **Any-time interpretation tree (IPT)** is a general-purpose method for arbitrary work spaces, in which feature associations are found in a given time window, processing a tree of fixed depth sector-wise[6]. For man-made work spaces such as buildings, city districts, or plants, a new method, the **Orthogonal surface assignment (OSA)**, has been designed exploiting the fact that usually some surface directions are orthogonal to each other, e.g. most wall normals are perpendicular to the vector of gravity. Both methods offer *several* pose hypotheses as output (k best local optima). Each solution comprises a correspondence relation z, pose parameters T (unit quaternion R and translation $\underline{t}$), a covariance matrix C (pose parameter uncertainty), and a rating g.

Fig. 1. Rotating laser scanner RoSi (left), 180° image (center), surface features (right)

4 Loop hypothesizing and closing

When a newly captured range view has been located, this event initiates further global actions: finding and closing cycles, assessing and possibly correcting errors, and merging range views to larger submaps (see fig. 2). A new range view is added to an *Elastic View Graph* (EVG). Each *node* contains the surface features of one or several merged range views, and the current transformation F taking it into global map coordinates. An *edge* represents the currently active pose hypothesis between the linked range views, as described in section 3.

Unless contradicting evidence is produced by loop closing, the best rated hypothesis is always followed from node to node. Usually, there will be little ambiguity, anyway. Ambiguity is considered high where widely separated poses with very similar ratingsare observed. We measure the ambiguity $\eta(T)$ of a pose hypothesis T with respect to T^*, the best rated hypothesis for the same node (global optimum), by

$$\eta(T) := d(T^*, T)/(g - g^*) \tag{1}$$

where $d(.,.)$ is a distance metric on transformations [8]. A global queue of fixed length stores the most ambiguous hypotheses sorted by decreasing values of η in order to restrict backtracking to finitely many of them (fig. 2).

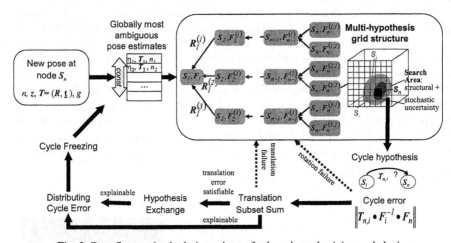

Fig. 2. Data flow and calculation scheme for loop hypothesizing and closing

When a range view has been linked to its predecessor and inserted into the graph, further links of correspondence and pose to previously mapped views are sought via overlapping volumes. In order to find them within bounded time, the current view's bounding box, amplified by an uncertainty region, is used as an index into a coarse voxel grid maintained of the provisional map (spatial hashing). This uncertainty derives from two sources, a stochastic one and a structural, set based, one.

Stochastic uncertainty is derived from the sensitivity of the pose estimating algorithms around particular solutions. The covariance matrices of 3D features, like surface directions and centers, are propagated from input to result views through the pose estimation and application. It is a first-order approximation

$$C_n = \sum_{j=1}^{m} \Im_{F_j}^T C_j \Im_{F_j} + \Im_{mri}^T C_{mri} \Im_{mri} \qquad (2)$$

where C_n and C_j denote the feature covariances at output node S_n resp. input nodes S_j, and C_{mri} the covariance matrix of the points in the *moving range image* captured for the current view Sn. Cmri depends on theerror of the laser scanner and the motion uncertainty of the sensor origin during the short time interval of image capture. $\Im_{mri}$ and $\Im_F$ denote Jacobian matrices of the respective function blocks extracting surface features (segmentation), estimating, and applying pose parameters. For example, when the rotation part, a quaternion, is estimated as an eigenvector of a positive definite matrix whose coefficients depend on uncertain input features, $\Im_F$ basically captures the eigenvector sensitivity (see [6] for details). *Structural uncertainty* accounts for different local optima of correspondence and pose and may result in totally different maps. In fig. 2, the hypotheses are arranged in a tree according to the different rotations, where for each choice there mayexist several translation hypotheses. If the distance between them is bounded by ε_t, the overall feature uncertainty is simply updated by $\delta^{(k+1)} \le \delta^{(k)} + \varepsilon_t$. This arrangement is efficient mainly for the OSA algorithm (section 3) where we get very few different rotations. Set-based and stochastic uncertainties may be combined since, from a covariance matrix, an uncertainty ellipsoid is obtained by specifying a desired significance level. If a view S_i intersects the combined uncertainty volume around the current view S_n, an extra transformation $T_{n,i}$ is estimated (by the IPT or OSA methods). Thus, a cycle of edundant transformations is formed whose product must equal the identity I. Since all transformations were estimated independently, actually, there will be a difference

$$d\left(I, T_{n,i} \cdot \prod_{j=i}^{n-1} T_j\right) \qquad (3)$$

which is used to measure the *cycle error*. Now we check if the feature errors resulting from a cycle error of this size lie within the accumulated stochastic uncertainty region. If the cycle error is *explainable* in that sense, it will be eliminated, by distributing the residual angular rotation and translation errors evenly among the nodes on the cycle, using a closed-form algorithm proposed by Sharp [11]. Otherwise, some pose hypotheses along the cycle are wrong. We check if the cycle error is *satisfiable*: an approximate and simplified *subset sum* problem is executed on the bounded set of hypotheses, to see if the cycle error becomes explainable by exchanging hypotheses. In that case, it can be distributed as before. After error distribution, all node and edge covariances are updated and all hypotheses on the cycle are deleted. This operation is called *cycle freezing* since the cycle nodes no longer remain mutually elastic. The merged cycle becomes a new node of the graph. If no rotation and no translation alternative is found under which the cycle error is explainable, no cycle is detected. (It does not guarantee that none exists: there might exist an inferior pose, zeroing the cycle error but not represented in the list). The complexity of the error distributing and cycle freezing operations is $O(N)$ in the total number N of views, since there is no bound on the cycle length. But any view may be involved only once, therefore the integrated cost remains linear. Maintaining and exchanging hypotheses and forming cycle errors have all constant complexity per view. The lookup time to find overlapping views in the grid is expected to be $O(1)$, too, but may be worse since it depends on the spatial distribution of the features. This point merits future research.

5 Preliminary experimental results

A mapping experiment with the *RoSi* sensor was carried out at *Technologiefabrik Karlsruhe*. The executed path went through several adjacent rooms of a laboratory building and back via a corridor; 18 still range views were taken. The motionstrategy was not to cover the complete unknown space, but to analyze the loop closing properties and cycle errors using very few, highly reduced observations. The scene character changes significantly within the sequence: cluttered rooms with office objects are followed by long, almost featureless corridors. Non-stationary objects such as door wings in different states, persons, and personal objects are also encountered. Most challenging, the portion of corresponding features in adjacent images is often very low, in particular when traversing a door. The pose tracking algorithms IPT and OSA from section 3 were thoroughly tested, at first with *no vehicle motion input*. On the average, the best rated hypothesis failed in ≈ 7 cases per sequence (IPT) compared to ≈ 0.4 (OSA). Analysis of the 12 best hypotheses for selected view pairs revealed that, even for OSA, registration remains brittle (fig. 3 left): quality values g are often similar but the translation components vary greatly. By comparing the poses with separate pose estimates from odometry input [3,6] and assigning probabilities according to Mahalanobis distance, only, the hypotheses became clearly disambiguated. For the best rated hypotheses, translation errors of 35-60 mm were measured in a cycle with diameter $\approx 17m$. The rotation error is close to $0°$, due to the successful tracking of a building coordinate system by OSA. The map obtained from 18 partialviews is shown in Fig. 3 right. It contains 196 merged or added surface patches and 890 surface relations. Color here corresponds to the surface membership to 21 automatically classified object classes (blue: floor, green: ceilings, grey: outer walls, lilac: tables,..). Unexplored areas are shown in white background color. A simple fuzzy membership

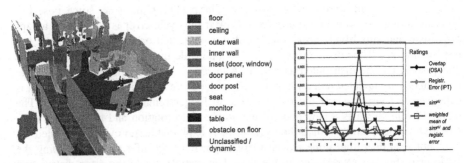

Fig. 3. Merged map with classified surfaces (left), 12 best rated hypotheses for one view pair (right)

classification was implemented to test and demonstrate that the map is indeed efficient for planning and decision making.

References

1. M. C. Bosse, P. M. Newman, J. J. Leonard, M. Soika, W. Feiten and S. Teller, An Atlas Framework for Scalable Mapping, IEEE Int'l Conf. on Robotics and Automation, Taipeh, Taiwan, 2003
2. J.A. Castellanos et.al., The SPmap: A Probabilistic Framework for Simultaneous Localization and Map Building, IEEE Trans. on Robotics and Automation 15(5), 1999, 948–953
3. J.Davison, N.Kita, 3D Simultaneous Localisation and Map-Building Using Active Vision for a Robot Moving on Undulating Terrain, IEEE Conf CVPR, Hawaii, 2001
4. C. Früh, A. Zakhor, Constructing 3D City Models by Merging Ground-Based and Airborne Views, IEEE Conf. CVPR 2003, Madison, USA, 562-69
5. X. Jiang, H. Bunke, Three-dimensional Computer Vision (in German), Springer Verlag 1997
6. P. Kohlhepp, P. Pozzo, M. Walther, R. Dillmann, Sequential 3D-SLAM for Mobile Action Planning, IEEE Conf. IROS, Sendai, Japan, 2004, 722-729
7. C. Kwok, D. Fox, M. Meila, Real-time Particle Filters, Advances in Neural Information Processing Systems 15, 2002
8. J.J. Kuffner, Effective Sampling and Distance Metrics for 3D Rigid Body Path Planning, IEEE Conf. ICRA, 2004.
9. Y.Liu et al, Using EM to Learn 3D Models of Indoor Environments with Mobile Robots, 18th Conf. on Machine Learning, Williams College, 2001
10. Nüchter, H. Surmann, K. Lingemann, J. Hertzberg, S. Thrun. 6D SLAM with Application in Autonomous Mine Mapping, Proc. IEEE 2004 Int'l Conf. Robotics and Automation (ICRA '04), New Orleans, USA, 2004
11. G. C. Sharp et al, Toward Multiview Registration in Frame Space, Proc. IEEE Int. Conference on Robotics & Automation, Seoul, Korea, 2001
12. P. Steinhaus, R. Dillmann, Construction and modeling of the RoSi scanner for 3D range image acquisition (in German), Workshop on Autonomous Mobile Systems (AMS), Karlsruhe, 2003
13. S. Thrun, D. Fox, W. Burgard, F. Dellaert, Robust Monte Carlo localization for mobile robots, Artificial Intelligence 128, 2001, 99-141
14. S. Thrun et al, A System for Volumetric Robotic Mapping of Abandoned Mines, Proc. ICRA, Taipei, Taiwan, 2003
15. S. Thrun et al, FastSLAM: An efficient solution to the simultaneous localization and mapping problem with unknown data association, Journal of Machine Learning, 2004

Selbstständige Erstellung einer abstrakten topologiebasierten Karte für die autonome Exploration

Kalle Kleinlützum, Tobias Luksch, Daniel Schmidt, Karsten Berns

Technische Universität Kaiserslautern, Fachbereich Informatik, AG Robotersysteme,

E-mail: luksch@informatik.uni-kl.de

Zusammenfassung. Mobile Roboter, die Tätigkeiten in einer Indoor-Umgebung ausführen sollen, benötigen zur Orientierung eine Karte ihrer Umgebung, die sie im besten Falle selbst erstellen sollten. Bei der Exploration unbekannter Umgebung hat sich die Verwendung topologiebasierter Karten als sinnvoll erwiesen, da diese die Lokalisierung und Navigation für mobile Roboter vereinfachen. Probleme gab es bisher häufig aufgrund ungünstiger Rasterung oder fehlender Autonomie. Der vorgestellte Ansatz legt schrittweise eine topologische Karte aus rechteckigen Räumen an, während sich der Roboter durch seine Umgebung bewegt, die mit Laserscannern erfasst wird. Die Leistung des Verfahrens wird in der Simulation und in ersten realen Testläufen untersucht.

1 Einleitung

Für mobile Roboter in strukturierter Umgebung wie Bürogebäuden, Krankenhäusern, Museen oder Fabrikgeländen sind eine Vielzahl von Tätigkeiten denkbar, zum Beispiel Transportaufgaben, interaktive Führungen oder Gebäudeüberwachung. Für diese Anwendungen ist es zumeist notwendig, dass der Roboter eine Karte der Umgebung besitzt, um sich zu orientieren, d.h. Lokalisation und Navigation durchführen zu können.

Soll dem Roboter zudem keine Karte der Umwelt vorgegeben werden und er sie stattdessen selbst erstellen, ist das Problem der autonomen Exploration zu lösen (*simultaneous localizaton and mapping, SLAM*). Die Wahl der Repräsentationsform spielt dabei eine entscheidende Rolle.

1.1 Kartierungsformen für die Exploration im Indoor-Bereich

Die in der Robotik verwendeten Kartierungsformen lassen sich grob in drei Gruppen einteilen: Rasterkarten (grid maps), geometrische Karten und topologische Karten.

In Grid Maps ist die Umgebung in gleichgroße Kacheln eingeteilt, die einen Belegungszustand besitzen. Hier gibt es bereits viele verschiedene Ansätze, die beispielsweise die gesamte Roboterumgebung in einer einzigen großen Karte erfassen [1] oder aus einer Grid Map eine topologische Karte generieren [2]. Darüber hinaus gibt es auch Möglichkeiten der Integration beider Kartentypen, um die Vorteile beider Verfahren zu nutzen [3].

Bei geometrischen Karten wird die Umwelt durch extrahierte geometrische Merkmale beschrieben, wie z.B. Polygonzüge. Auch hier gibt es in der Literatur vielfältige

Bezeichnungen und Ausprägungen: Sie reichen von *Maps containing geometrical beacons* [4] über *Shape-based maps* [5] bis hin zu *Polygonal maps* [6].

In topologischen Repräsentationen ist die Karte zusätzlich in Regionen (Knoten) eingeteilt, die durch ihre Position und Konnektivität die topologische Struktur der Roboterumgebung wiedergeben. Existierende Ansätze, die topologische Karten bei der Exploration fortlaufend anlegen, generieren diese meist in festen Abständen [7][8]. Im vorgestellten Ansatz dagegen werden die einzelnen Räume (d.h. Regionen) der Topologie als Rechtecke modelliert und orientieren sich soweit wie möglich an den räumlichen Gegebenheiten; eine Darstellung, die sich z.B. in den Karten von [2] oder [9] findet und eher dem menschlichen Verständnis von Kartenelementen entspricht. Ein Grund für dieses vereinfachte Raummodell war die Intention, die Kartierung so abstrakt wie möglich zu halten und damit die Robustheit zu verbessern. Kleinere Strukturen der Umgebung, die in einer solchen Karte nicht verzeichnet sind, werden von der verhaltensbasierten Basissteuerung behandelt.

Die Lokalisierung mit derart vereinfachten Raummodellen wird in [9] und [10] untersucht. Dieses einfache Raummodell setzt allerdings restriktive Annahmen wie rechtwinklige Gebäudearchitektur voraus und ist lediglich in strukturierter Indoor-Umgebung einsetzbar, die man jedoch z.B. in Bürogebäuden häufig vorfindet.

2 Systemübersicht

Der Systemaufbau des vorgestellten Ansatzes gliedert sich in drei wesentliche Komponenten, die *Raumerkennung*, die *Kartenerstellung* und die *Lokalisierung* (siehe Abb. 1), die in die vorhandene verhaltensbasierte Basisarchitektur eingebunden werden. Die schon umgesetzten Komponenten werden im Folgenden beschrieben.

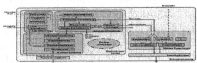

Abb. 1. Aufbau des (geplanten) Gesamtsystems

2.1 Raumerkennung

Die Raumerkennung extrahiert aus den Daten zweier planarer Laserscanner einen den Roboter umgebenden rechteckigen Raum. Die Raum-Extraktion wird in sechs Einzelschritten durchgeführt:

Zuerst werden die Rohdaten der Scanner (Abb. 3(a)) geglättet, dann werden Kannten mittels *recursive line splitting* extrahiert (Abb. 3(b)). In diesen Kanten werden nun Kandidaten für die vier Raumwände bestimmt(Abb. 2). Aufgrund des rechteckigen Raummodells wird dann ein Raum um den Roboter festgelegt (Abb. 3(c)). Im letzten Schritt werden Öffnungen und Verdeckungen des rechteckigen Raumumrisses bestimmt (Abb. 2). Dazu werden die vier Raumwände in so genannte Segmente eingeteilt, die angeben, wie die extrahierten Scankanten zu den aufgespannten Wänden liegen. Liegen Scankanten vom Roboter aus gesehen hinter der Wand, so werden an diesen Stellen Segmente vom Typ „Öffnung" eingefügt. Ab einer gewissen Mindestgröße werden diese Öffnungen als Durchgänge (Türen) in die topologische Karte eingetragen. Umgekehrt werden Segmente vom Typ „Verdeckt" eingesetzt, wenn Scankanten vor der Wand liegen. Das ist der Fall, wenn Teile der Wand durch Objekte verdeckt sind [11].

```
Eingabe aller realen Kanten k aus Kantenextraktion in Menge K
Aussortieren aller Kanten k mit Länge l(k) < l_min aus Menge K
FOR ALL k_i IN K DO
    Erzeuge leeren Wandkandidaten w_i mit Längenmaß L(w_i) = 0 in Menge
W
    FOR ALL k_j IN K DO
        IF k_j ELEMENT OF w-Umgebung von k_i
        AND Winkelabweichung alpha_ij < alpha_max
            Füge k_j dem von k_i aufgespannten Wandkandidaten w_i hinzu
            Addiere die Länge l(k_j) zu der Längenmaß L(w_i)
```

Abb. 2. Algorithmus zur Bestimmung von Wandkandidaten

2.2 Raumaktualisierung

Zwei weitere wichtige Aspekte sind die Erweiterung und das Aktualisieren der bereits erkannten Räume. Die Aktualisierung ist besonders hinsichtlich dynamischer Türen interessant, da diese geöffnet und geschlossen werden können. Dazu werden alle Wandsegmente um einen Zuverlässigkeitswert erweitert, der angibt, wie sicher diese erkannt wurden. Bewegt sich der Roboter durch einen erkannten Raum, erzeugt er regelmäßig eine zweite Raumkarte, die er mit der bereits gespeicherten vergleicht. Wenn neue oder besser erkannte Segmente gefunden werden, so wird die vorhandene Karte aktualisiert. Abbildung 4(a) zeigt einen langen Gang direkt nach der Initialisierung, d.h. die Raumdimensionen wurden erstmals ermittelt und die vier Raumwände indie bereits genannten Segmente unterteilt und Türen ermittelt. Dabei wurden die rechten und linken Bereiche des Raumes noch relativ unzuverlässig erkannt. Bewegt sich der Roboter nun durch den Raum, werden nach und nach die unzuverlässigen Segmente durchbesser erkannte Segmente ersetzt (Abb. 4(b)). Auf diese Weise erscheinen auch neue Türen, die in die topologische Karte eingetragen werden.

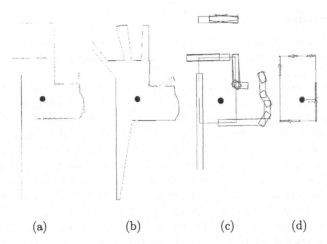

(a)   (b)   (c)   (d)

Abb. 3. (a) Die Rohdaten der Scanner, (b) Ergebnis der Kantenextraktion, (c) Aufgespannter
Raum, (d) Türen semantisch erfasst und Raumkoordinaten festgelegt

Neben der Aktualisierung ist noch die Erweiterung der Raumdimensionen notwendig: Mit zunehmender Entfernung werden die Laserscannerdaten unzuverlässig, weswegen in langen Gängen oder großen Räumen bei der hier vorgestellten Raumerkennung zuerst ein kleinerer Raum angelegt wird. Mit Hilfe der zweiten aktuellen Raumkarte kann der vorhandene Raum erweitert werden, wenn eine Wand undefiniert ist (d.h. die Scannerdaten in dieser Richtung zu schlecht waren) oder sie aus einer einzigen großen Öffnung besteht (Abb. 4(c)). Auf diese Weise können auch größere Räume vom Roboter erschlossen werden, die er nicht sicher mit einem Scan erfassen kann (Abb. 4(d)).

Bei dieser Aktualisierung können auch bereits erkannte Türen wieder aus der Raumkarte ausgetragen werden (wie beim Wechsel von Abb. 4(b) nach 4(c) unten links). Sie werden aber fest eingetragen, sobald sie vom Roboter durchfahren wurden. Auf diese Weise bleiben Türen in der topologischen Karte gespeichert, auch wenn sie zur Zeit geschlossen sind, der Roboter aber bereits hindurchgefahren ist.

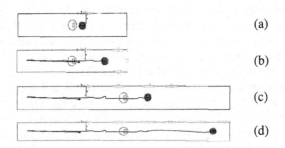

(a)

(b)

(c)

(d)

Abb. 4. (a) Raumdarstellung direkt nach der Initialisierung, (b) Aktualisierter Raum, (c) Erweiterter Raum, (d) Komplett erkannter Raum

2.3 Kartenerstellung

Wenn sich der Roboter durch seine Umgebung bewegt, werden, ausgehend vom Startraum, neue, noch unbekannte rechteckige Räume zur topologischen Karte hinzugefügt. Abbildung 5(a) zeigt eine Beispielfahrt durch den Flur und einige benachbarte Räume sowie die geometrische Darstellung der dabei erstellten topologischen Karte aus der Simulation. Die Topologie selbst beschreibt die Verbindung einzelner Räume. Diese Verbindungen entsprechen den Türen, die, ausgehend vom jeweiligen Raum, angeben, in welchen weiteren Raum sie führen. Führt eine Tür in einen unbekannten Raum, so wird beim Betreten ein neuer Raum entsprechend dem vorgestellten Verfahren angelegt und gespeichert. Ist der Raum hingegen bereits bekannt, wird er aus der topologischenKarte geladen. Dabei ist das Auftreten von Zyklen möglich, d.h. eine Tür führt in einen bereits bekannten Raum, den der Roboter jedoch durch eine andere Tür verlassen hat. Derartige Fälle werden bemerkt, indem für jede Tür des aktuellen Raumes überprüft wird, ob sie auf einer Wand eines Nachbarraumes liegt.

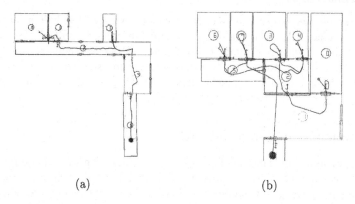

(a)   (b)

Abb. 5. (a) Kartenerstellung in langen Fluren, (b) Ablauf mit vielen Räumen

3 Ergebnisse

3.1 Simulationsergebnisse

Die Kartierung wurde in der Simulation getestet, in der keine Odometriefehler auftreten. Dazu wurden Abstandsdaten der Laserscanner entsprechend einer vorgegebenen Büroumgebung (Abb. 6(a)) generiert und mit Rauschen behaftet. Bei der Fahrt durch die simulierten Räumlichkeiten konnte das schrittweise Anlegen einer Karte überprüft werden (Abb. 5(b)). Da zur Zeit noch Explorations- und Navigationsstrategien fehlen, muss der Roboter manuell durch die Umgebung gefahren werden. Die topologische Karte wird aber ausschließlich anhand der vom Roboter gemessenen Umgebungsdaten erstellt und erweitert.

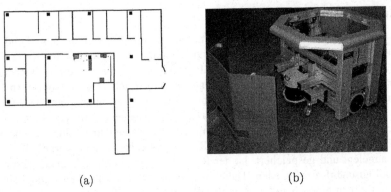

(a)   (b)

Abb. 6. (a) Simulierte Büroumgebung, (b) Roboter *Marvin* (offen)

3.2 Der Roboter *Marvin*

Der Roboter *marvin* (**m**obile **a**utonomous **r**obotic **v**ehicle for **in**door **n**avigation, Abb. 6(b)) ist eine AMR-Plattform mit Differentialantrieb, zwei SICK-Laserscannern und einer verhaltensbasierten Steuerungsarchitektur. Erste Versuche zeigen, dass der Ansatz mit dem realen Roboter vielversprechend ist und sich wie in der Simulation verhält, jedoch sind weitere Versuche über einen längeren Zeitraum nur mit einer vollständig umgesetzten Positionskorrektur sinnvoll.

4 Zusammenfassung und Ausblick

Es wurde ein Ansatz zur selbstständigen Erstellung einer Umgebungskarte für Indoor-Roboter vorgestellt. Da die lokale Navigation durch die verhaltensbasierte Basissteue-rung durchgeführt wird, fiel die Entscheidung auf eine abstrakte, topologiebasierte Repräsentationsform. Es konnte in der Simulation und eingeschränkt in der Realität gezeigt werden, dass der Roboter eine solche Karte während einer Explorationsfahrt in einer Umgebung mit rechtwinkligen Räumen autonom anlegen und erweitern kann. Die darin enthaltene ungefähre Position der Türen jedes Raumes reicht aus, um eine verhaltensbasierte Navigation von Raum zu Raum zu erlauben.

Nächste Schritte bei der Weiterentwicklung sind die Erweiterung der Positionsbe-stimmung des Roboters, die bislang nur zum Teil implementiert ist. Das folgende Ziel ist die autonome Exploration und Kartierung, die mittels einer verhaltensbasierten Steuerung realisiert werden soll. Dabei soll der Roboter nicht unbedingt den kürzesten, wohl aber den für ihn interessantesten Weg wählen. Weiterhin ist angedacht, inner-halb der lokalen Räume zusätzliche Detailinformationen zu sammeln und beispiels-weise in Form von grid maps oder geometrischer Primitive einzufügen. Das ist be-sonders dann wichtig, wenn die Räume eine semantische Bedeutung erhalten und der Roboter verschiedene Punkte anfahren oder manipulieren soll. Dafür muss aber vor allem auch die Sensorik des Roboters erweitert werden. In zugestellten Räumen kann es bei dem planaren Laserscanner vorkommen, dass bei der Raumerkennung keine oder falsche Wände gefunden werden. Hier kann beispielsweise ein 3D-Laserscanner Abhilfe schaffen. In Zukunft ist auch eine Erweiterung des Raummodells denkbar, welches auch nicht-rechteckige Räume zulässt.

Literaturverzeichnis

1. Yamauchi E, Schultz A, Adams W: Mobile Robot Exploration and Map-Building with continuous Localization. IEEE/ICRA, 1998
2. Fabrizi E, Saffiotti A: Augmenting Topology-Based Maps with Geometric Information. Robotics and Autonomous Systems: 91–97, 2002
3. Thrun S: Learning Metric-Topological Maps for Indoor Mobile Robot Navigation. Artificial Intelligence, Bd. 1: 21–71, 1998
4. Althaus P, Christensen H I: Automatic Map Acquisition for Navigation in Domestic Environments. IEEE/ICRA: 1551–1556, 2003
5. Wolter D, Latecki L J, Lakämper R, Sun X: Shape-Based Robot Mapping. 27th German Conference on Artificial Intelligence: 439–452, 2004
6. Latecki L J, Lakämper R, Sun X, Wolter D: Building Polygonal Maps from Laser Range Data. ECAI International Cognitive Robotics Workshop, 2004
7. Bosse M, Newman P, Leonard J, et al.: An Atlas Framework for Scalable Mapping. IEEE/ICRA: 1899–1906, 2003
8. Yamauchi B, Langley P: Place learning in dynamic real-world environments. RoboLearn 96:, 123–129, 1996
9. Jensfelt P, Austin D. J, et al.: Feature Based Condensation for Mobile Robot Localization. IEEE/ICRA: 2531–2537, 2000
10. Jensfelt P, Christensen H. J.: Laser Based Pose Tracking. IEEE/ICRA: (4), 2994–2998, 1999
11. Kleinlützum K: Selbständige Kartierung mit einem autonomen Indoor-Roboter. Diplomarbeit, Technische Universität Kaiserslautern, 2005

Integration of a Sound Source Detection into a Probabilistic-based Multimodal Approach for Person Detection and Tracking

Robert Brückmann, Andrea Scheidig, Christian Martin, Horst-Michael Gross

Ilmenau Technical University, Department of Neuroinformatics and Cognitive Robotics

Abstract. Dealing with methods of Human-Robot-Interaction and using a real mobile robot, stable methods for people detection and tracking are fundamental features of such a system and require information from different sources. Based on an existing probability-based and multimodal approach for person detection and tracking, in this paper, we discuss the integration of a further sensory cue. This sensory cue is a sound source detection emerged from auditory information. Firstly, we discuss a newly developed approach for a sound source detection applied for a real world problem, dealing with the difficulty of reverberant environments. Secondly, we show a possible solution to integrate the sound source detection into the already existing person detection and tracking system applied for the mobile interaction robot HOROS working in a real office environment.

1 Introduction

Dealing with Human-Robot-Interaction (HRI) especially in real-world environments, one of the general tasks is the realization of a stable people detection and the respective tracking functions. Depending on the specific application that integrates a person detection, different approaches are possible. For real world problems, most promising approaches combine different sensory channels like visual cues and the scan of a laser-range-finder. Beside these sensory cues in the context of HRI also the auditory cue yields important information of the position of an interaction partner. Exemplary approaches which combine such different sensory cues are the SIG robot (auditory and visual cues) [7] or the BIRON project [2] (laser-range-finder, visual and auditory cues). The drawback of these approaches is the sequential processing of the sensory cues. For instance, people are detected by the laser information only and are subsequently verified by visual cues. Problems occur, when the laser-range-finder yields no information, for instance in situations when only the face of a person is perceivable.

To overcome this drawback, in [6] we propose a multimodal approach to realize the detection of people and the respective tracking functions. As sensory channels in [6] we use the following sensory modalities of our mobile interaction robot HOROS: the omnidirectional camera, the sonar sensors, and the laser-range-finder. A main advantage of our approach is the simple integration of further sensory channels, like sound sources because of the used aggregation scheme. So in this paper, we firstly present an approach for sound source detection (see section 4), that will be integrated in the whole tracker system (see section 3). In result people can be detected by their legs, their faces and also by their speech based interaction or by only one of these features respectively. Respective results for the sound source detection in the context of a real world application will be shown in section 4.

2 Robot System HOROS

To investigate respective methods, we use the mobile interaction robot HOROS as an information system for employees, students and guests of our institute. The system's task includes that HOROS autonomously moves in the institute, detects persons as possible interaction partners and interacts with them, for example, to answer questions like the current whereabouts of specific persons.

The hardware platform for HOROS is a Pioneer-II-based robot from ActiveMedia. For the purpose of HRI, this platform was extended with different modalities. This includes a Tablet PC running under Windows XP for touch-based interaction, speech recognition and speech generation. It was further extended by a robot face which includes an omnidirectional fisheye camera, two webcams, and two microphones.

Laser-based Information: The laser-range-finder is a very precise sensor with a resolution of one degree, perceiving the frontal 180 degree field of HOROS. The laser-range-finder is fixed on the robot approximately 30 cm above the ground. Therefore it can only perceive the legs of people. Based on the approach presented in [1], we also analyze the scan of the laser-range-finder for leg-pairs using a heuristic method.

Sonar Information: Information from the sonar tends to be very noisy, imprecise und unreliable. Therefore, the variances are large and the impact on the certainty of a hypothesis is minimal. Nevertheless, the sonar is included to support people tracking behind the robot. So we are able to form an estimate of the distance in vision-based hypotheses.

Fisheye Camera: For HOROS we use an omnidirectional camera with a fisheye lens yielding a 360 degree view around the robot. Because of the task of person detection, the usage of such a camera requires that the position of the camera is lower than the position of the faces. To detect people in the omnidirectional camera image a skin-color-based multi-target-tracker[8] is used. This tracker is based on the condensation algorithm[3] which has been extended, so that the visual tracking of multiple people at the same time is now possible. A person detection using omnidirectional camera images yields hypotheses about the direction of a person but not about the distance.

Sound Source Detection: There are two electret-microphones attached to the head of HOROS which are used to detect acoustic sources. The distance between them is approximately 27 cm. With the detection algorithm described in section 4 the angle between the sound source and the robot can be calculated by using the time delay of the sound. The possible resolution of the angle is up to two degrees for sources right in front of the robot. The two microphones don't allow for a full 360-degree-detection. Only sources between -90 and +90 degrees can be detected. Thus the combination with other sensory cues is necessary to avoid wrong detections.

The integration of the information from the camera and the sound source detection with the information from the laser-range-finder and the sonar sensors results in a powerful person detection system. Subsequently the developed method for the combination of the sensory systems will be discussed.

3 Generation of User Models

At first, a suitable data representation for the aggregation of the multimodal hypotheses resulting from the different sensor readings has to be choosen. The possibilities range from simple central point representation to probability distributions approximated by particles. The aggregation scheme we use is based on Gaussian distributions. Because of the unknown correlations between the different sensor readings, we did not use a Kalman Filter based approach to combine these hypotheses. Instead Covariance Intersection is applied [6].

First for the purpose of tracking, the sensory information about detected humans is converted into Gaussian distributions. The mean of each Gaussian distribution equals the position of the detection and the covariance matrix represents the uncertainty about this position. The form of the covariance matrix is sensor dependent due to different sensor characteristics, like their accuracy. Furthermore, the sensors have different error rates of misdetections that have to be taken into account.

Tracking based on probabilistic methods attempts to improve the estimate of the position of a human at each time. These estimates are integrated into a local map that contains all hypotheses around the robot. This map is also used to aggregate the sensor hypotheses from the current sensor readings. A sensor reading and a hypothesis with a minimum distance are merged. This update is done via the *Covariance Intersection* rule [4]. Sensor readings not matched with a hypothesis of the local map are introduced as a new hypothesis.

4 Integration of Sound Source Detection as a further Hypothesis

Besides the other sensory cues of the multimodal tracker, information gathered from sound sources is another important input for an interaction system. When implementing a sound source detection, especially reverberation often leads to wrong detection results. Therefore, we present a new approach for dealing with this issue.

Detection of sound sources can be achieved by using at least two spatially separated microphones. These receive the sound with a time delay which can be used to calculate the angle between the microphone array and the sound source. Hence our approach of detecting a speaker is based on the time delay of arrival (TDOA) between two microphones. In absence of noise and reverberation, the cross-correlation is a good method to measure the TDOA value:

$$r_{ij}(t) = \int_{-\infty}^{+\infty} x_i(\tau) x_j(t+\tau) d\tau \qquad (1)$$

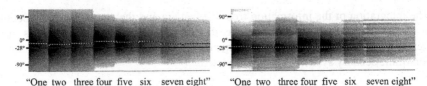

"One two three four five six seven eight"   "One two three four five six seven eight"

Fig. 1. Localization results compared to an unmodified cross-correlation for a speaker counting from 1 to 8 in a reverberant room. Distance to the robot was approx. 6 m, angle was approx. -28 degrees. Left: Results using only standard cross-correlation. The correct angle to be detected is marked with a dark-gray line at -28 degrees, the actual result is shown by the dotted line. Right: Results using the proposed localization approach. The dotted line shows the maximum of the correlation function at -28 degrees.

where x_i is the signal of microphone i. The position of the maximum in the cross-correlation represents the delay between the two signals. The cross power spectrum

$$R_{ij}(\omega) = X_i(\omega) \cdot X_j^*(\omega) \qquad (2)$$

is the Fourier transform of the cross-correlation, where $X_i(\omega)$ is the Fourier transform of $x_i(t)$. Using the cross power spectrum, the influence of each frequency component can be weighted. The phase transform proposed in [5] uses the cross power spectrum enhanced by such a weighting function:

$$r_{ij}^{(g)}(t) = \int_{-\infty}^{\infty} \Psi_{ij}(\omega) X_i(\omega) X_j^*(\omega) e^{j\omega t} d\omega \qquad (3)$$

where $\Psi_{ij}(\omega)$ is defined as

$$\Psi_{ij}(\omega) = \frac{1}{\left| X_i(\omega) X_j^*(\omega) \right|} \qquad (4)$$

This weighting function normalizes the Fourier spectrum by setting the absolute values for all ω to 1. Thus only the phase of each frequency component remains. This whitening of the data narrows the resulting peak in the cross correlation function $r_{ij}^{(g)}(t)$ making the detection of the TDOA value easier.

A drawback of this transform is that every frequency bin of the Fourier spectrum will have the same influence to the resulting cross correlation, even if it is dominated by noise or contains reverberation. We added another weight to the phase transform which provides the possibility to weight different frequencies according to their probability to contain reverberation.

$$\Psi_{ij}^{(e)}(\omega) = \frac{w(\omega)}{\left| X_i(\omega) X_j^*(\omega) \right|} \qquad (5)$$

The function $w(\omega)$ can be used to decrease the influence of a frequency bin if it contains reverberation.

It is assumed that reverberation is received after the direct sound with a room-specific delay because the echo always has to cover a longer distance. By applying the cross-correlation only to the beginning of a perceived sound we can improve the results of the phase transform. A kind of onset-filter is used to implement this behaviour. The digital audio data of the two microphones is processed using windows of 1024 samples at a sample frequency of 44.1 kHz. Using the Fast Fourier Transform (FFT) we calculate the Fourier coefficients $X_i(k)$ of the windows for each microphone i. The weighting function for the discrete spectrum is expressed by

$$\Psi_{ij}^{(e)}(k) = \frac{w(k)}{\left|X_i(k)X_j^*(k)\right|} \tag{6}$$

We use thresholds for each frequency component to calculate the weights. The threshold values and the weights are adapted after the processing of each window.

$$w^t(k) = \min\left(0, X(k) - o^t(k)\right) \tag{7}$$

$$o^{t+1}(k) = \begin{cases} X^t(k) & , o^t(k) < X^t(k) \\ \alpha \cdot o^t(k) & , o^t(k) \geq X^t(k) \end{cases} \tag{8}$$

with $X(k)$ being the mean power spectral density of the two microphone channels. If a peak in one frequency bin has been found, then $X(k)$ is usually much larger than $o(k)$. The weight won't suppress the influence of this frequency for the current window. Subsequently this frequency band is inhibited for some time by raising the threshold $o(k)$. This ensures that the weight will be nearly 0 for the following windows. Therefore the reverb following the direct sound will not be evaluated in the cross correlation. It will decay over time and the threshold values will be decreased with each window using the decay factor $\alpha \in (0...1)$. α determines how fast a frequency band will be available for the detection of new onsets. For our implementation, we empirically set α to 0.95. Such an onset detection is used separately for each frequency component. For speech signals, where the spectrum of the sound changes over time, it is possible to detect new onsets in different frequency bands while other bands are inhibited due to reverberation.

The angle between the microphone array and the sound source can be computed if the TDOA is known. We use a model to estimate this angle which assumes that the distance to the sound source is much larger than the distance between the two microphones.

$$\Theta = \arcsin\frac{v_{sound} \cdot \Delta t_{TDOA}}{\Delta s_m} \tag{9}$$

The distance between the two microphones is described by Δs_m, v_{sound} is the sound velocity.

To test the localization algorithm, we placed a loudspeaker at different positions around the robot at a distance of approximately 75 cm and presented a recording of a male speaker. The mean angular error for different angles is shown in Fig. 2. Because of the non-linear relation between the TDOA and the computed angle, the resolution of the detection decreases for sources located on the sides of the robot. Practically, angles over 70 degrees only result in a rough directionestimation, whereas source right in front of the robot can be located quite accurately. Since our system is developed for

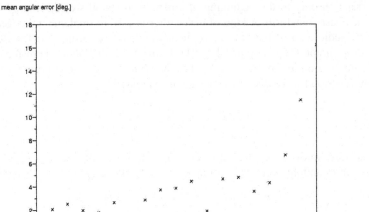

mean angular error [deg.]

angle [deg.]

Fig. 2. Mean angular error of the sound source detection as a function of the actual angle between the robot and the sound source.

Human-Robot-Interaction, the robot will turn towards the speaker. So the sound source will always be in front of the robot after therotation, making the detection of the speaker easier.

Integration into a multimodal tracker

The integration of the found angle as a new hypothesis for the multimodal tracker is done by adding a Gaussian distribution representing the perceived speaker. It turned out that different types of audio signals yield different results with respect to localization robustness. A wide-band signal like a hand clap is generally easier to detect than a narrow-band signal, eg. a sine tone. Additionally, louder signals yield better localization results because of their higher signal-to-noise ratio. To take such details into account, the covariance of the Gaussian distribution illustrates the uncertainty depending on the broadness of the spectrum in conjunction with the overall loudness of the perceived audio signal. The following uncertainty value can be used for the determination of the covariance:

$$Q_{angle} = \sum_{k=1}^{N} (w(k) \cdot X(k)) \tag{10}$$

This quality value increases if there are many frequency bins which contain loud signals and which are not inhibited by small weights. High values of Q_{angle} result in a narrow covariance while low values lead to wider covariances.

5 Summary and Conclusions

In this paper we have shown an approach for detecting a sound source using two microphones. A new method has been described allowing detection even in reverberant environments. The result has been integrated with a multimodal sensor tracker supporting thedetection of people by adding a hypothesis of a possible speaker position. In our future work, we will adjust the balancing between the already existing sensory cues and the newly integrated tracker hypothesis generated by the sound source localization.

References

1. J. Fritsch, M. Kleinehagenbrock, S. Lang, T. Ploetz, G.A. Fink, and G. Sagerer. Multimodal anchoring for human-robot-interaction. *Robotics and Autonomous Systems, Special issue on Anchoring Symbols to Sensor Data in Single and Multiple Robot Systems*, 43(2–3):133–147, 2003.

2. A. Haasch, S. Hohenner, S. Huewel, M. Kleinehagenbrock, S. Lang, I. Toptsis, G.A. Fink, J. Fritsch, B. Wrede, and G. Sagerer. Biron – the bielefeld robot companion. In *International Workshop on Advances in Service Robots*, pages 898–906, May 2004.

3. M. Isard and A. Blake. Condensation – conditional density propagation for visual tracking. *International Journal on Computer Vision*, 29:5–28, 1998.

4. S. Julier and J. Uhlmann. A nondivergent estimation algorithm in the presence of unknown correlations. In *Proceedings of the 1997 American Control Conference*, pages 2369–2373 vol.4. IEEE, June 1997.

5. C.H. Knapp and C. Carter. The generalized correlation method for estimation of time delay. *IEEE Transactions on Acoustics, Speech and Signal Processing*, 24(4):320–327, 1976.

6. C. Martin, E. Schaernicht, A. Scheidig, and H.-M. Gross. Sensor fusion using a probabilistic aggregation scheme for people detection and tracking. In *Proc. of ECMR*, 2005.

7. K. Nakadai, H.G. Okuno, and H. Kitano. Auditory fovea based speech separation and its application to dialog system. In *IEEE/RSJ International Conference on Intelligent Robots and Systems (IROS-2002)*, volume 2, pages 1320–1325, 2002.

8. T. Wilhelm, H.-J. Boehme, and H.-M. Gross. A multi-modal system for tracking and analyzing faces on a mobile robot. In *Robotics and Autonomous Systems*, volume 48, pages 31–40, 2004.

Using Descriptive Image Features
for Global Localization of Mobile Robots

Hashem Tamimi, Alaa Halawani, Hans Burkhardt, Andreas Zell

Computer Science Dept.,University of Tübingen, Sand 1, 72076 Tübingen, Germany
Albert-Ludwigs-University of Freiburg, Chair of Pattern Recognition and Image Processing,
79110 Freiburg, Germany

E-mail: {tamimi,zell}@informatik.uni-tuebingen.de
{halawani,burkhardt}@informatik.uni-freiburg.de

Abstract. In this paper descriptive visual features based on integral invariants are proposed to solve the global localization of indoor mobile robots. These descriptive features are locally extracted by applying a set of non-linear kernel functions around a set of interest points in the image. To investigate the approach thoroughly, we use a set of images taken by re-assigning the robot position many times near a set of reference locations. Also, the presence of illumination variations is encountered many times in the images. Compared to a well-known approach, our approach has better localization rate with moderate computational overhead.

1 Introduction

Vision-based robot localization demands image features with many properties. On one hand the features should exhibit invariance to scale and rotation as well as robustness against noise and changes in illumination. On the other hand they should be extracted very quickly so as not to hinder other tasks that the robot plans to perform. Although both global and local features are used to solve the robot localization problem, local features are more commonly employed because they can be computed efficiently, are resistant to partial occlusion, and are relatively insensitive to changes in viewpoint. There are two considerations when using local features [4]: First, the interest points should be localized in position and scale. Interest points are positioned at local peaks in a scale-space search, and filtered to preserve only those that are likely to remain stable over transformations. Second, a signature of the interest point is built. This signature should be distinctive and invariant over transformations caused by changes in camera pose as well as illumination changes. While point localization and signature aspects of interest point algorithms are often designed together, they can be considered independently [6].

In this paper we propose the application of the integral invariants to the robot localization problem on a local basis. First, our approach detects a set of interest points in the image based on a Difference of Gaussian (DoG)-based interest point detectordeveloped by Lowe [5]. Then, it finds a set of descriptive features based on the integral invariants around each of the interest points. These features are invariant to similarity transformation (translation, rotation, and scale). Our approach proves to lead to significant localization rates and outperforms a previous work that is described in Section 4.

2 Integral invariants

Following is a brief description of the calculation of the rotation- and translation-invariant features based on integration. The idea of constructing invariant features is to apply a nonlinear kernel function $f(\mathbf{I})$ to a gray-valued image, $\mathbf{hbfI}$, and to integrate the result over all possible rotations and translations (Haar integral over the Euclidean motion):

$$\mathbf{T}[f](\mathbf{I}) = \frac{1}{PMN} \sum_{n_0=0}^{M-1} \sum_{n_1=0}^{N-1} \sum_{p=0}^{P-1} f(g(n_0, n_1, p\frac{2\pi}{P})\mathbf{I}) \tag{1}$$

where $\mathbf{T}[f](\mathbf{I})$ is the invariant feature of the image, M, N are the dimensions of the image, and g is an element in the transformation group G (which consists here of rotations and translations). Bilinear interpolation isapplied when the samples do not fall onto the image grid. The above equation suggests that invariant features are computed by applying a nonlinear function, f on the neighborhood of each pixel in the image, then summing up all the results to get a single value representing the invariant feature. Using severaldifferent functions finally builds up a feature space.

To preserve more local information we remove the summation over all translations. This results in a map $\mathbf{T}$ that has the same dimensions of $\mathbf{I}$:

$$\left(\mathbf{T}[f]\mathbf{I}\right)(n_0, n_1) = \frac{1}{P} \sum_{p=0}^{P-1} f\left(g\left(n_0, n_1, p\frac{2\pi}{P}\right)\mathbf{I}\right) \tag{2}$$

Applying a set of different fs will result in a set of maps. A global multi-dimensional feature histogram is then constructed from the elements of these maps. The choice of the non-linear kernel function f can vary. For example, invariant features can be computed by applying the monomial kernel, which has the form:

$$f(\mathbf{I}) = \left(\prod_{p=0}^{P-1} \mathbf{I}(x_p, y_p)\right)^{\frac{1}{p}} \tag{3}$$

One disadvantage of this type of kernels is that it is sensitive to illumination changes. The work in [7] defines another kind of kernels that are robust to illumination changes. These kernels are called relational kernel functions and have the form:

$$f(\mathbf{I}) = rel\left(\mathbf{I}(x_1, y_1) - \mathbf{I}(x_2, y_2)\right) \tag{4}$$

with the ramp function

$$rel(\gamma) = \begin{cases} 1 & \text{if} \gamma < -\varepsilon \\ \dfrac{\varepsilon - \gamma}{2\varepsilon} & \text{if} -\varepsilon \leq \gamma \leq \varepsilon \\ 0 & \text{if} \varepsilon < \gamma \end{cases} \tag{5}$$

centered at the origin and $0 < \varepsilon < 1$ is chosen by experiment. Please refer to [?] for detailed theory.

3 DoG-based point detector

The interest points, which are used in our work, were first proposed as a part of the Scale Invariant Feature Transform (SIFT) developed by Lowe [5]. These features have been widely used in the robot localization field [8] [11].The advantage of this detector is its stability under similarity transformations, illumination changes and presence of noise.

The interest points are found as scale-space extrema located in the Difference of Gaussians (DoG) function, $D(x, y, \sigma)$, which can be computed from the difference of two nearby scaled images separated by a multiplicative factor k:

$$D(x,y,\sigma) = (G(x,y,k\sigma) - G(x,y,\sigma)) * I(x,y)$$
$$= L(x,y,k\sigma) - L(x,y,\sigma) \tag{6}$$

where $L(x, y, \sigma)$ defined the scale space of an image, built by convolving the image $I(x, y)$ with the Gaussian kernel $G(x, y, \sigma)$. Points in the DoG function which are local extrema in their own scale and one scale above and below are extracted as interest points. The interest points are then filtered for more stable matches, and more accurately localized to scale and subpixel image location using methods described in [2].

4 Using global integral invariants for robot localization

In [12], integral invariants are used to extract global features for solving the robot localization problem by applying Equation 2 to each pixel (n_0, n_1) in the image **I**. The calculation of the matrix **T** involves finding an invariant value around each pixel in the image which is time consuming. Instead of this, Monte-Carlo approximation is used to estimate the overall calculation [10]. This approximation involves applying the nonlinear kernelfunctions to a set of randomly chosen locations and directions rather than to all locations and directions.

Global features achieve robustness mainly because of their histogram nature. On the other hand, local features, extracted from areas of high relevance in the image under consideration, are more robust in situations where the objects in images are scaled or presented in different views [3]. Such situations are often encountered by the robot during its navigation. In the next sections we modify the global approach by applying the integral invariants locally around a set of interest points.

5 Extracting local integral invariants

Unlike the existing approach, explained in Section 2, the features that we propose are not globally extracted; they are extracted only around a set of interest points. Our approach can be described in the following steps:

1. **Interest point detection:**
   The first stage is to apply the DoG-based detector to the image in order to identify potential interest points. The location and scale of each candidate point are determined and the interest points are selected based on measures of stability described in [5].

2. **Invariant features initial construction:** For each interest point located at (n_0, n_1) we determine the set of all points which lie on the circumference of a circle of radius r_1. We use bilinear interpolation for sub-pixel calculation. Another set of points that lie on a circumference of a circle of radius r_2 are determined in the same manner. Both circles have their origin at (n_0, n_1). To make the features invariant to scale changes, the radii are adapted linearly to the local scale of each interest point. This way the patch that is used for feature extraction always covers the same details of the image independent of the scale.

3. **Nonlinear kernel application:** A non-linear kernel function is applied to the values of the points of the two circles. Each point located at (x_0, x_0) on the circumference of the first circle is tackled with another point located at (x_1, x_1) on the circumference of the second circle. taking into consideration a phase shift θ between the corresponding points. This step is repeated together with step 2 for a set of V kernel functions f_i, $i = 1, 2, ..., V$. The kernels differ from each other by changing r_1, r_2 and θ Finally we apply Equation 7 for each interest point located at (n_0, n_1).

$$F_i(n_0, n_1) = \frac{1}{P} \sum_{p=0}^{P-1} f_i\left(g\left(n_0, n_1, p\frac{2\pi}{P} \right) \mathbf{I} \right), i = 1, 2, \cdots, V. \tag{7}$$

We end up with a V-dimensional feature vector, F, for each single interest point.

6 Experimental results

In this section we present the experimental results of our local integral invariants compared with the global integral invariants reviewed in Section 4.

6.1 The database of images

To simulate the robot localization we use a set of 264 gray scale images taken at 33 different reference locations. Each has a resolution of 330×240. In each reference location we apply the following scenario capturing an image after each step: (1) The robot stops. (2) It translates 50 cm to the right. (3) It rotates 20 degrees to the left. (4) It moves 50 cm ahead. (5) It rotates 40 degrees to the right. (6) It rotates 20 degrees to the left. (7) It moves the camera up 5 degrees using the pan-tilt unit. (8) It moves the camera down 10 degrees using the pan-tilt unit.

The database is divided into two equal parts. 132 images are selected for training and 132 for testing. This partitioning is repeated 50 times with different combinations for training and testing images. The average localization rate is computed for each combination. We assume that optimal localization results are obtained when each input image from one of the reference locations matches another image that belongs the same reference location.

6.2 Global integral invariants

In order to compare our work with the work of [12] which involves calculating the global features, a set of 40×10 random samples is used for the Monte-Carlo approximation, which is also suggested in [10] for best performance. For each sample we apply a set of three different kernels. Both monomial and texture kernels functions are investigated for best localization accuracy using the above images. For each image a single D-dimensional histogram is build with $D=3$. Each dimension contains 8 bins which has experimentally led to best results. The histograms are compared using the $l-Norm$ measure.

Fig. 1. Matching two different images using the proposed approach.

6.3 Local integral invariants

We use the following parameters when implementing our descriptive features: For each interest point we set $V=12$ which gives us a 12-dimensional feature vector that is generated using a set of either relational kernel functions or monomial kernel functions. Best results were obtained with $\varepsilon = 0.098$ in Equation 5.

When evaluating the localization approach we first compare each individual feature vector from the image in the query with all the other feature vectors, extracted from the training set of images, using the $l-Norm$ measure. Correspondences between the feature vectors are found based on the method described in [1] which leads to robust matching. Then we apply a voting mechanism to find the corresponding image to the one in query. The voting is basically performed by finding the image that has the maximum number of matches. Figure 1 gives an example of the correspondences found between two different images using the proposed approach.

6.4 Results

Figures 2(a) and 2(b) both demonstrate the localization rate of our approach and the existing approach. Figure s(a) shows the results that we get using the relational kernels whereas Fig. 2(b) shows the results using the monomial kernels. It can be seen that our approach performers better than the global integral invariants approach using any of the two kernel types but gives best results using the relational kernel based features.

Table 1 shows the best results of the two approaches. The overall average localization rate and the computation time of the proposed approach (using relational kernels), compared with the global integral invariants (using monomial kernels), are depicted.

Table. 1. The overall localization rate and localization time of both approaches.

	Overall average localization rate	Average localization time (seconds)
Local integral invariants (relational)	96.51%	0.86
Global integral invariants (monomial)	74.79%	0.42

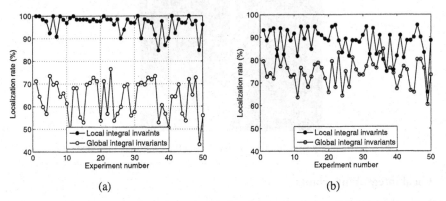

(a)   (b)

Fig. 2. Localization rate of the of the descriptive features and the global features using (a) relational kernels (b) monomial kernels.

7 Conclusion

In this paper we have proposed new descriptive features for robot localization based on local integral invariants. The descriptive features have a compact size but are capable of matching images with high accuracy. In comparison with global features, our descriptive features show better localization results with a moderate computational overhead.

8 Acknowledgment

The first and second author would like to acknowledge the financial support by the German Academic Exchange Service (DAAD) of their PhD. scholarship at the Universities of Tübingen and Freiburg in Germany.

References

1. P. Biber and W. Straßer. Solving the correspondence problem by finding unique features. In *16th International Conference on Vision Interface*, 2003.
2. M. Brown and D. Lowe. Invariant features from interest point groups. In *British Machine Vision Conference*, BMVC, Cardiff, Wales, September 2002.
3. A. Halawani and H. Burkhardt. Image retrieval by local evaluation of nonlinear kernel functions around salient points. In *Proceedings of the 17th International Conference on Pattern Recognition (ICPR)*, volume 2, pages 955–960, Cambridge, United Kingdom, August 2004.

4. Y. Ke and R. Sukthankar. PCA-SIFT: A more distinctive representation for local image descriptors. In *CVPR (2)*, pages 506–513, 2004.
5. D. Lowe. Distinctive image features from scale-invariant keypoints. *Int. J. Comput. Vision*, pages 91–110, 2004.
6. K. Mikolajczyk and C. Schmid. A performance evaluation of local descriptors. In *International Conference on Computer Vision & Pattern Recognition*, pages 257–263, June 2003.
7. M. Schael. Texture defect detection using invariant textural features. *Lecture Notes in Computer Science*, 2191:17–24, 2001.
8. S. Se, D. Lowe, and J. Little. Vision-based mobile robot localization and mapping using scale-invariant features. In *Proceedings of the IEEE International Conference on Robotics and Automation (ICRA)*, pages 2051–2058, Seoul, Korea, May 2001.
9. S. Siggelkow. *Feature Histograms for Content-Based Image Retrieval*. PhD thesis, Albert-Ludwigs-Universität Freiburg, Fakultät für Angewandte Wissenschaften, Germany, December 2002.
10. S. Siggelkow and M. Schael. Fast estimation of invariant features. In W. Förstner, J. M. Buhmann, A. Faber, and P. Faber, editors, *Mustererkennung, DAGM*, pages 181–188, Bonn, Germany, September 1999.
11. H. Tamimi, H. Andreasson, A. Treptow, T. Duckett, and A. Zell. Localization of mobile robots with omnidirectional vision using particle flter and iterative SIFT. In *Proceedings of the 2005 European Conference on Mobile Robots (ECMR05)*, Ancona, Italy, 2005.
12. J. Wolf, W. Burgard, and H. Burkhardt. Robust vision-based localization by combining an image retrieval system with monte carlo localization. *IEEE Transactions on Robotics*, 21(2):208–216, 2005.

Outdoor-Systeme

Extension Approach for the Behaviour-Based Control System of the Outdoor Robot RAVON

Bernd Helge Schäfer, Martin Proetzsch, Karsten Berns

Kaiserslautern University of Technology, Department of Informatics, Gottlieb-Daimler-Straße, 67663 Kaiserslautern, Germany

b_schaef@informatik.uni-kl.de, proetzsch@informatik.uni-kl.de

Abstract. This paper describes the extension of a behaviour-based control system for autonomous outdoor navigation. To perform robust obstacle avoidance the existing stereo vision system is complemented with a 2D laser scanner. Instead of combining raw sensor data, the additional information is integrated exploiting behaviour fusion. The scanner data processing is wrapped into new behaviours, which are added to the existing network. That way no modifications to existing behaviours and interconnections are necessary. Furthermore sensors can be evaluated individually augmenting traceability and testability. The performance of the resulting system is shown in a real world scenario.

1 Introduction

Recent development in robotics has revealed a strong demand for autonomous vehicles in unmanned space travel, autonomous farming, civil protection and humanitarian demining. These applications require robots to navigate in unstructured natural terrain. The diversity of outdoor environments calls for a flexible and extensible control architecture. In order to achieve both robust and predictable performance an integrative paradigm is required. This paper presents the behaviour-based control system of the mobile outdoor platform RAVON (Robust Autonomous Vehicle for Off-road Navigation, see Fig. 1) with a strong focus on the extensibility aspect. In this context new sensor equipment is integrated into a working system. New behaviours reacting

Abb. 1. The four-wheeled outdoor vehicle RAVON

on the additional sensor data are designed and interconnected with the existing behaviour structure. The resulting system represents a robust upgrade to the previous behaviour network. That way only functionality which is directly affected by the behaviours introduced has to be tested again.

Using multiple sensors for obtaining information about the environment has been an issue in many fields of robotic research (see [1] for an overview). Contributions dealing with outdoor environments often introduce explicit mechanisms for sensor fusion. In [2] for example the learning aspect of classifier fusion is presented while [3] proposes a traversability grid for joining data yielded from different sensors. The high complexity of such approaches is in particular problematic on reactive control layers. In [4] a behaviour-based control system employing multiple sensors is described. Yet no overlapping sensor data is considered in this work neglecting certain aspects covered here.

2 Mobile robot platform RAVON

The platform used in experiments is the all-terrain vehicle RAVON depicted in Fig. 1. With 2.4 m length and 1.4 m width the 350 kg heavy robot can climb slopes of 100 % inclination. The vehicle features a four wheel drive with independent motors yielding maximal velocities of 3 m/s. Front and rear axis can be steered separately.

Self localisation is realised employing odometry and an inertial measurement system (IMS). Fusion with a GPS-receiver is under development. As visual sensors a planar laser range finder from SICK (field of vision: 180°, angular resolution: 0.5, distance resolution: about 0.5 cm) as well as a custom design stereo camera head have been mounted to support obstacle detection. The stereo vision system (see Fig. 1) has been built up using two Sony DFW V500 ccd cameras equipped with 6 mm lenses (horizontal field of vision: 43°). A detailed description of the setup can be found in [5].

3 Control software extension exploiting behaviour fusion

The hierarchical behaviour architecture examined in this work is described in [6]. Each behaviour computes meta output signals *activity* and *target rating*. The impact of behaviours can be influenced using their meta inputs *activation* and *inhibition*. The meta signals allow the arrangement of behaviours in a hierarchical network. If more than one behaviour tries to influence another, a fusion node is inserted, where the weighting of inputs is deduced from the influencing behaviours' activities. This architecture has been used in [7] to implement a stereo vision based obstacle avoidance system.

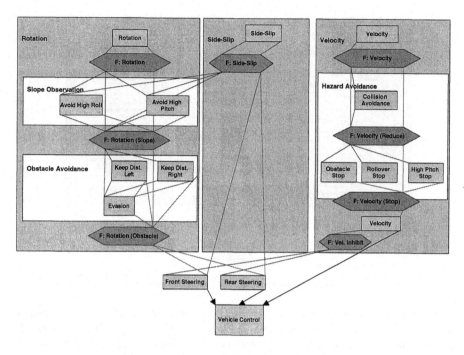

Fig. 2. Behaviour network before extension

The resulting behaviour network (see Fig. 2) comprises three control chains affecting desired rotation, side-slip, and velocity of the vehicle. The rotational component is influenced by slope observing behaviours and obstacle avoidance facilities using the stereo vision system mentioned before. The desired side-slip is forwarded as is at the moment. For safety reasons the velocity is adjusted according to obstacle proximity and critical vehicle inclination. This control system has proven robust and suitable for locally flat outdoor terrain. Tests in urban environments and hilly grassland featuring low obstacle densities have been performed successfully in [7]. Due to the cameras' limited field of vision higher obstacle densities are critical. Furthermore the interaction with target driven navigation behaviours (reach a global goal position) revealed deficits.

To avoid these problems an additional sensor system shall be called into service. The large field of vision and the accuracy of laser range finders has appeared suitable to complement the existing configuration as indicated in Fig. 3.

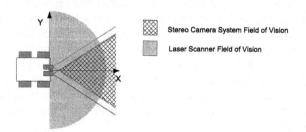

Fig. 3. Laser scanner and camera system field of vision overlap

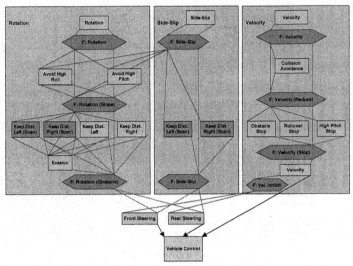

Fig. 4. Behaviour network after extension

One possibility of integrating the mentioned sensor system is to carry out fusion on the level of raw data and feed the combined output to the existing behaviour network. However, in this case all affected behaviours would require modification and recalibration to be able to cope with the different characteristics of the new sensor input. As a result, the effort for testing rises and traceability becomes more difficult. The approach chosen supports separate evaluation of each sensor and integration on the behaviour level. The resulting behaviour network is depicted in Fig. 4 where new behaviours have been indicated as darker boxes. Besides rotational motion the laser scanner additionally supports parallel steering commands. The integration itself requires little effort. New behaviours are implemented according to the characteristics of the laser scanner. Afterwards they are added to the control structure by connecting them to existing fusion nodes which do not require modification.

3.1 Experiments

The extended obstacle detection and avoidance system has been evaluated in various test runs on hilly grassland. In an exploration mode the vehicle was supposed to move forward without colliding with obstacles. This way several hundred metres have been covered autonomously by the vehicle. In this chapter one representative run will be explained in detail. Figure 5 shows the course the robot has taken from a bird's eye perspective based on the fused data from odometry and the IMS. Prominent spots, called *checkpoints* in the following, have been marked with numbers in order to simplify the correlation with the robot's control actions which are depicted in Fig. 6. In this scenario, the robot passes along a conformation of bushes to its right, indicated with green ellipses in the pose trace. Note that the bushes have been marked by the authors without precise measurement to help the reader to bring together the robot's point of view with the bird's eye perspective.

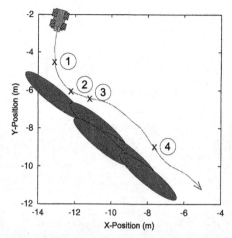

Fig. 5. Robot pose trace of outdoor experiment

In order to document the results as comprehensive as possible, the activity traces of the important behaviours have been plotted in Fig. 6. At the checkpoints a visualisation of the scanner data, the left camera image and the depth map yielded from the stereo vision system are associated with the traces. In the depth map the brightness (dark spots are closer, light spots further away) of gray indicates the distance to the robot while the detected obstacles are marked with crosshairs.

Scanner data is presented as a 2D plot. Dark points indicate the raw data yielded by the scanner hardware. In order to avoid vehicle reaction on sensor noise, clusters are formed according to the distance between particular data points. These clusters are highlighted by red boxes. Note that the clustering of out of range values has no impact on the behaviour activity. The scanner's field of vision is divided into sectors each of which provides an input for the behaviours. A green line indicates the selected distance value for each of these sectors.

At checkpoint (1), the robot heads directly for the bushes. The stereo camera system reconstructs a large obstacle and selects one closest point to the right and one to the left of the robot. That way `KeepDistRotLeft` (`Camera`) and `KeepDistRotRight` (`Camera`) which compete with each other become active trying to guide the robot into the respectively other direction. In contrast the laser scanner only detects close clusters to the right due to the bushes' loose structure. This explains the jitter which can be observed in the activity traces of `KeepDistRotRight` (`Scanner`) and `KeepDistTransRight` (`Scanner`). As the obstacles detected are already very close to the robot, behaviour `Collision Avoidance` slows the robot down until the bushes have been passed. Additionally behaviour `Evasion` observes if obstacles reside in a centre corridor in front of the vehicle and arbitrates between the `KeepDist` behaviours. The target ratings of the `KeepDistRight` behaviours (not shown here) indicate that these are less content than the `KeepDistLeft` behaviours. For that reason these are supported by behaviour `Evasion` and the robot turns to its left away from the bushes. The result of this manoeuvre can be observed at checkpoint (2) where the vehicle has turned enough to see only bushes

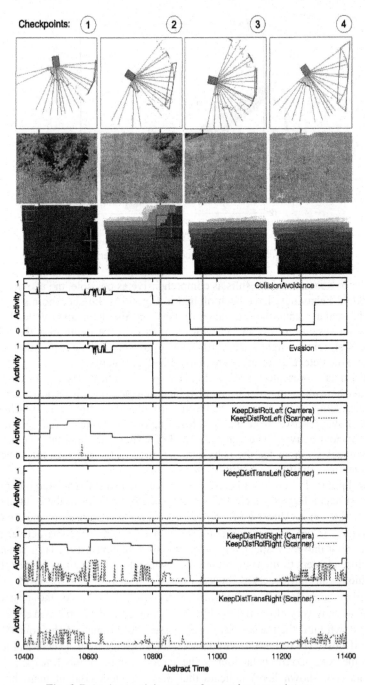

Fig. 6. Behaviour activity trace of an outdoor experiment

to its right. The `KeepDistRight` behaviours accomplish the following of the bushes. Between checkpoints (3) and (4) no noteworthy behaviour activity can be observed as both sensors detect no solid obstacle. At Checkpoint (4) the `KeepDistRight` behaviours gradually become active again as the robot has drifted closer to the bushes. Clearly the scanner perceives the bushes earlier than the cameras due to its larger field of vision.

Note that despite the jitter in the behaviour activities reacting to the laser scanner, the robot pursuits a smooth course fusing rotatory and translatory steering commands. Evidently laser scanner and stereo camera system complement each other in a synergetic way. The scanner compensates the cameras' narrow horizontal field of vision while the cameras perceive a larger terrain patch at once, which makes this sensor more reliable when facing unsolid structures like the bushes in the scenario described above. Furthermore the translatory elements fused into the originally only rotatory steering enhances the robot's manoeuvrability. Finally the deficits in interaction with target driven navigation behaviours could be eliminated successfully.

4 Conclusion and outlook

The presented behaviour-level extension approach has successfully been applied during the integration of sensor hardware. New behaviours processing the additional information have been interlinked with the existing behaviour network without touching present structures. That way individual sensors could easily be calibrated, tested and integrated by incremental behaviour activation. Experiments have shown that the additional laser scanner together with the new set of behaviours represents a robust upgrade of the former system.

In the future the scanner shall be mounted on a tilting unit in order to better consider the three dimensional nature of outdoor terrain. Additionally a local memory for keeping track of obstacles leaving the field of vision is planned.

References

1. Novick D.: Implementation of a Sensor Fusion-Based Object-Detection Component for an Autonomous Outdoor Vehicle. Ph.D. Dissertation, unpublished, 2002.
2. Dima C.S., Vandapel N. and Herbert M.: Classifier Fusion for Outdoor Obstacle Detection. IEEE ICRA 2004
3. Crane C., Armstrong D., Ahmed M., et al.: Development of an integrated sensor system for obstacle detection and terrain evaluation for application to unmanned ground vehicles. in Proc. of the SPIE, vol. 5804, pp. 156-165, 2005
4. Huntsberger T.L. and Rose J.: Behavior-based control for autonomous mobile robots. in Proc. ROBOTICS2000, pp. 299-305
5. Schäfer B.H.: Security Aspects of Motion Execution in Outdoor Terrain. Master's Thesis, University of Kaiserslautern, unpublished, 2005
6. Proetzsch M., Luksch T. and Berns K.: Fault-Tolerant Behavior-Based Motion Control for Offroad Navigation. 20th IEEE ICRA 2005
7. Schäfer B.H., Luksch T. and Berns K.: Obstacle Detection and Avoidance for Mobile Outdoor Robotics. EOS 2005

Visual Odometry Using Sparse Bundle Adjustment on an Autonomous Outdoor Vehicle

Niko Sünderhauf, Kurt Konolige, Simon Lacroix, Peter Protzel

Technische Universität Chemnitz Fakulät für Elektrotechnik und Informationstechnik
SRI International, Menlo Park, USA
LAAS/CNRS, Toulouse, Frankreich

E-mail: niko.suenderhauf@informatik.de, peter.protzel@etit.tu-chemnitz.de,
konolige@ai.sri.com, simon.lacroix@laas.fr,

Abstract. Visual Odometry is the process of estimating the movement of a (stereo) camera through its environment by matching point features between pairs of consecutive image frames. No prior knowledge of the scene nor the motion is necessary. In this work, we present a visual odometry approach using a specialized method of Sparse Bundle Adjustment. We show experimental results that proof our approach to be a feasible method for estimating motion in unstructured outdoor environments.

1 Introduction

Estimating the motion of a mobile robot is one of the crucial issues in the SLAM problem. Besides odometry, inertia sensors, DGPS, laser range finders and so on, vision based algorithms can contribute a lot of information. Visual Odometry approaches have recently been used in different ways [1] [2]. However, the basic algorithm behind these approaches is always the same: The first step is detecting feature points (or interest points) in the images. This is usually done using the *Harris Corner Detector* [3] which has profen to be a very stable operator in the sense of robustness and invariance against image noise [4]. The second step is to match interest points between the left and right images of a single stereo frame and between two consecutive frames. This matching can be done using a correlation based method as in [1].

After interest point detection and successful matching we can go on and determine the motion the camera undertook between two pairs of images. A variety of methods is available, but in this paper we will concentrate on a method called Sparse Bundle Adjustment.

2 Sparse Bundle Adjustment

Bundle Adjustment provides a solution to the following problem: Consider a set of world points X_j is seen from a set of cameras with camera matrices P_i. Each camera projects X_j to $X_{ij}=P_iX_j$, so that X_{ij} are the image coordinates of the j-th world point in the i-th image.

What are the "optimal" projection matrices P_i and world coordinates X_j so that the summed squared *reprojection error* is minimal?

Thus we want to solve

$$\min_{P_i, \mathbf{X}_j} \sum_{ij} d(P_i \mathbf{X}_j, \mathbf{x}_{ij})^2 \tag{1}$$

where $d(\mathbf{x}, \mathbf{y})$ is the Euclidean distance between image points $\mathbf{x}$ and $\mathbf{y}$.

Bundle Adjustment is a non-linear minimization problem which can be solved using iterative non-linear least squares methods such as Levenberg-Marquardt. A very efficient solution to the problem has been proposed by [5] and implemented by [6].

Solving (1) with Levenberg-Marquardt involves iterative solving of normal equations of the form

$$\mathbf{J}^T \mathbf{J} \delta = \mathbf{J}^T \varepsilon \tag{2}$$

where $\mathbf{J}$ is the Jacobian of the *reprojection function* $f_{(\mathbf{a}, \mathbf{b})} = \tilde{\mathbf{x}}$. f takes $\mathbf{a} = (\mathbf{a}_1^T, \mathbf{a}_2^T, \dots \mathbf{a}_m^T)^T$ and $\mathbf{b} = (\mathbf{b}_1^T, \mathbf{b}_2^T, \dots \mathbf{b}_n^T)^T$ as parameters and returns $\tilde{\mathbf{x}} = (\tilde{\mathbf{x}}_{11}^T, \tilde{\mathbf{x}}_{12}^T, \dots \tilde{\mathbf{x}}_{mn}^T)^T$. Here $\mathbf{a}_i$ is the 6-Vector of the currently estimated parameters of the i-th camera, $\mathbf{b}_j$ is the 3-vector with the parameters of the j-th world point respectively. The projected image coordinates of world point j in the i-th image (according to $\mathbf{a}_i$ and $\mathbf{b}_j$) are given by $\tilde{\mathbf{x}}_{ij}$.

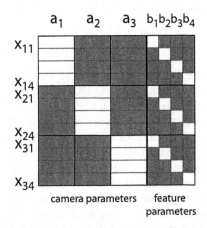

Fig. 1. Structure of a sparse Jacobian matrix for a bundle adjustment problem consisting of 3 cameras and 4 feature points. The gray entries are all zero.

The Jacobian $\mathbf{J}$ of f is made up of entries $\partial \tilde{\mathbf{x}}_{ij}/\partial a_k$ and $\partial \tilde{\mathbf{x}}_{ij}/\partial b_k$. One may notice, that $\partial \tilde{\mathbf{x}}_{ij}/\partial a_k = 0$ unless $i = k$ and similar $\partial ldex_{ij}/\partial b_k = 0$ unless $j = k$. This is simply because the projected coordinates of world point j in the i-th image are not dependent on any camera's parameters but the i-th and they neither depend on any other world point but the j-th.

Given this, one verifies that $\mathbf{J}$ contains large blocks with 0–entries (see Fig. 1). In other words, $\mathbf{J}$ has a sparse structure. The *Sparse Bundle Adjustment* (SBA) implementation as presented by [6] takes advantage of that very structure and thus enables SBA to solve huge minimization problems over many thousands of variables within seconds on a standard PC. The C source code is freely available (under the terms of the GNU General Public License) on the author's website http://www.ics.forth.gr/lourakis/sba.

3 Visual Motion Estimation using Sparse Bundle Adjustment

Sparse Bundle Adjustment (SBA) as implemented in [6] solves the minimization problem as stated in (1), not considering any uncertainty of whatever kind, assuming calibrated cameras and monocular image information.

We extended the original SBA implementation in a way that it can handle stereo data input directly. We therefore had to change some internal algorithms and datastructures, so that the routines take 4–vectors $\mathbf{X}_{ij}$ instead of 2–vectors. The additional two entries to these vectors are the image coordinates in the (right) stereo image. This way, we do not necessarily have to provide proper initial estimates for the world points $\mathbf{X}_j$ or the camera poses P_i. Although this impressively demonstrates the power of Bundle Adjustment, it is of course not an reasonable approach and should not be used in any seriously motivated application. Instead, we should provide SBA with initial estimates for the camera pose acquired by odometry or other sensor systems like inertial navigation systems, compass, or GPS and initialize the 3D coordinates of the feature points by triangulation. Starting with reasonable initial estimates speeds up the optimization process significantly and may detain the optimization from stepping into a false local minimum.

Besides the extension to stereo input, we also implemented a simple iterative outlier rejection method. After SBA finished its minimization loop with the solutions $\mathbf{P}^*$ and $\mathbf{X}^*$, there is still an *residual error* $\varepsilon_{ij} = d(P_i^* \mathbf{X}_j^*, \mathbf{x}_{ij})^2$ left for every $\mathbf{X}_{ij}$. A simple outlier rejection is to discard all $\mathbf{X}_{ij}$ where $\left| \varepsilon_{ij} - \overline{\varepsilon} \right| > k\sigma_\varepsilon$ for any k > 0 where $\overline{\varepsilon}$ is the mean of all residual errors and σ_ε is the corresponding standard deviation. $k = 1,5$ was chosen empirically. After the so determined outliers have been removed from the input data, SBA is restarted using $\mathbf{P}^*$ and $\mathbf{X}^*$ as initial estimates. The procedure is repeated for a fixed number of iterations or until no more outliers are found. This process helps decreasing the mean residual error and is in many cases sufficient to discard outliers arising from false or bad matches. However, as the experiments showed, it is a fairly naive approach, computationally quite expansive, and may fail sometimes. We therefore favor a more robust outlier rejection method, based on RANSAC.

After all these modifications, were able to use SBA for visual motion estimation in two different ways:

1. Sliding Window SBA
2. full SBA

Sliding Window SBA The simplest and fastest estimation method is estimating structure and motion parameters between two consecutive stereo frames only. The overall motion is obtained by simple concatenation or "chaining" of the single estimates. Intuitively one will expect this to be fairly inaccurate, as possible small errors will accumulate quickly.

To avoid the problems of simple chaining, we implemented a sliding window SBA approach. Instead of optimizing for two consecutive images only, we choose a *n-window*, e.g. a subset of *n* images which we perform SBA upon. The pose and structure parameters estimated in this way are used as initial estimates in the next run, where we slide the window further one frame. With that basic sliding window approach, we would bundle adjust every consecutive *n* -window in the sequence of the obtained images, even if the robot has not or only very little moved while the images of the window were taken. Therefore another idea is to only include those images into the window, which pose are more than a certain threshold away from the pose of their respective predecessor in the window. The final algorithm can be summarized as follows:

1. Starting from image I_i find the closest image I_{i+k} so that the motion between I_i and I_{i+k} exceeds a certain threshold. This can be determined by pairwise SBA between I_i and I_{i+k} or, of course, using odometry data.
2. Add I_{i+k} to the window
3. Set $i = i + k$ and repeat from 1. until there are sufficient many (*n*) images in the window
4. bundle adjust the window using the poses obtained in step 1 as initial estimates

In this way a window size of two corresponds to the simple chaining approach.

Full SBA Full SBA optimizes the whole bundle of obtained images at once. It determines camera poses and structure parameters for all recorded frames in one big optimization loop. Although this should intuitively yield the best results, it is, due to its complexity, an off line (batch) method not usable to continuously update the robot's position as he moves along.

4 Experimental Results

During all experiments we used the triangulated world coordinates as initial estimates for the world points, but did not initialize the camera poses. Instead, we assumed the cameras did not move at all. The outlier rejection was enabled and limited to 10 iterations. We tested our algorithms on a dataset of 80 images acquired in an outdoor environment using a stereovision bench made of two 640x480 greyscale cameras with 2.8mm lenses and a baseline of approximately 8cm. We would like to acknowledge Max Bajracharya fromJPL for providing us with the images and data. Ground truth data was extracted from using a Leica "total station" surveying instrument. Between 18 and 308 feature points were visible in each image, with 106 on average. Each point was visible in only 4 consecutive images on average.

Sliding Window SBA The dataset was tested with several window sizes and motion thresholds.

The average deviation from the ground truth position for all tested methods was approximately 23 cm (2.3 % of traveled distance). The error tends to increase slightly with increasing motion threshold above 20 cm because fewer feature points can be matched between two consecutive images and the quality of the matches drops with increasing movement between the images. However, the differences between the different parameter settings are not significant. They are far below 1% of the traveled distance. The camera was mounted very close to the ground and was tilted down, so only few feature points were identified (and succesfully tracked) in a feasible distance from the robot to compare the different window sizes and motion thresholds.

Table 1. Distance from the ground-truth end pose in cm for sliding window SBA

motion threshold	min.	max.	mean error
0	22.30	22.61	22.37
10	22.37	23.35	22.67
15	22.03	22.33	22.18
17	22.49	22.97	22.68
20	22.07	23.01	22.53
25	22.48	24.05	23.11
30	23.19	24.37	23.76
35	23.04	24.27	23.29
40	23.57	24.76	24.32

Full SBA The position estimated by full SBA was 20.48 cm away from the ground truth position which is slightly better than the above results, as we expected. The estimation involved 2129 3D points and 8462 image projections. 6947 parameters had to be estimated (3 for each 3D point and 7 for every camera pose). Our implementation took 6 iterations to detect and remove outliers. On a 1GHz P3 machine running Linux, the algorithm finished in 4.5 minutes. This rather long time is caused by the computationally expansive simple outlier rejection method and the stereo-extension forcing us to use SBA in the inefficient "simple driver" mode. The required runtime may be reduced if more care is taken for runtime efficiency.

We also tested the algorithms with a sequence of indoor images. The results there were even better than for the outdoor sequence, mainly because feature points were visible longer and could be matched more accurately.

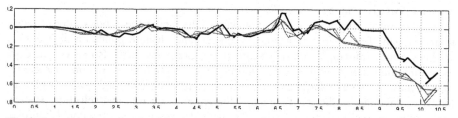

Fig. 2. Several estimated trajectories for the outdoor dataset. The thick black line is the ground truth.

5 Conclusions and Further Work

The error, 2.3 % of the traveled distance for the outdoor data, proves SBA to be a feasible method for visual motion estimation. The tested datasets did not show significant differences among the different parameter settings. However, as a rule of thumb one should not use simple chaining (resp. window size 2) as the error tends to be higher than with window sizes 3 or 4. Window sizes above 4 or 5 do in general not help to improve the result, as feature points may in average not be tracked for more than 5 images. This, of course, is highly dependent on your matching and tracking algorithm, image quality and the environment. Larger windows increase computation time drastically, so a tradeoff between computational costs and accuracy has to be considered.

In further work, the methods will be refined to make them more robust against outliers arising from false matches. This will help decreasing the error and, if implemented efficiently, can even decrease overall computation time.

A possible way to achieve this is to use RANSAC [7] to estimate hypotheses for motion between two consecutive frames.

We can use three 3D point correspondences (from before and after the motion) to determine a least squares estimate of the 6D motion parameters as shown in [8] or [9]. A RANSAC approach will then be able to find a consistent estimateof the motion parameters, preventing the negative influence of outliers in the 3D point data. SBA can then be used to refine the solution on the resulting inlier sets of several consecutive frames at once. The techniques we presented here (sliding window, motion threshold) should of course be used further in such an approach.

Visual Odometry should also be fused with other motion-estimating sensors or methods, such as DGPS or laser-scanmatching. To achieve this (using an Extended Kalman Filter) it is necessary to acquire the covariance matrix of SBA's results. This, in return, can be achieved by propagating the covariance matrix of the initial estimates (structure and motion) through SBA. In its current implementation, SBA uses the identity matrix as covariance matrix for the world points. It should be possibleto use any arbitrary covariance matrix instead (see [6]). As pointed out in [5] in Algorithm A6.4, the covariance matrix for the camera parameters is the pseudo-inverse of SBA's internal matrix S. One can use the matrices V and U in the implementation to calculate S as given in [5].

Another interesting idea is to exploit knowledge about the robot's current environment in the feature selection process. For instance feature points on objects that are known to be moving could be discarded. Feature points may also be weighted in the estimation process by the type of terrain they are seen on. Points on concrete or asphalt may be more reliable and useful than points found in high grass (which is likely to move due to the wind). Finally, the feature detection and selection process itself could be modified. For instance the methods presented in [10] could improve feature quality significantly.

References

1. David Nister, Oleg Naroditsky, and James Bergen. Visual odometry. In *Proc. IEEE Computer Society Conference on Computer Vision and Pattern Recognition (CVPR 2004)*, pages 652–659, 2004.
2. C. Olson, L. Matthies, M. Schoppers, and Maimone Maimone. Robust stereo egomotion for long distance navigation. In *Proceedings of the IEEE Conference on Computer Vision and Pattern Recognition (CVPR-00)*, pages 453–458, Los Alamitos, June 13–15 2000. IEEE.
3. M. Stephens C. Harris. A combined corner and edge detector, 1988.
4. C. Bauckhage C. Schmid, R. Mohr. Evaluation of interest point detectors, 2000.
5. R. I. Hartley and A. Zisserman. *Multiple View Geometry in Computer Vision*. Cambridge University Press, ISBN: 0521540518, second edition, 2004.
6. M.I.A. Lourakis and A.A. Argyros. The design and implementation of ageneric sparse bundle adjustment software package based on the levenbergmarquardt algorithm. Technical Report 340, Institute of Computer Science–FORTH, Heraklion, Crete, Greece, Aug. 2004. Available from http://www.ics.forth.gr/~lourakis/sba.
7. Martin A. Fischler and Robert C. Bolles. Random sample consensus: a paradigm for model _tting with applications to image analysis and automated cartography. *Communications of the ACM*, 24(6):381–395, June 1981.
8. Shinji Umeyama. Least-squares estimation of transformation parameters between two point patterns. *IEEE Trans. Pattern Anal.* Mach. Intell., 13(4):376–380, 1991.
9. K.S. Arun, T.S. Huang, and S.D. Blostein. Least-squares _tting of two 3–d point sets. *IEEE Trans. Pattern Anal.* Mach. Intell., 9(5):698–700, 1987.
10. David G. Lowe. Distinctive Image Features from Scale-Invariant Keypoints. In *International Journal of Computer Vision*, 60, 2, pages 91–110, 2004.

Verbesserte GPS-Positionsschätzung mit IP-transportierten Korrekturdaten für autonome Systeme im Outdoor-Bereich

Johannes Pellenz, Sabine Bauer, Tobias Hebel, Sebastian Spiekermann,
Gerd Tillmann, Dietrich Paulus

Universität Koblenz-Landau, Universitätsstr. 1, 56070 Koblenz

E-mail: pellenz@uni-koblenz.de

Zusammenfassung. Für autonome Systeme im Outdoor-Bereich können zur groben Selbstlokalisation GPS-Positionsdaten genutzt werden. Leider sind diese Positionsdaten bedingt durch verschiedene Fehlerquellen recht ungenau. Eine Möglichkeit zur Verbesserung der Genauigkeit bieten Korrekturdaten, die von einer festen Referenzstation berechnet und ausgesendet werden. Für den Empfang der Referenzdaten ist jedoch normalerweise ein eigener Empfänger erforderlich, der auf dem autonomen, mobilen System mitgeführt werden muss. Diese Arbeit untersucht als alternativen Transportweg für Korrekturdaten das (W)LAN und stellt die Verbesserung der Positionsschätzung nach Anwendung der Korrekturdaten dar.

1 GPS

GPS (Global Positioning System) ist ein satellitengestütztes Navigationssystem. Mit Hilfe dieses Systems ist es möglich eine dreidimensionale Positionsbestimmung zu bekommen, d.h. den Längen- und Breitengrad und die Höhe des Ortes, an dem sich die GPS-Antenne befindet. Dazu muss der GPS-Empfänger die Signale von mindestens vier Satelliten auswerten und dabei eine Laufzeitmessung der Signale von jedem einzelnen Satelliten zum Empfänger durchführen. Die Ungenauigkeiten in der Positionsbestimmung sind auf Ereignisse zurückzuführen, die die Laufzeit der Signale beeinflussen, wie z. B. atmosphärische Störungen [1]. Weitere Ursachen für Ungenauigkeiten sind ungünstige Satellitenkonstellationen sowie Ungenauigkeiten der Satellitenuhren. Im Durchschnitt kann man bei einer solchen Positionsbestimmung ohne Einbezug von Korrekturdaten nur eine Genauigkeit von 10 bis 15 Metern erreichen. Die vom GPS-Empfänger berechnete Position kann mittels NMEA-Protokoll über eine serielle Verbindung an einen PC übertragen werden, der zur Steuerung des Roboters verwendet wird. In Abb. 1 ist unser mobiles System „Robbie 5" zu sehen, auf dem der GPS-Empfänger installiert ist.

Abb. 1. Unser mobiles System „Robbie 5" mit GPS-Empfänger

2 Differential GPS

Die verschiedenen Störungen resultieren in einer falschen Abschätzung des Abstands des Empfängers zu einem Satelliten; diese geschätzten Abstände werden Pseudostrecken genannt. DGPS (Differential GPS) ist eine Ergänzung zum GPS System, bei dem mit Hilfe von Korrekturdaten eine präzisere Ermittlung der Position des GPS-Empfängers ermöglicht wird, indem Korrekturdaten für die Pseudostrecken zu jedem Satelliten berücksichtigt werden. Dazu positioniert man einen GPS-Empfänger an einem Ort, dessen Position zuvor exakt vermessen wurde. Aus der Differenz der bekannten und der momentan durch das GPS ermittelten Entfernungen lassen sich dann die Pseudostreckenkorrekturen berechnen. Ein solches System nennt man Referenzstation. Da sich die Fehlerquellen bei benachbarten Empfängern zum gleichen Zeitpunkt in etwa gleich auswirken, kann die Referenzstation die Pseudostreckenkorrekturen einem nahe gelegenen GPS-Empfänger übermitteln und diesem somit eine bessere Positionsbestimmung ermöglichen. In der Echtzeitübertragung solcher Korrekturdaten hat sich das RTCM-SC 104 Format durchgesetzt [4]. Dieses Format wird in erster Linie über Funk ausgestrahlt, z.B. über einen Mittelwellensender. Auf der Nutzerseite muss ein passender Empfänger zur Verfügung stehen, der die Signale auffängt und an das GPS-Gerät weiterleitet. Dieses kann dann die Korrekturwerte bei der Positionsbestimmung berücksichtigen.

3 NTRIP

Neben der Übertragung über Radiowellen kann auch das Internet dazu verwendet werden, Pseudostreckenkorrekturen zu übertragen. NTRIP (Networked Transport of RTCM via Internet Protocol) ist eine im September 2004 vorgestellte Technik, um

Satellitennavigationsdaten über ein TCP/IP-basiertes Netzwerk zu übertragen. NTRIP ermöglicht Systemen auch ohne eigenen Radioempfänger RTCM Korrekturdaten zu nutzen [2]. Es basiert auf dem HTTP 1.1 Standard, so dass als Übertragungswege zu einem mobilen System neben GSM- oder UMTS-Netzen auch WLAN (Wireless LAN) genutzt werden kann. WLAN ist für mobile Systeme besonders interessant, da oft eine WLAN-Verbindung bereits vorhanden ist und Korrekturdaten daher ohne zusätzlichen Hardwareaufwand und Stromverbraucher verwendet werden können.

4 Versuchsaufbau

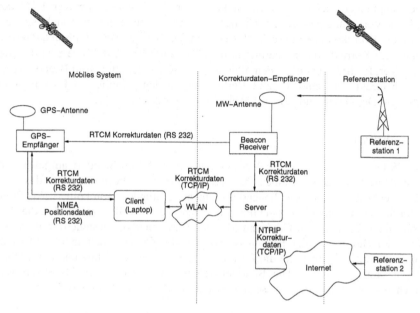

Abb. 2. Versuchsaufbau

Ziel der Arbeit ist es zu untersuchen, welche Auswirkung die Einspeisung von Korrekturdaten auf die Positionsbestimmung hat und wie sich der alternative Übertragungsweg über TCP/IP auf die Genauigkeit der Positionsschätzung auswirkt. Dazu wurden vier verschiedene GPS Empfänger mit unterschiedlichen Korrektursignalen über unterschiedliche Wege versorgt. Als Empfänger wurden verwendet:

- *Trimble 4000 DS*, ein professioneller GPS-Empfänger mit externer Antenne. Dieses Gerät wird u.a. für die Vermessung eingesetzt.
- *Garmin eTrex*, ein sehr verbreitetes Consumer GPS Handgerät, das über eine RS 232 Schnittstelle mit einem Rechner verbunden werden kann.
- *Garmin GPS 12*, ebenfalls ein Consumer GPS Handgerät älterer Bauart, das mit einer RS 232 Schnittstelle ausgestattet ist.
- *u-blox RCB-LJ Receiver Board*, ein professionelles GPS Receiver Board, das als Platine geliefert wird und in eigene Systeme integriert werden kann.

Folgende Quellen für Korrekturdaten wurden verwendet:

– Referenzstation der Wasser- und Schifffahrtsverwaltung des Bundes (WSV) in Koblenz. Die Referenzstation ist nur ca. 2 km vom GPS-Empfänger entfernt. Die Signale werden mit einem *Differential Beacon Receiver MBX-3* der Firma *csi* empfangen.
– NTRIP Datenstrom des Broadcasters www.euref-ip.net aus Frankfurt (Kennung FFMJ2). Die Referenzstation ist ca. 84 km vom GPS-Empfänger entfernt. Die Korrekturdaten werden über das Internet empfangen.

Sowohl die NTRIP-Daten als auch die Daten der WSV werden zunächst von einem Server empfangen und dort an einem Port über die Software dgpsipd [5] dem Client zur Verfügung gestellt. Durch diesen Aufbau ist für den Client nicht sichtbar, von wo die Korrekturdaten tatsächlich stammen. Das Auslesen des GPS und das Weiterleiten der Korrekturdaten übernimmt das Programm gpsd (GPS service daemon) [3]. Zu Vergleichszwecken wurden alternativ die Korrekturdaten vom Beacon Receiver auch direkt über eine RS 232 Verbindung an den jeweiligen GPS-Empfänger übertragen. Der Gesamtaufbau ist in Abb. 2 wiedergegeben.

5 Ergebnisse

Die GPS-Geräte wurden für die Versuche an einem festen Standort aufgebaut und über ein serielles Kabel an einen PC angeschlossen. Die Positionen, die das GPS-Gerät berechnete, wurde anschließend über mehrere Stunden protokolliert. Als Protokoll für die Kommunikation vom GPS zum Client wurde NMEA verwendet; die Positionsangaben darin sind in Grad und Minuten (mit Nachkommastellen) kodiert. Zur einfacheren Bewertung der Ergebnisse wurden die Gradangaben in Gauß-Krüger-Koordinaten umgerechnet, die die Positionen in Form eines Rechts- und eines Hochwerts in Metern angeben. Die Beurteilung der Abweichung vom Mittelwert wird dadurch erleichtert. Die Ergebnisse sind in den Tabellen 1 bis 4 wiedergegeben.

Tabelle. 1. Trimble 4000 DS: Standardabweichungen der Messungen vom Mittelwert (in Metern)

Trimble 4000 DS	(ohne DGPS)	RTCM	RTCM über IP	NTRIP
$\sigma_{Rechtswert}$	1,1193	0,7631	0,6804	0,7949
$\sigma_{Hochwert}$	1,7204	1,2748	1,1361	1,1489
Anz. Messungen	62867	56708	54596	50333

Tabelle. 2. Garmin eTrex: Standardabweichungen der Messungen vom Mittelwert (in Metern)

Garmin eTrex	(ohne DGPS)	RTCM	RTCM über IP	NTRIP
$\sigma_{Rechtswert}$	2,5502	4,3400	4,8082	2,3542
$\sigma_{Hochwert}$	2,9501	5,1217	4,7685	5,1078
Anz. Messungen	16842	10624	4099	6030

Tabelle. 3. Garmin GPS 12: Standardabweichungen der Messungen vom Mittelwert (in Metern)

Garmin GPS 12	(ohne DGPS)	RTCM	RTCM über IP	NTRIP
$\sigma_{Rechtswert}$	3,0202	–	2,9210	–
$\sigma_{Hochwert}$	4,4464	–	4,9631	–
Anz. Messungen	16635	–	5378	–

Tabelle. 4. u-blox RCB-LJ: Standardabweichungen der Messungen vom Mittelwert (in Metern)

u-blox RCB-LJ	(ohne DGPS)	RTCM	RTCM über IP	NTRIP
$\sigma_{Rechtswert}$	2,0941	1,6848	1,8741	2,0442
$\sigma_{Hochwert}$	2,5074	2,7444	3,1590	2,7955
Anz. Messungen	31645	26602	48610	23174

5.1 Unterschiede zwischen den GPS-Geräten

Die Ergebnisse zeigen, dass die verschiedenen GPS-Empfänger sehr unterschiedlich auf die Einspeisung von Korrekturdaten reagieren. Während das Trimble die Genauigkeit der Positionsmessung deutlich verbessern kann (die Standardabweichung sinkt auf bis zu unter 70 cm), beeinflussen die Korrekturdaten beim Garmin GPS 12 die Positionsmessungen kaum. Beim Garmin eTrex wurde die Positionsschätzung sogar unstabiler; die Standardabweichung stieg auf fast das Doppelte an. Grund dafür könnte sein, dass bei ausgeschaltetem DGPS interne Filter die Daten stark glätten, und damit die Positionsschätzung stabiler erscheinen lassen.

5.2 Unterschiede durch die Art der Übertragung

Die Art der Übertragung der Korrekturdaten – ob über eine direkte RS 232-Verbindung oder über ein IP-Netzerk – hat keine signifikante Auswirkung auf die Genauigkeit der Positionsschätzung. Dieses Ergebnis wurde erwartet, da die Korrekturdaten bei ausgeschalteter SA (Selective Availability; künstliche Verschlechterung der Daten; seit 1. Mai 2000 normalerweise deaktiviert) mehrere Minuten gültig sind, und eine Verzögerung durch den Transport dadurch nicht ins Gewicht fällt.

5.3 Unterschiede durch verschiedene Referenzquellen

Die Genauigkeit des Trimble-Empfängers lässt sich sowohl durch die Einspeisung der Korrekturdaten von der nahe gelegenen Referenzstation in Koblenz, also auch von der weiter weg liegenden Referenzstation in Frankfurt (über NTRIP) deutlich verbessern. Dies überrascht, da die atmosphärischen Störungen lokal unterschiedlich sein können und durch Daten von einer näher gelegenen Referenzstation besser kompensiert werden sollten.

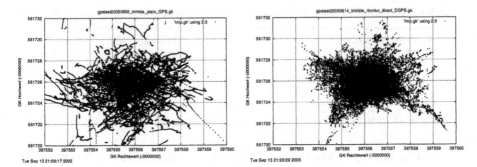

Abb. 3. Plot zweier Messung mit dem Trimple 4000 DS: Ohne (links) und mit (rechts) Korrekturdaten

6 Ausblick

Die vorliegende Arbeit zeigt, wie ein Diffential GPS zur Groblokalisation auf einem mobilen System eingesetzt werden kann, ohne dass zusätzliche Hardware auf dem Roboter mitgeführt werden muss. Folgende Aspekte werden in der weiteren Arbeit betrachtet:

Verhalten von Consumer GPS-Empfängern erklären. Wir werden weitere Experimente mit Low-cost-GPS-Empfängern durchführen um festzustellen, ob die Anwendung von Differenzdaten bei solchen Geräten überhaupt gerechtfertigt ist.

Eigene Referenzstation. Wir planen den Aufbau einer eigenen Referenzstation, um damit auch sehr lokale Störungen kompensieren zu können und Differenzdaten auch in Gebieten zur Verfügung zu haben, in denen keine Referenzstation in der Nähe ist.

7 Danksagung

Wir danken Herrn Michael Hoppe und Herrn Mario Walterfang von der Wasser- und Schifffahrtsverwaltung (WSV) des Bundes in Koblenz für die freundliche Leihgabe von Hardware und die fundierte fachliche Unterstützung. Dank geht auch an das Projektteam von Robbie 5: Richard Arndt, Peter Decker, Christian Delis, Andreas Klöber, Alexander Kubias, Ken McManus, Felix Nagel, Sarah Steinmetz und Thorsten Tillack.

Literaturverzeichnis

1. Bauer, M.: Vermessung und Ortung mit Satelliten. Wichmann Verlag, 2003.
2. Lenz, E.: Networked Transport of RTCM via Internet Protocol (NTRIP): Application and Benefit in Modern Surveying Systems. FIG Working Week, 2004.
3. Raymond, E: gpsd - a GPS service daemon. http://gpsd.berlios.de/index.html.
4. Radio Technical Commission For Maritime Services: RTCM recommended standards for differential GNSS (Global Navigation Satellite Systems) service, Version 2.2. Alexandria, Virginia, 1998.
5. Rupprecht, W.: DGPS corrections over the Internet. FIG Working Week, 2004. http://www.wsrcc.com/wolfgang/gps/dgps-ip.html.

Fahrerassistenzsysteme

Videobasierte Fahrspurerkennung zur Umfelderfassung bei Straßenfahrzeugen

Stephan Neumaier, Georg Färber

Lehrstuhl für Realzeit-Computersysteme RCS, TU München

E-mail: {neumaier,faerber}@rcs.ei.tum.de

Zusammenfassung. Basierend auf dem 4D-Ansatz zur schritthaltenden Bildfolgeverarbeitung wird ein Verfahren zur Fahrspurerkennung vorgestellt. Eine erweiterte Dynamikmodellierung ermöglicht die verallgemeinerte Betrachtung sowohl ortsfester als auch mit dem Eigenfahrzeug mitbewegter Spursegmente unter beliebiger Ablage. Eine Klassifikation des Spurtyps durch explizite Vermessung der Markierungszeichen dient bei zugleich reduziertem Rechenzeitbedarf der Steigerung der Robustheit des rekursiven Schätzprozesses sowie ersten Ansätzen zur Situationsanalyse. Bei Testfahrten mit einem Versuchsträger der Audi AG konnte eine beeindruckende Systemleistung demonstriert werden.

1 Einleitung

Die folgend vorgestellten Arbeiten entstanden im Rahmen des Forschungsverbundes FORBIAS [1] „Bioanaloge Sensomotorische Assistenz" der Bayerischen Forschungsstiftung. Eines der Ziele ist dabei die Realisierung einer bioanalogen Fahrzeugkamera [2] zur Umfeldwahrnehmung, welche den Anspruch einer multifokalen Lösung inkl. Blickrichtungssteuerung/-stablisierung erfüllt.

Motivation hierfür ist der zunehmende Einzug von Assistenzfunktionen ins Fahrzeug und die damit verbundenen steigenden Anforderungen an eine maschinelle Umfeldwahrnehmung. Während herkömmliche Funktionen wie ABS (Anti-Lock Braking System) oder ESP (Electronic Stability Program) lediglich mit Sensordaten bzgl. des Eigenfahrzeugs (Raddrehzahlen, Lenkwinkel, Querbeschleunigung) auskommen, sind neuere Systeme auch auf Informationen aus dem Fahrzeugumfeld angewiesen. Als Beispiel sei hier die Funktion ACC (Adaptive Cruise Control) genannt, bei welcher eine Vorgabe der Eigengeschwindigkeit auf Basis von Abstands- und Geschwindigkeitsdaten voraus fahrender Fahrzeuge erfolgt. Dabei stützt sich die Umfeldwahrnehmung derzeit in erster Linie auf Radar. Nachteil dieses Sensors ist seine geringe Winkelauflösung, der schmale Erfassungsbereich in der Ferne sowie eine durch die hohe Fehldetektionsrate bedingte Beschränkung auf bewegte Objekte. Zunehmendes Interesse richtet sich daher auf den Bereich der Videosensorik. Als die dem menschlichen Sehen verwandteste Art der Umfelderfassung ermöglicht sie insbesondere auch die Wahrnehmung explizit auf den Fahrer ausgerichteter Reize, wie z.B. Verkehrszeichen, Warnlichter oder auch Spurmarkierungen.

Auf letztere stützt sich der im Folgenden vorgestellte Ansatz einer videobasierten Fahrspurerkennung, welcher in Zusammenarbeit mit dem Industriepartner Audi AG der Umsetzung einer erweiterten ACC-Funktionalität dient. Zusätzliches Wissen über den Verlauf sowohl der Eigen- als auch benachbarter Fahrspuren ermöglicht eine räumliche Spurzuordung der mittels Radar generierten Objekthypothesen. Dies ist essentiell für eine sichere Auswahl des zur Abstandsregelung relevanten voraus fahrenden Fahrzeugs. Zudem dient die räumliche Spurzuordnung als Ausgangskriterium für eine videobasierte Verifikation der Radar-Hypothesen. Auch nicht bewegte Objekte wie z.B. das stehende Ende eines Staus – bislang von der Radarsensorik ausgeblendet – sollen somit erfassbar werden. Ferner kann die Spurkrümmung als weiteres Kriterium zur Anpassung der Eigengeschwindigkeit herangezogen werden. In Assistenzsystemen der nächsten Generation, welche neben der Längsführung zusätzlich in die Fahrzeugquerführung eingreifen, wird basierend auf einer videobasierten Fahrspurerkennung sogar eine autonome Spurhaltung realisierbar sein.

In Absatz 2 wird nun zuerst ein allgemeiner Überblick über das Verfahren zur Spurerkennung auf Basis des 4D-Ansatzes gegeben um dann detailliert auf die Dynamikmodellierung sowie den Nutzen einer expliziten Vermessung der Spurmarkierungen und Klassifikation des Spurtyps einzugehen.

2 Fahrspurerkennung auf Basis des 4D-Ansatzes

Ausgangsbasis für das vorgestellte Verfahren zur Fahrspurerkennung stellt eine Bildfolgeverarbeitung nach dem sog. 4D-Ansatz [4, 4] dar. Klassische Verfahren des maschinellen Sehens verwenden viel Energie auf Einzelbildverarbeitung sowie eine Invertierung der perspektivischen Abbildung, d.h. dem versuchten Rückschluß von 2D-Bildinformation auf eine 3D-Szene. Diese Ansätze ziehen jedoch keinen Vorteil aus der zeitlichen Kontinuität, welcher alle Objekte in der physikalischen Welt unterworfen sind, da das Objektverhalten im Bild aufgrund der nichtlinearen perspektivischen Abbildung nicht mehr durch einfache Modelle beschreibbar ist. Dies gilt sowohl für Bewegung als auch Form von Objekten. Nach dem 4D-Ansatz hat daher die Beschreibung von Objekten zugleich in Zeit und Raum zu erfolgen, mittels dynamischer Zustandsmodelle sowie 3D-Formmodellen. Letztere definieren die Lage visuell vermeßbarer Objektmerkmale im Raum, welche über ein Abbildungsmodell ins Bild projiziert werden. Basierend auf diesen Erwartungen lassen sich mittels speziell auf die jeweiligen Objektmerkmale zugeschnittener Bildverarbeitungsoperatoren äußerst effektive, selektive Meßprozesse zur Ermittlung der tatsächlichen Merkmalslage im Bild aufsetzen. Dies garantiert einen auch für komplexere Szenen beherrschbaren Rechenaufwand und somit eine im Videotakt schritthaltende Bildverarbeitung. Die Rückkopplung der Vorhersagefehler der Merkmalslage erfolgt über ein rekursives Schätzverfahren. Die so aktualisierte Modellvorstellung dient in Verbindung mit den hinterlegten Dynamikmodellen wiederum als Ausgangsbasis zur Prädiktion des nächsten Videotakts.

Für die konkrete Aufgabe der Fahrspurerkennung haben sich nun richtungsselektive, ternäre Korrelationsmasken als Bildverarbeitungsoperator zur Kantenextraktion als bestens geeignet erwiesen (vgl. Abb. 1). Bestehend aus je einem Block positiv bzw. negativ besetzter Spalten, deren Orientierung der erwarteten Kantenrichtung entspricht, werden diese entlang eines Suchpfades auf den interessierenden Bildbereich angewandt. Der Verlauf der Maskenantwort gibt Aufschluß über Hell-Dunkel- bzw. Dunkel-Hell-Übergänge im Intensitätsverlauf, d.h. die Lage potentieller Kanten im Bild. Dabei zu beachten ist die explizite Modellierung des für diesen Operator

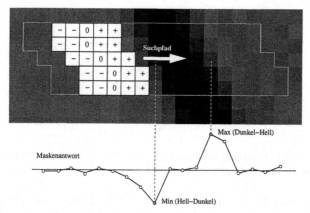

Abb. 1. Schematische Darstellung einer ternären Korrelationsmaske zur richtungsselektiven Kantenextraktion inkl. Verlauf der Maskenantwort entlang eines Suchpfades.

charakteristischen sog. Aperturfehlers, d.h. einer nicht meßbaren Translationskomponente tangentiell zur Kantenrichtung.

Als rekursives Schätzverfahren dient ein erweitertes Kalmanfilter, wobei insbesondere auf eine recheneffiziente Umsetzung als sequentielles Filter [5] Wert gelegt wurde. Die Formmodellierung von Fahrspursegmenten in 3D erfolgt mittels Klothoidennäherung, d.h. einer sich linear über der Lauflänge ändernden Krümmung.

$$c(l) = c_0 + c_1 \cdot l \tag{1}$$

Die Dynamikmodellierung besteht derzeit aus drei voneinander entkoppelten Teilsystemen zur Schätzung des Fahrzeugnickwinkels, der Fahrspurbreite sowie der gemittelten horizontalen Spurkrümmung (die vertikale Krümmung der Fahrbahn sei vorerst vernachlässigt). Auf letzteres Teilsystem zur Krümmungsschätzung, in welchem zugleich die Relativlage des Eigenfahrzeug zur Fahrspur modelliert wird, soll nun detaillierter eingegangen werden.

2.1 Erweiterte Modellierung der Dynamik

Das Teilmodell zur Spurkrümmungsschätzung basiert auf den Vorarbeiten von Mysliwetz, welche unter [6] vollständig nachzulesen sind. Dabei wird ein mit dem Eigenfahrzeug mitbewegtes Spursegment mit definierter Vorausschauweite betrachtet. Das Modell stützt sich neben den gemittelten horizontalen Krümmungsparametern c_0 und c_1 (vgl. Gl. 1) im Wesentlichen auf Zustandsgrößen die Querdynamik des Eigenfahrzeugs betreffend, d.h. laterale Spurablage y, rel. Gierwinkel ψ bzgl. Spur, Schwimmwinkel β, sowie die Steuergrößen Lenkwinkel λ und Eigengeschwindigkeit v (vgl. Abb. 2). Die Querdynamikmodellierung entspricht mit einigen Vereinfachungen weitgehend dem linearen ebenen Einspurmodell [7].

$$\dot{x} = -v\cos(\beta + \psi) - yv_S c_0 + v_S \tag{2}$$

$$\dot{y} = +v\sin(\beta + \psi) + xv_S c_0 \tag{3}$$

$$mit\, c_0 = v_S c_1$$

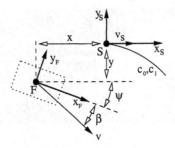

Abb. 2. Relative Lage Fahrzeugkoordinaten (F) zu Spursegmentkoordinaten (S) sowie zur Spurkrümmungsschätzung relevante Zustandsgrößen

Das ursprüngliche Modell wurde durch zusätzliche Hinzunahme der Ablage x als variable Zustandsgröße verallgemeinert, was u.a. in den beiden neuen Beziehungen (Gl. 2, 3) resultiert. Ohne die bisherige Nebenbedingung $x_{const} = 0$ ist es nun nicht mehr zwingend die gesamte Vorausschauweite über ein einziges Spursegment mit gemittelter Krümmung abzudecken (Ein realer Fahrspurverlauf besteht i.A. aus einer verketteten Abfolge mehrerer Klothoidensegmente unterschiedlicher Krümmung). Spursegmente können mit beliebiger Ablage x relativ zum Eigenfahrzeug definiert und somit auch verkettet werden, z.B. bei Verwendung eines bifokalen Kamerasystems bestehend aus Weitwinkel- und Teleoptik (vgl. Abs. 1 zu FORBIAS). Zudem ist nun mit ein und derselben Systemdynamik zugleich die Modellierung eines mit dem Eigenfahrzeug mitbewegten oder alternativ ortsfesten Spursegments möglich, wie z.B. in Behringer [8] zur Betrachtung ortsfester Segmentübergänge vorgeschlagen. Das Umschalten zwischen beiden Alternativen erfolgt über eine jeweils entsprechende Definition der Segmentgeschwindigkeit v_S (Gl. 4).

$$v_S = \begin{cases} \dfrac{v\cos(\beta+\psi)}{1-yc_0} & mitbewegtes Segment, d.h. \dot{x} = 0 \\ 0 & ortsfestes Segment \end{cases} \quad (4)$$

Die Verwendung einer ortsfesten Spurmodellierung ist immer dann sinnvoll, wenn innerhalb der Vorausschau Segmentübergänge mit signifikant unterschiedlichem Krümmungsverhalten liegen und eine Möglichkeit zur Detektion dieser Übergänge besteht. Diese Übergangsdetektion kann in Spezialfällen, beispielsweise bei der Annäherung an eine Autobahnausfahrt, im Rahmen einer Situationsanalyse stattfinden. Eine universelle Lösung zur Fahrspursegmentierung hat über eine Fehlerdetektion innerhalb des rekursiven Schätzprozesses zu erfolgen, welche i.A. äußerst rechenaufwendig ist und noch Gegenstand aktueller Arbeiten darstellt.

Die Linearisierung zur Überführung in ein für den Schätzprozeß zeitdiskretes System [9] mittels Kleinwinkelnäherung ist jedoch für $x \gg 0$ nicht mehr adäquat. Hier hat sich eine Taylorentwicklung um den aktuellen Schätzwert bewährt.

2.2 Klassifikation der Spurmarkierung

Eine weitere Neuerung des vorgestellten Ansatzes zur Fahrspurerkennung ist die direkte Vermessung einzelner Markierungszeichen (vgl. Abb. 3). Während die hor.

und vert. Meßfenster zur Kantenextraktion den rekursiven Schätzprozeß stützen, dienen bei Ausfall mehrerer benachbarter Meßfenster zusätzlich spurtangentiale Suchfenster der Detektion sowie anschließenden Verfolgung der Enden einzelner Markierungszeichen. Dies ermöglicht im Vergleich zum bisherigen lediglich auf einer Verteilungsfunktion der Meßausfälle basierenden Verfahren eine zuverlässige Spurklassifizierung, welche aktuell zur selbständigen Aktivierung bzw. Deaktivierung der Nebenspurerkennung genutzt wird. Die Nebenspurerkennung wird lediglich bei Vorliegen *gestrichelter* Markierungen aktiviert, umgekehrt wird bei *durchgezogener* Markierung von nicht vorhandenen Nebenspuren ausgegangen.

Abb. 3. Fahrspurerkennung während einer Autobahn Testfahrt mit Audi-Versuchsträger: Zu erkennen sind die hor. und vert. Meßfenster der Kantenextraktion zur Spurkrümmungsschätzung sowie spurtangentiale Suchfenster zur Vermessung der Enden einzelner Markierungszeichen. Zusätzlich sind mit Index die Objekthypothesen der Radarsensorik eingetragen.

Ein weiterer Vorteil besteht in der Möglichkeit die Kantenextraktion für den rekursiven Schätzprozeß in den freien Markierungszwischenräumen bewußt auszublenden (vgl. Mittelspur in Abb. 3). Dies spart nicht nur Rechenzeit sondern steigert auch nachweisbar die Robustheit der Spurschätzung, da es die Wahrscheinlichkeit für potentielle Fehlkorrelationen deutlich reduziert. Einer ansonsten gerne in den Markierungszwischenräumen erfolgten Interpretation von Fahrbahnschäden/-verschmutzungen, Schlagschatten oder auch von durch die Hinderniserkennung fälschlicher Weise nicht ausgeblendeten Objektkanten als Spurverlauf wird somit vorgebeugt.

Mit den vermessenen Enden der Markierungszeichen liegen zudem bzgl. der Fahrbahn und somit der Welt ortsfeste 3D-Merkmale vor. Diese können einerseits bei Verwendung einer ortsfesten Spursegmentmodellierung (vgl. Abs. 2.1) zur Stabilisierung der geschätzten Fahrzeugablage x herangezogen werden. Zum anderen, bei Nutzung als Stützpunkte zur weiteren Meßfensterverteilung entlang der Vorausschau, ermöglichen sie auch bei mitbewegtem Spurmodell eine eindeutige Merkmalsassoziation [10]. D.h. einzelne vermessene Kantenmerkmale, deren Platzierung entlang des Kantenverlaufs bisher unbestimmt war (sog. Aperturproblem), können nun präzise einem definierten Kantenpunkt zugeordnet und über mehrere Videotakte verfolgt werden. Dies ermöglicht eine adaptive, bzgl. des jeweiligen Kantenpunkts optimierte Parametrierung des Bildverarbeitungsoperators (vgl. Abb. 1) sowie die Einführung von Assoziationskriterien, z.B. aufgrund charakteristischer photometrischer Eigenschaften, im Falle von Mehrdeutigkeiten (sog. Korrespondenzproblem).

Letztlich erlaubt dieses Verfahren bei unerwartetem Merkmalsausfall sogar die Generierung von Objekthypothesen parallel zur Radarsensorik (vgl. LKW rechts in Abb. 3). Im Nahbereich ist dies insbesondere zur Detektion überholender oder seitlich einscherender Fremdfahrzeuge von Interesse, im Fernbereich zur Erfassung neu in den Vorausschaubereich kommender Hindernisse.

3 Zusammenfassung und Ausblick

Es wurde ein Verfahren zur videobasierten Fahrspurerkennung basierend auf dem 4D-Ansatz vorgestellt. Insbesondere wurde dabei auf eine erweiterte Dynamikmodellierung sowie die Vorteile einer Klassifikation der Spurmarkierung durch direkte Vermessung der Markierungszeichen eingegangen. Das bestehende System zeigte im Rahmen von Autobahn Testfahrten mit einem Versuchsträger der Audi AG bereits eine beeindruckende Robustheit im Nahbereich (ca. 80m Vorausschauweite bei monokularer Kamerakonfiguration mit 9mm Brennweite). Zum Ausbau auf den Fernbereich wird derzeit der Versuchtsträger um eine zweite Kamera mit Teleoptik inkl. Blickrichtungssteuerung/-stablisierung erweitert.

In diesem Kontext sollen Strategien zur Kombination mehrerer mit dem Eigenfahrzeug mitbewegter bzw. ortsfester Spursegmente untersucht werden (vgl. Abs. 2.1 zur erweiterten Dynamikmodellierung). Eine zusätzliche Steigerung der Systemrobustheit wird durch die Einführung weiterer Assoziationskriterien für ortsfeste Kantenmerkmale erwartet (vgl. Abs. 2.2 zum Nutzen der Vermessung von Spurmarkierungen).

Literaturverzeichnis

1. FORBIAS–Forschungsverbund Bioanaloge Sensomotorische Assistenz.
2. www.abayfor.de/abayfor/presse_print/infomaterial
3. Harms, P., Neumaier, S.: Bioanaloge Ansätze zur Steigerung der Robustheit maschinellen Sehens bei Fahrzeugen. Workshop Fahrerassistenzsysteme FAS2005, Walting im Altmühltal (2005)
4. Dickmanns, E.D.: 4D-Szenenanalyse mit integralen raumzeitlichen Modellen. DAGM87, Informatik Fachberichte Nummer 149 (1987)
5. Dickmanns, E.D.: The 4D-Approach to Dynamic Machine Vision. Proc. 33rd IEEE Conf. on Decision and Control (1994)
6. Bierman, G. J.: Factorization Methods for Discrete Sequential Estimation. Mathematics in Science and Engineering Nr. 128, Academic Press (1977)
7. Mysliwetz, B.: Parallelrechner-basierte Bildfolgen-Interpretation zur autonomen Fahrzeugsteuerung. Universität der Bundeswehr München (1990)
8. Zapp, A.: Automatische Straßenfahrzeugführung durch Rechnersehen. Universität der Bundeswehr München (1988)
9. Behringer, R.: Visuelle Erkennung und Interpretation des Fahrspurverlaufes durch Rechnersehen für ein autonomes Straßenfahrzeug. Universität der Bundeswehr München (1996)
10. Wünsche, H.-J.: Bewegungssteuerung durch Rechnersehen. Fachberichte Messen-Steuern-Regeln, Springer-Verlag (1988)
11. von Holt, V.: Integrale Multisensorielle Fahrumgebungserfassung nach dem 4D-Ansatz. Universität der Bundeswehr München (2004)

Introduction of a Full Redundant Architecture into a Vehicle by Integration of a Virtual Driver

Frédéric Holzmann, Frank Flemisch, Roland Siegwart, Heiner Bubb

DaimlerChrysler AG
German Aerospace Center, Braunschweig, Germany
Federal Institute of Technology (EPFL), Lausanne, Switzerland
University of Technology, Munich, Germany

Abstract. This paper introduces a concept for advanced driver assistance by means of a redundant architecture including all system components spanning from environment perception to vehicle controllers. Concepts like Dickmanns et.al. show that automation can be sophisticated enough to drive fully autonomously in certain environments. The concept described here uses automation to introduce a virtual driver redundant to, but not excluding, the human driver. The first part of this paper is an overview of the project framework based on a model of the driver cognition. After that the use of a multiagent technology for the generation of a virtual driver will be described into more details. A decision control and its use to optimize the driver's command depending on his fitness and the confidence of the vehicle. The final part shows some preliminary results and concludes towards future work and research issues.

1 Introduction

According to [1], 95% of the accidents are due to a human mistake because of a wrong interpretation of the environment or lack of information. According to [1], about 40% of those accidents could be avoided by some warnings or a preventive and helpful influence of the driver's command. The consequences of the other accidents could be drastically decreased by an automatic reaction when the driver has no chance to react anymore.

Up to now, as only the driver can control the vehicle, his command is realized even if it is not adequate to the situation. By integrating an additional driver within a dedicated electronical control unit, it will be possible to use its output to improve the reliability of the driver's command within some compatibility tests. The confidence of the system and the driver's fitness are fused first to understand the conflicts and second to give advices to the driver.

Every element of the short time loop from the environment sensing to the command control will be integrated in the virtual driver. As defined by Lunenfeld in [2], the necessary tasks can be serialized as shown in Fig. 1: environment sensing, planning and the stabilization of the vehicle by coordinating the aggregates. The choice of the maneuver still stay with the driver, who is responsible for the vehicle action.

The stabilization is made depending on the output of this new element – the Decision Control. That task is transferred from the driver to the vehicle in order to get the possibility to modify his command. Furthermore the actuators are modifed in order to give the possibility to act depending in the corrected driver's command.

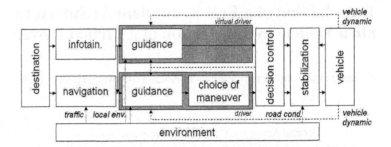

Fig. 1. Redundant data flow integrated into the vehicle

2 Modeling a virtual driver

An improved safety technology has to combine comfort range (latency time about 50 ms) and a safety range (reaction time lower than 10 ms) with both actions accurate enough for the situation. The changing from a comfort modus to a safety modus depends in practice on a time to collision (*TTC*, [4]) to the closest source of danger. It will be extremely short if a pedestrian suddenly appears or extremely long if the vehicle follows another vehicle with nearly the same speed.

2.1 Architecture of the multiagent system

A multiagent system (like on Fig. 2) can be extracted from the driver model for managing the different situations around the vehicle. It analyzes continuously the information about the environment stored in its *blackboard* : *E**. The agent *emergency analysis* determines the riskiness with the corresponding *TTC*. Depending on this maximal time, the agents integrated into the Contract Net will reply to the query. The different bids will be sorted and thebest agent will receive the contract and make the analysis.

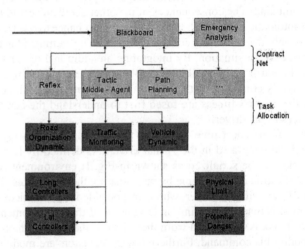

Fig. 2. Architecture of the system

Several agents have already been integrated. They need a validity range V:

- The occlusion rules: they describe what is necessary to allow the agent to take the initiative. For example the agent *vehicle following* starts automatically as soon as there is a vehicle ahead on the same lane.
- The exclusion rules: they describe the events that disable automatically the agent. For example the agent *lane changing* is switched off if any crossing road is detected.

Each of the agents can deliver different types of information :

- A matching rate D, which describes the adequation of the validity range (V) on the model of the environment needed ($E^{**}\ subsetE^*$).
- A safety envelope SE for the motion command of the vehicle.

$$SE : \begin{cases} SE_{long.} : E^{**} \longrightarrow \gamma_{min}, \gamma_{max} \in R \\ SE_{lat.} : E^{**} \longrightarrow \theta_{min}, \theta_{max} \in R \end{cases}$$

- An optimum of command OPT: ($\gamma_{opt}, \theta_{opt}$)

2.2 Definition of the tactic agents

A middle tactic agent splits automatically the work between road organization, traffic monitoring and vehicle dynamic. The agents analyzing the traffic flow will extract the possible maneuvers and set the adequate safety envelope for them. This theory will be downsized to the physical possiblities of the vehicle on the road in two steps. The agents working on road organization will analyze the maximal acceleration allowed because of speed limits (if there is no emergency case) and road curvature.After that the agents working on the vehicle dynamic will set the maximal capacities of the vehicle for the acceleration, braking, and steering.

Some conflicts may appear between differents agents triggered by the tactic agent. For example if a vehicle ahead is breaking abruptly and the vehicle behind still accelerate, the agents *vehicle following* and *vehicle leading* willget incompatible results. The conflict between them will be avoided by the middle agent by asking if there is one chance to change to another lane.

2.3 Towards a fault-tolerant technology

The resoures available into an automotive electronic control unit (ECU) are limited in terms of computation capacities per cycle. Therefore the fault-tolerance methodology cannot use a full N-voting algorithm for the different agents.

The agents are duplicated two or three times only if they are working at the current cycle. Depending on the current computation capacities, the duplications of the tactic agents start with the agents working for the analysis of the current lane, than of left lane and finally of the right lane.

3 Integration of a decision control into the vehicle

The Decision Control analyzes first the driver's command to understand which maneuver the driver wants to realize. If the driver's command stays between two possible maneuvers some doubts are legitimate. Once the maneuver extracted, its safety range and its optimum can be fused with the driver's command.

3.1 Understanding the driver's maneuver

Like every stochastic phenomenon, the driver's choice of maneuver can be determined only by selecting the maneuver with the best probability. In practice the evolution of a dynamic Markov Chain, depending on an-priori choice and the current command, will be computed over the time. The position of the driver's command and of the optimums of maneuvers will be extrapolated until the matrix stabilizes.

First the history of the driver is used to define the a-priori probabilities of the change of a maneuver for another (P_{ij}^*) or to itself (P_{ii}^*) like in Fig. 3. As not all scenarios are possible at the same time, only a sub-set of the connections will be enabled depending on the situation.

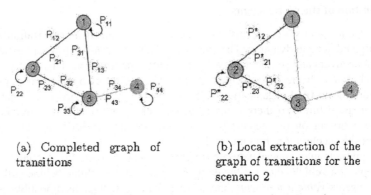

(a) Completed graph of transitions

(b) Local extraction of the graph of transitions for the scenario 2

Fig. 3. Use of the static graph of the maneuvers for a scenario

But the probabilities of the transition are weighted with other parameters. The most important parameters to extract the choice of the driver are the distance ($\Delta_{acc.}$) and ($\Delta_{steer.}$) to the optimums of the maneuvers and the variation of these distances. The additional parametera are the quality of each maneuver and the use of the blinkers. Indeed the higher the quality is, the more probable is the driver's choice because the driver tends to realize an easy manoeuver.

The posed problem can be explained as a probabilistic equation to set the probability of a maneuver depending on the former probability and the properties of the manoeuver:

$$P(M_n \otimes M_{n-1} \otimes H_{n-1} \otimes Q \otimes \Delta_{acc.} \otimes \Delta_{steer.} \otimes \dot{\Delta}_{acc.} \otimes \dot{\Delta}_{steer.} \otimes \delta \otimes \Pi) \quad (1)$$

The equation is split between three independent parts because all the parameters are not correlated. Therefore it is possible to get the probability for each maneuver (i) by using the computation for the Markov chain by using the matrix P and adapting it depending on the extrapolated positions of the optimums and the qualities like on the next equation :

$$\forall_i, P(M_n^i) = P(M_n^i \mid M_{n-1} \cdot H) \otimes P(M_n^i \mid Q^i) \otimes P(M_n^i \mid \Delta^i) \otimes P(M_n^i \mid \dot{\Delta}^i) ((2)$$

4 Fusion of the two commands

On one hand the Decision Control gets the driver's command and an information about his fitness that can be extracted from the steering angles over the time like in [5] or the eyes lids like in [6]. On the other hand the system gets the confidence of the computation, the safety envelope and sometimes the corresponding optimum.

If there is an optimum available, the distance to the command will describe the confidence of the driver's command. Otherwise only the inclusion of the command into the safety envelope helps to determine the confidence of the driver's command.

Let define the fitness of the driver as $x\%$ and the confidence of the co-pilot as $y\%$ to fuse of the driver's command (MC) and the optimum (OPT) together. A method based on a Markov Chain will be used, that can also beextended to n commands (like two pilots and an assistant system in an airplane). Each element will sort the group depending on his quality. For the driver, the a-priori quality is x and for the other an equal probability : $(1-x)/(n-1)$. This delivers here the a-priori matrix (M), that is always transient. Therefore the real weights of each element can be computed as the stabilization (generally obtained after 7 iterations) of the markov chain:

$$\forall i \in \mathbb{N} \times [1, n], P(E_i) = \lim_{x \to \infty} [0_1 ... 1_i ... 0_n]^T \cdot M^x \tag{3}$$

5 Implications for the roles and interaction between human and virtual driver

Redundant architectures like the one described here, with a virtual driver able to asses the situation and determine meaningful maneuvers, open up a whole spectrum of use, from light assistance, semi-automation to full autonomous vehicles. The point of choice on this spectrum depends at the first glance on the quality of sensors, sensor fusion, navigation and guidance. How good is the virtual driver, how reliable? On the second glance the degree of automation will also dependent on the interactionquality and understandability of the virtual driver: The better human and machine can interact and understand each other, the better the chance, that a higher level of automation also leads towards an overall improvement.

It will be wise to use this technology as assistance at first, and increase the level of assistance or automation, as shown in Fig. 4, only if the severity of the situation, the quality of the virtual driver and interaction quality/understandability is high enough. With this technology becoming more capable over time, the discussion might shift from a "how shall we", towards a conscious choice of roles between humans and machines, where "virtual driver" is one of several role models for this kind of automation.

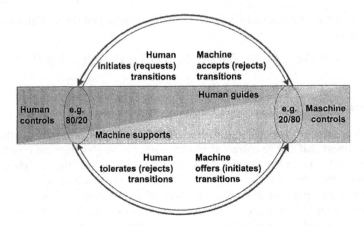

Fig. 4. Non-exclusive levels of automation adapted from [3]

6 Conclusion and next research

The concept presented here is heading towards a full redundant architecture from the environment sensing down to the vehicle by integrating a virtual driver in order to improve the safety on the roads.

A virtual driver determines all the possible maneuvers by using a multi agent system, in which different agents are working in parallel depending on the time to collision. After that the choice of the maneuver is extracted from this list and from the driver's command by using a Bayesian approach and then fuse with the driver's command depending on the fitness of the driver and the confidence of the virtual driver to improve the reliability of the command.

Contrary to other concepts, this system does not work binarily with the driver's command. It gets a continuous mix of the driver's command with the output of virtual driver depending on the driver's fitness and the accuracy of the certainty of theprediction. This method permits to compensate a lack of alertness of the driver's lack of accuracy of the assistant system by modifying their weights on the command of the vehicle.

But the most important part is still under investigation. The driver has to feel the fusion of his/her command with the output of the co-pilot in term of degradation or strengthening of the feedback. Therefore during the next years the interactionof the assistant systems with the driver will have to be improved. Otherwise the driver will not understand what the vehicle really does and even will be taken out of to the control loop. The hope is that this integration of new concept of interaction will also lead to a drastic modification of the ergonomy of the vehicle.

References

1. Jürgen GRENDEL. *Sicherheitsanalyse im Strassenverkehr*. Bundesanstalt für Strassenwesen, Mensch und Sicherheit, 1993
2. Nathan GARTNER et al.. *Traffic Flow Theory*. Technical report, Transportation Research Board, 1997
3. Frank O. FLEMISCH, Catherine A. ADAMS et al.. *The H-Metapher as a Guideline for Vehicle Automation and Interaction*. NASA/TM-2003-212672, Dec. 2003
4. Matthias KOPF et al.. Milestones for the development of a full active safety concept. FTM-UTM Safety Technology Symposium, 2004
5. Richard GRACE, V. E. BYRNE et al.. *A drowsy driver detection system for heavy vehicles*. Proceedings of the 17th Digital Avionics Systems Conference, 2001, pp. I36/1–I36/8
6. Paul SMITH, Mubarak SHAH, and N. DA VITORIA LOBO. *Monitoring Head/Eye Motion for Driver Alertness with One Camera*. IEEE International Conference on Pattern Recognition, 2000

Systemplattform für videobasierte Fahrerassistenzsysteme

Matthias Goebl, Sebastian Drössler, Georg Färber

Lehrstuhl für Realzeit-Computersysteme, Technische Universität München

E-mail: {goebl,droessler,faerber}@rcs.ei.tum.de

Zusammenfassung: Videobasierte Fahrerassistenzsysteme müssen große Bildatenmengen verarbeiten und gleichzeitig eine hohe Reaktivität für regelungstechnische Aufgaben wie die Kamerastabilisierung aufweisen. Die hier vorgestellte Systemplattform orientiert sich am menschlichen Bildstabilisierungs- und –verarbeitungssystem. Sie besteht aus einer starren und einer in der Horizontalen und Vertikalen schwenkbaren Kamera, einem 6–DOF Beschleunigungssensor zur Erfassung der Fahrzeugeigenbewegung sowie einem Hochleistungsrechner und einem eingebetteten System. Die enge Kopplung von Bildverarbeitung und Regelung erfordert es, die Systemplattform als Ganzes zu betrachten.

1 Einleitung

Der zunehmende Umfang von aktiven Fahrerassistenzsystemen lässt die Anforderungen an eine maschinelle Umfeldwahrnehmung rapide ansteigen. Im Bereich der Komfortsysteme können die Autokäufer bereits heute aus einer Vielzahl von neuen Funktionen auswählen, wie z.B. adaptive Abstandsregelung (ACC). Sicherheitskritische Systeme wie Spurhaltung oder Notbremsung, die stärker in die Fahrzeugaktorik eingreifen, sind in Deutschland trotz jahrzehntelanger Forschung [1] noch nicht in Serie erhältlich. Ein Grund dafür ist das unzureichende Vertrauen in die wahrgenommene Verkehrssituation. Diese wird in Serienfahrzeugen mittels Umfeldsensoren wie z.B. Radar und Informationen über eigene Bewegungsgrößen (Raddrehzahlen und Lenkwinkel) errechnet.

Für den Durchbruch aktiver Fahrerassistenzsysteme ist daher eine robuste Umfelderfassung unumgänglich. Eine mögliche Lösung zeigt die Natur auf, sie hat den meisten Lebewesen nach einer jahrmillionenlangen Evolution die Augen zum primären Organ der Umfelderfassung gemacht. Der bioanaloge Ansatz, Videosensorik auch im Fahrzeug einzusetzen, wird aber durch die, im Vergleich zu klassischen Sensoren wie Radar, aufwändigere Handhabung von Kamerasystemen erschwert: Sie benötigen eine breitbandige Hardwareanbindung, rechenintensive Auswertung und für größtmögliche Robustheit eine aktive Stabilisierung und Regelung.

Die im folgenden vorgestellte Systemplattform erfüllt diese Anforderungen und orientiert sich im Design am biologischen Vorbild: Abschnitt 2 gibt einen Einblick in das menschliche Verarbeitungssystem, Abschnitt 3 zeigt die Hard- und Software der Systemplattform und Abschnitt 4 gibt einen Ausblick.

2 Menschliches Vorbild

Im Folgenden werden die beim Menschen für das Sehen zuständigen sinnesverarbeitenden Organe bzw. Areale im Gehirn kurz erläutert [2,3].

Die Bilderfassung erfolgt über die Rezeptoren des Auges. Diese sind so angeordnet, dass sie eine sehr hohe Auflösung im Zentrum (*Fovea*) besitzen und nach außen hin in ihrer Anzahl abnehmen. Dies ermöglicht eine klare Erfassung in der Sichtachse bei gleichzeitig breitem Blickwinkel für das periphere Umfeld. Drei Muskelpaare erlauben dem Auge neben langsamen Folgebewegungen auch *Sakkaden* (schnelle Blicksprünge mit bis zu 800°/sec) in horizontaler und vertikaler Richtung. Bis zu einem bestimmten Grad wird sogar eine Torsion um die Blickachse durchgeführt, um das Bild auch bei einer Kopfneigung auszugleichen.

Für die Blickstabilisierung ist u.a. der sog. *vestibulo-okuläre Reflex* (VOR) zuständig. Das *Vestibularorgan* (Gleichgewichtsorgan) im Innenohr erfasst die drei rotatorischen und drei translatorischen Freiheitsgrade und liefert über den sog. 3-Neuronen-Bogen direkte Steuersignale an die Okulo-Motorik. Durch diese enge Kopplung wird eine Latenzzeit von 10 ms erreicht. Da diese Information alleine für eine zuverlässige Stabilisierung nicht ausreicht, erfolgt eine Rückkopplung visueller Informationen, aus denen der *retinal slip* (Bildverschiebung auf der Netzhaut) berechnet wird und dem 3-Neuronen-Bogen entsprechende Korrektursignale aufgeschaltet werden.

Bei bewegten Szenen tritt ein weiterer Stabilisierungsmechanismus auf: Der *optokinetische Nystagmus* (OKN). Dabei folgt das Auge einem bewegten Punkt, indem versucht wird, den retinal slip für diesen Punkt möglichst gering zu halten. Verlässt der anvisierte Punkt den Sichtbereich erfolgt eine Sakkade zur Rückstellung der Augen. So wird z.B. beim seitlichen Blick aus einem fahrenden Auto ein scharfes Bild der Umwelt garantiert.

Der *visuelle Kortex* ist für eine erste Bildanalyse zuständig. Mit massiver Parallelität und dementsprechend hoher Verarbeitungsleistung werden dort Kanten und teilweise bereits einfache Formen mit deren Orientierung sowie Bewegungsrichtung erkannt. Auffällig dabei ist, dass die Erkennung von Kanten unterschiedlicher Orientierung und auch die Bewegungsrichtung in gesonderten Bereichen des Gehirns erfolgt, was die parallele Verarbeitung von Reizen deutlich aufzeigt.

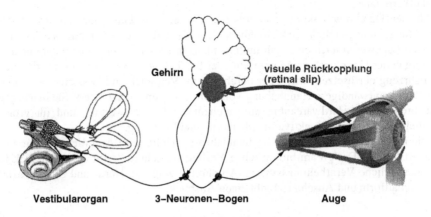

Abb. 1. Am Sehen beteiligte Organe

Für eine Situationsanalyse mit z.b. Gefahrenabschätzung und anschließender We-geplanung sind derart viele Bereiche des Gehirns zuständig, dass eine genaue Lokali-sation nicht mehr möglich ist; die dafür aufgewendete Verarbeitungsleistung ist je-doch enorm. Als Beispiel der Leistungsfähigkeit kann ein Versuch herangezogen werden, der die Zeit von Bildaufnahme bis Erkennung analysiert. Dabei werden ei-nem Probanden Bilder (Naturaufnahmen) präsentiert, wobei sich in manchen Tiere befinden können. Falls der Proband ein Tier detektiert, soll er dies mit einem Tasten-druck deutlich machen. Im Durchschnitt lag die Latenzzeit bei etwa 300 ms. Ohne die Latenzzeit in die Sehrinde (ca. 50–80 ms) und ohne die motorische Reaktion (ca. 150 ms) benötigt das Gehirn lediglich 70–100 ms für die Erkennung [3].

3 Systemplattform

Bei der technischen Umsetzung dieses leistungsstarken biologischen Systems wurde bei Sensorik und Aktorik auf eine platzsparende Umsetzung geachtet. Bei der Aus-wahl der Hardwareplattform standen eine einfache, schnelle Erweiterbarkeit und die funktionale Aufteilung im Vordergrund, die zur Entwicklung von videobasierten Fahrerassistenzsystemen hervorragend geeignet ist; für Serienfahrzeuge muss eine entsprechende Skalierung vorgenommen werden.

3.1 Kameraplattform

Die junge HDRC-Technologie (High Dynamic Range CMOS) bietet eine dem menschlichen Auge ebenbürtige Helligkeitsdynamik von 120 dB. Ihre RAM-artige Zugriffsmöglichkeit z.B. zur direkten Adressierung von ROIs (region of interest) macht sie zu einem optimal geeigneten Bildsensor.

Für die gleichzeitige Erfassung sowohl von detaillierten Bildinformationen in der Ferne als auch des peripheren Umfelds kommt eine Plattform mit je einer Kamera für den Fern- und den Nahbereich zum Einsatz. Aus Platzgründen wurde auf eine zweite Weitwinkelkamera für Stereo-Sehen zunächst verzichtet; zudem versprechen bioana-loge Algorithmen – wie z.B. der Merkmalsfluss oder die Nutzung von Vorwissen – eine ausreichende Tiefenwahrnehmung. Die Telekamera ist zusätzlich mit einer Be-wegungssteuerung zur Stabilisierung und Blickrichtungssteuerung ausgestattet. Ein 6–DOF Beschleunigungssensor liefert die Eingangssignale für einen Regler zur Ka-merastabilisierung.

Zur Steigerung der Robustheit können weitere Sensoren (Gierrate, Radar, Lidar, etc.) eingebunden werden; dies wird hier jedoch nicht weiter betrachtet.

3.2 Hardwareplattform

Die Hardwareplattform dient zum einen der rechenintensiven aber langsameren Bild-auswertung, wie sie hochparallel im menschlichen Gehirn geschieht, zum anderen der schnellen aber weniger rechenlastigen Blickrichtungsstabilisierung, wie sie beim Menschen über den VOR erfolgt.

Zur Realisierung realzeitfähiger Bildverarbeitungsalgorithmen kann zum einen ei-gens zu diesem Zweck konstruierte Spezialhardware dienen, die erst nach einer zeit-aufwendigen und teuren Entwicklungsphase verfügbar wird. In der Forschung wird

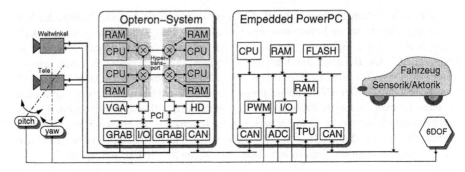

Abb. 2. Hardwareplattform mit Sensorik und Aktorik

daher Bildverarbeitung hauptsächlich in Software auf Mikroprozessoren durchgeführt, eine schritthaltende Verarbeitung ist dann aber auf Hardwareseite nicht mehr gegeben und muss durch Software garantiert werden.

Das Design der Systemplattform zeigt Abb. 2. Ein Hochleistungs-Opteron-System (hier mir vier Prozessoren) liefert die nötige Verarbeitungsleistung für die Bildverarbeitung, ein embedded PowerPC dient der Blickrichtungsstabilisierung, der Sensordatenvorverarbeitung und der Kommunikation mit den Fahrzeugsystemen. Für die Fahrzeuganbindung wird der in der Automobilindustrie etablierte und in seinem Zeitverhalten hinreichend untersuchte CAN–Bus verwendet. Im Folgenden werden die Kriterien für das Design der Plattform erörtert.

Hochleistungs-Rechner Hier werden handelsübliche standardisierte PC-Industriekomponenten (COTS, commercial-off-the-shelf) eingesetzt. Diese zeichneten sich in der Vergangenheit aufgrund der Anforderungen des PC-Markts durch konsequente Rückwärtskompatibilität aus, so dass auch in Zukunft ein problemloser Austausch durch leistungsfähigere Komponenten möglich sein wird. Analog wird bei konstanter Leistung eine Miniaturisierung erwartet, wie sie PC/104-Systeme und DIMM-PCs heute zeigen, so dass die Plattform dadurch in einigen Jahren eine fahrzeugtaugliche Größe erreichen kann. Bei Auswahl einer zukunftsfähigen Architektur rechtfertigen diese Erfahrungen den Einsatz eines heutigen Hochleistungs-PCs. Hierbei können Multiprozessorsysteme zum Einsatz kommen, durch die Multicore-Technologie werden diese Prozessorkerne später auf einem Chip integriert sein.

Es werden nur 64 Bit–fähige Architekturen berücksichtigt, die auch 32 Bit–Software nach dem IA-32 Standard (Intel Architecture-32) ausführen können, um existierende Applikationen ohne Modifikation einsetzen zu können. 64 Bit–Architekturen können problemlos Speichergrößen jenseits von 4GB verwalten, wie es für die Zwischenspeicherung längerer Bildsequenzen notwendig ist. Durch breitere 64–Bit Worte und den größeren Registersatz können neue Bildverarbeitungsalgorithmen mit höherer Parallelität entwickelt werden. Die Anforderungen erfüllen sowohl die AMD64-Architektur, als auch die befehlssatzkompatible EM64T-Erweiterung von Intel. Folgende Systeme wurden daher betrachtet:

Intel Itanium: Bei der von Intel eingeführten Hochleistungsarchitektur IA-64 (Intel Architecture-64) des Itanium Prozessors zeichnet sich keine große Marktdurchdringung ab, so dass mit einer zukünftigen Miniaturisierung nicht gerechnet werden kann. Ein Grund dafür ist auch die mangelhafte Verarbeitungsleistung bei 32 Bit–Software.

Intel Pentium/Xeon, AMD Athlon: Diese Mehrprozessor-PCs weisen eine sehr große Verbreitung auf, es existieren bereits zahlreiche eingebettete Lösungen. Allerdings entsprechen sie der SMP–Architektur (symmetric multiprocessing): Alle Prozessoren greifen über einen gemeinsamen Speichercontroller (Host–Bridge) auf den gleichen Speicher zu, was bei gleichzeitigen Zugriffen die beteiligten Prozessoren ausbremst [4].

AMD Opteron: Opteron–Systeme sind nach der NUMA–Architektur (non uniform memory access) aufgebaut, wie sie Abb. 2 zeigt: Jeder Prozessor verfügt über einen eigenen lokalen Speicher, auf dem er über einen eigenen Speichercontroller arbeiten kann. Die Prozessoren sind untereinander mit schnellen Hypertransport–Links verbunden, über die sie auch Zugriff auf den Speicher anderer Prozessoren haben. Crossbar–Switches in jedem Prozessor verbinden die Hypertransport–Links mit dem Prozessorkern und dem Speicher, so dass Laufzeiteinflüsse minimiert werden. Dadurch skalieren mit steigender Prozessoranzahl NUMA–Systeme besser als vergleichbare SMP–Systeme [5].

Für diese Hardwareplattform wurden daher Mehrprozessor–PCs mit AMD Opteron–Prozessoren gewählt, da deren breite Speicheranbindung eine wichtige Voraussetzung für die gleichzeitige Auswertung großer Bilddatenmengen erfüllt. Durch zwei Framegrabber an verschiedenen PCI–Bussen ist sogar der parallele Bildeinzug und die gleichzeitige Verarbeitung auf unterschiedlichen CPUs möglich, so wie es auch beim Menschen in beiden Gehirnhälften geschieht. Die gute Skalierbarkeit ermöglicht eine signifikante Leistungssteigerung durch Hinzufügen weiterer Prozessoren ohne großen Softwareaufwand. Ein Verbund mehrerer PCs, deren Netzwerkverbindungen entsprechende Latenzzeiten bei der Kommunikation erzeugen, kann so vermieden werden.

Eingebettetes System Die zeitkritischen Regelungsalgorithmen der Blickrichtungssteuerung stellen harte Realzeitanforderungen, die moderne Standard–PCs aufgrund der Latenzzeiten ihrer auf hohen Durchsatz optimierten Bussysteme nur schwer erfüllen können [6,7].

Für die Steuerung der Kameraplattform wird daher ein embedded Power-PC 555/565 (MPC) von Freescale Semiconductor (ehemals Motorola) eingesetzt. Dieser zeichnet sich durch die notwendigen kurzen Interrupt–Reaktionszeiten aus, bei einer für eingebettete Systeme vergleichbar hohen Rechenleistung. Der MPC enthält bereits zahlreiche Peripherie–Module wie I/O–Ports, A/D–Wandler, Schnittstellen, CAN und Speicher „on-chip", so dass durch diese Plattform unnötiger Hardwareentwicklungsaufwand bei der Anbindung weiterer Sensorik und Aktorik eingespart werden kann.

Der MPC besitzt zusätzlich zwei Time-Processing-Units (TPU): Diese I/O-Koprozessoren verfügen über einen eigenen Speicher und können unabhängig von der CPU komplexe Echtzeitsteueraufgaben durchführen, wie sie bei der Anbindung weiterer Aktorik anfallen werden.

3.3 Softwareplattform

Um dem Anspruch einfacher zukünftiger Erweiterbarkeit bei maximaler Flexibilität zu erfüllen, werden nur Standard Echtzeit–Betriebssysteme eingesetzt, deren Quellcode öffentlich verfügbar ist.

Auf PC–Seite ist dies das freie Unix–Derivat Linux. Darauf setzt das Applikations–Framework auf, das verschiedene Module zur Sensordatenerfassung, Verarbeitung, Wahrnehmung, Situations-Interpretation und Aktoransteuerung aufnimmt und ihnen eine einheitliche Softwareschnittstelle (API) zur Verfügung stellt (Abb. 3). Zentraler Bestandteil ist dabei eine *dynamische Objektdatenbasis* (DOB) [1], in der alle relevanten Daten in Objekten als kleinste Organisationseinheit zusammengefasst und konsistent gehalten werden.

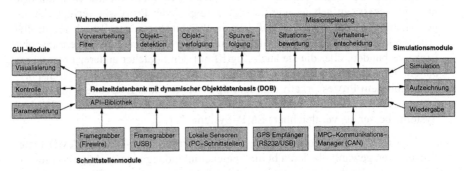

Abb. 3. Applikationsframework (PC) mit Beispielmodulen

Auf dem MPC–System läuft das Realzeit–Betriebssystem eCos. Ein Realzeitframework nimmt Algorithmen zur Kamerastabilisierung, Sensordatenverarbeitung und Ansteuerung des Fahrzeugs auf. Es sorgt zudem für eine transparente Kommunikation mit dem Hochleistungs–System.

4 Ergebnisse und Ausblick

Die vorgestellte Systemplattform orientiert sich in ihrer Struktur an den Verarbeitungseinheiten des Menschen. Die enge Kopplung von Sensorik und Aktorik über ein eingebettetes Realzeitsystem garantiert die kurzen Reaktionszeiten, um menschliche Bildstabilisierungsreflexe nachzubilden. Die enorme Rechenleistung für die Bildverarbeitung sowie Szenenanalyse, -interpretation, Missionsplanung etc. stellt ein Hochleistungsrechner bereit.

Derzeit wird das Applikationsframework erweitert, um das Zeitverhalten der Module, insbesondere der Bildverarbeitung, zu kontrollieren und die Echtzeitfähigkeit der Datenbasis sicherzustellen. Dazu wird die Realzeiterweiterung RTAI-Linux eingesetzt, die bereits auf die 64–Bit Architektur portiert wurde.

5 Danksagung

Diese Arbeiten entstanden in Kooperation des *Forschungsverbundes Bioanaloge Sensomotorische Assistenz* (FORBIAS), der von der Bayerischen Forschungsstiftung gefördert wird, und dem Projekt *Architektur und Schnittstellen für kognitive Funktionen in Fahrzeugen* (ASKOF), das durch die Deutsche Forschungsgemeinschaft unter dem Förderzeichen FA 109/17-1 unterstützt wurde.

Literaturverzeichnis

1. Ernst D. Dickmanns. Dynamic vision-based intelligence. In *AI Magazine 25(2)*, pp. 10–30, 2004.
2. Andrew H. Clarke. Neuere Aspekte des vestibulo-okulären Reflexes. Europ Arch ORL (Suppl.1995/I), pp. 117–153. 1995.
3. Karl R. Gegenfurtner. Unterlagen zur Vorlesung *Sinnesphysiologie und Wahrnehmung*, 2005.
4. Jürgen Stohr, Alexander von Bülow, und Georg Färber. Bounding worst-case access times in modern multiprocessor systems. In *Proceedings of the 17th Euromicro Conference on Real-Time Systems*, Palma de Mallorca, Balearic Islands, Spain, July 2005.
5. C. Guiang, K. Milfeld, A. Purkayastha, J. Boisseau. Memory Performance on Dual-Processor Nodes: Comparison of Intel Xeon and AMD Opteron Memory Subsystem Architectures. In *Proceedings of the ClusterWorld Conference and Expo*, San Jose, CA, June 24-26, 2003.
6. Jürgen Stohr, Alexander von Bülow, und Matthias Goebl. Einflüsse des PCI-Busses auf das Laufzeitverhalten von Realzeitsoftware. *Technischer Bericht*, Lehrstuhl für Realzeit-Computersysteme, TU München, Dezember 2003.
7. Jürgen Stohr, Alexander von Bülow, und Georg Färber. Controlling the Influence of PCI DMA Transfers on Worst Case Execution Times of Real–Time Software. In *Proc. of the 4th Intl. Workshop on Worst Case Execution Time Analysis in conjunction with 16th ECRTS*, Catania, Italy, June 2004.

Kognitive Sensordatenverarbeitung

Sequential Parameter Estimation for Fault Diagnosis in Mobile Robots Using Particle Filters

Christian Plagemann, Wolfram Burgard

Department of Computer Science, University of Freiburg, Georges-Koehler-Allee 79, 79110 Freiburg, Germany

E-mail: {plagem,burgard}@informatik.uni-freiburg.de

Abstract. The autonomous detection and handling of faults is an important skill for mobile robot systems. Faults in the motion-control system can strongly decrease the robots' performance or compromise its mission completely. In this paper, we demonstrate how a mobile robot system can, in case of a fault, switch to a richer internal system model and estimate the newly introduced parameters to reliably diagnose its state and possibly continue its operation. We discuss three methods for sequential parameter estimation using particle filters and evaluate their performance in physically accurate simulation runs.

1 Introduction

Mobile robots act in their environment using an internal system model that defines possible actions and their effects. When failures like broken wheels or deflating tires occur, this model is not suitable anymore. In such cases, we propose to switch to a higher dimensional model and estimate the newly introduced parameters in a sequential manner.

Many works have dealt with the fault diagnosis problem for technical systems. Classic approaches in the literature include model-based methods, where the system is assumed to transition between discrete steady states [1]. Here, diagnosis is often based on system snapshots without a history. For highly dynamic systems like mobile robots, these approaches are limited in their capability to distinguish between normal operational modes and faulty ones. Recently, probabilistic state tracking techniques have been applied to this problem. Adopted paradigms range from Kalman filters [2] over sampling in factored dynamic systems [3,4] to particle filters in various modelings [5,6,7]. Particle filters represent the belief about the state of the system by a set of state samples that are updated based on sensory input and based on issued controls [8]. In particle filter based approaches to failure diagnosis, the system is typically modeled by a dynamic mixture of linear processes [9] or a non-linear markov jump process [10].

In this work, we assume a non-linear system that is appropriately represented by a low complexity model under normal operational conditions, but has to be analyzed using a more complex and higher dimensional model after a failure has occurred. We describe three methods for the sequential estimation of the newly introduced model parameters and evaluate their performance in realistic simulation runs.

2 Sequential Estimation of the System State and Model Parameters

Partical filters can be used to efficiently estimate the variables of dynamic systems, if some mild preconditions are met [8]. For the normal operation of mobile robots, a simple planar motion model is assumed. Obviously, when a serious failure such as tire deflation occurs, this model is not applicable anymore. In this case one needs to switch to a richer, more complex model that can more accurately track the behavior of the faulty system. One key problem in this context is how to initialize and quickly estimate the additional parameters of these richer models.

Particle filters represent the probability distribution of the system states by a discrete set of weighted samples. By using this form of representation, new sensor measurements and action commands can be integrated in a direct way. Algorithm 1 formulates the particle filtering approach for the case that state variables x as well as model parameters η are to be estimated. This algorithm differs from the standard particle filtering algorithm [11] by explicitly including η in Lines 3,4, and 6.

Algorithm 1 Parameter_Estimating_Particle_filter(χ_{t-1}, u_t, z_t)

1: $\overline{\chi}_t = \chi_t = \emptyset$
2: **for** $m = 1$ to M **do**
3:     sample $\eta_t^{[m]} \sim p(\eta_t | \chi_{t-1})$
4:     sample $x_t^{[m]} \sim p(x_t | u_t, x_{t-1}^{[m]}, \eta_t^{[m]})$
5:     $w_t^{[m]} = p(z_t | x_t^{[m]})$
6:     $\overline{\chi}_t = \overline{\chi}_t + \langle x_t^{[m]}, \eta_t^{[m]}, w_t^{[m]} \rangle$
7: **end for**
8: **for** $m = 1$ to M **do**
9:     draw i with probability $\propto w_t^{[i]}$
10:    add $x_t^{[m]}$ to χ_t
11: **end for**
12: return χ_t

In Algorithm 1, the state of the system at time t is represented by a set χ_t of state samples $x_t^{[m]}$ and parameter samples $\eta_t^{[m]}$. In Lines 1 and 2, we perform a state prediction step using the external motion command u_t and the motion model $p(x_t | u_t, x_{t-1}^{[m]}, \eta_t^{[m]})$. Line 5 incorporates the current sensor measurement z_t by reweighting the state samples according to the measurement model $p(z_t | x_t^{[m]})$. From Line 8 to 11, a resampling step is performed to concentrate the samples on high-probability regions of the state space. We refer to [11] for details about the resampling step and its efficient implementation.

The estimation process for the model parameters η is guided by the transition model $p(\eta_t | \chi_{t-1})$ as applied in Line 3. We specify and evaluate three possible transition models for the evolution of these parameters. Method 1 views η as a fixed model constant and continuously samples possible values from an uniform distribution. Method 2, as proposed by Gordon [12], adds Gaussian noise ("roughening penalties") to the parameter samples in each iteration. Finally, as a motivation for Method 3, Andrieu et al. [13] as well as Liu and West [14] point out, that constants cannot be estimated by just augmenting the state space accordingly. They argue, that the repeated addition of noise artificially increases the variance and therefore results in too diffuse posterior distributions. They propose to apply kernel smoothing techniques to this problem domain. In the following, we specify the parameter transition models for the alternative methods and evaluate their practical value in the next section.

Method 1: Sampling from a Uniform Prior Distribution
For Method 1, we sample new values for η from a uniform distribution in an interval $[\eta\_min, \eta\_max]$ for a certain fraction p of samples from the particle set

$$\eta_t^{[m]} \sim U_{[\eta\_min, \eta\_max]}.$$

Method 2: Applying Independent Random Shocks
An alternative way of evolving the parameter estimates is the application of small independent random shocks to the parameter samples. This method has originally been proposed by Gordon [12]. Here, independent Gaussian noise of variance σ^2 is added to all $\eta_t^{[m]}$ values in the particle set

$$\eta_t^{[m]} \sim N(\eta_{t-1}^{[m]}, \sigma^2).$$

Method 3: Kernel Smoothing
Liu and West [14] also propose to add Gaussian noise to evolve the parameters, but additionally apply a kernel smoothing step to reduce the variance of the distribution. Here, a smoothing parameter h has to be chosen between 0 and 1 in addition to the noise parameter σ^2

$$a = \sqrt{1-h^2}$$
$$k_t^{[m]} = a\,\eta_{t-1}^{[m]} + (1-a)\,\overline{\eta}_{t-1}$$
$$\eta_t^{[m]} \sim N(k_t^{[m]}, h^2\sigma^2).$$

3 Evaluation of the Estimation Methods in a Fault Diagnosis Scenario

For the experimental comparison of the three proposed methods to parameter estimation, we consider a failure diagnosis scenario where a mobile robot is losing air in one of its tires. For the faultless operational mode, the standard odometry-based motion model as described in [11] is used. In case of a failure, the state predictions based on this motion model become inaccurate and the observation likelihood drops. Now, we switch to a more complex model that includes an additional parameter $\overline{\eta}_t$ for the difference in the air pressure of the tires.

We compared the estimation quality of the three methods described above using the high-fidelity simulator Gazebo [15], where physics and motion dynamics are simulated accurately using the Open Dynamics Engine [16]. The modeled scenario includes an Active Media Pioneer 2DX robot executing motion commands and losing

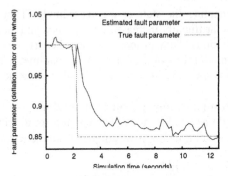

Fig. 1. A typical simulation run with estimated fault parameter and ground truth.

air pressure in the left tire after a predefined period of time. The followed trajectory contains sharp turns and strong acceleration changes. For each estimation method, 720 simulation runs have been performed. In these simulation runs, the meta parameters p, σ, and h have been varied to find their optimal values. As a performance measure for the different methods, we used the average error of the parameter estimates against the true values. This measures how quickly the correct value is attained and how stable it remains.

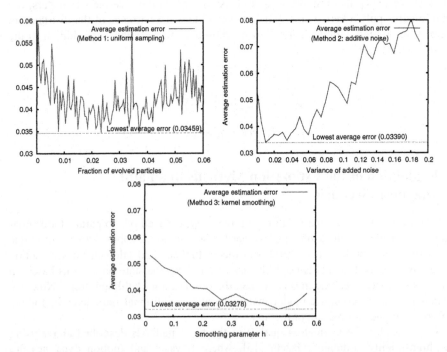

Fig. 2. The average estimation error of Method 1 (top left), Method 2 (top right), and Method 3 (bottom). The results are plotted for varying meta parameters p, σ and h respectively.

Figure 1 shows a typical estimation sequence of the fault parameter $\bar{\eta}_t$ and the underlying ground truth against the simulation time t. In this run, the average deviation of the estimate from the ground truth was 0.023 (tire deflation factor), which corresponds to an overestimation of the diameter of the left wheel by only 0.36 cm.

The achievable estimation accuracies for varying meta parameters are depicted in Fig. 2. Method 1 shows a high sensitivity to the choice of p (the fraction of particles which are redrawn from a uniform distribution). The performance of Method 2 changes more smoothly and results in better estimates when σ is tuned towards the optimal value. Method 3 is even less sensitive to the choice of smoothing parameter h and yields better results than the other methods.

According to these experiments, the kernel smoothing approach appears to be best suited for the problem of tracking failure states in the domain considered here.

4 Conclusions and Further Research

In this paper, we dealt with the problem of online failure diagnosis for mobile robots. We formulated the diagnosis task as as an online parameter estimation problem using particle filters. We described three methods for the continuous adaption of model parameters and evaluated their performance using simulation runs. In this comparison, the kernel smoothing technique archieved the most accurate results and was least sensitive to the choise of its meta-parameters.

As a continuation of this work, we are exploring how models of different abstractions can be integrated into a common framework to allow smooth transitions between them based on available evidence from measurements. In the context of the particle filter, we believe that this will allow for an adaptive distribution of computational resources over competing system models and at the same time avoid the need for introducing fixed thresholds.

Another area for future work is motivated by the observation that the parameter estimation quality is correlated to the performed actions and the structure of the environment. For example, a deflating tire can easily be diagnosed when the robot is moving fast and straight. In a slow rotational motion, however, the fault may not change the observable behavior greatly. A promising area for further reasearch would be the quantification of this influence.

References

1. Williams, B.C., Nayak, P.P.: A model-based approach to reactive self-conguring systems. In Minker, J., ed.: Workshop on Logic-Based Articial Intelligence, Washington, DC, June 14–16, 1999, College Park, Maryland, Computer Science Department, University of Maryland (1999)
2. Washington, R.: on-board real-time state and fault identication for rovers (2000)
3. Koller, D., Lerner, U.: Sampling in Factored Dynamic Systems. In: Sequential Monte Carlo Methods in Practice. Springer (2001) 445–464
4. Ng, B., Peshkin, L., Pfeer, A.: Factored particles for scalable monitoring. In: Proceedings of the Eighteenth Conf. on Uncertainty in Articial Intelligence. (2002)
5. Verma, V., Thrun, S., Simmons, R.G.: Variable resolution particle lter. In: IJCAI. (2003) 976–984

6. Dearden, R., Clancy, D.: Particle lters for real-time fault detection in planetary rovers. In: Proceedings of the Thirteenth International Workshop on Principles of Diagnosis. (2002) 1–6
7. Ng, B., Pfeer, A., Dearden, R.: Continuous time particle ltering. In: Proceedings of the 19th International Joint Conference on Articial Intelligence (IJCAI), Edinburgh. (2005)
8. Thrun, S.: Particle lters in robotics (2002)
9. de Freitas, N., Dearden, R., Hutter, F., MoralesMenendez, R., Mutch, J., Poole, D.: Diagnosis by a waiter and a mars explorer (2003)
10. Driessen, J., Boers, Y.: An ecient particle lter for nonlinear jump markov systems. In: IEE Seminar Target Tracking: Algorithms and Applications, Sussex, UK, March 23-24. (2004)
11. Thrun, S., Burgard, W., Fox, D. In: Probabilistic Robotics. MIT Press (2005)
12. Gordon, N.J.: Bayesian methods for tracking. PhD thesis, Imperial College, University of London (1993)
13. Andrieu, C., de Freitas, J., Doucet, A.: Sequential mcmc for bayesian model selection (1999)
14. Liu, J., West, M.: Combined parameter and state estimation in simulation-based ltering. In A. Doucet, J.F.G.D.F., Gordon, N.J., eds.: Sequential Monte Carlo Methods in Practice. New York, Springer-Verlag, New York (2000)
15. Koenig, N., Howard, A.: Design and use paradigms for gazebo, an open-source multi-robot simulator. technical report. Technical report, USC Center for Robotics and Embedded Systems, CRES-04-002 (2004)
16. Smith, R.: Open dynamics engine. http://www.q12.org/ode/ode.html (2002)

Ermittlung von Linienkorrespondenzen mittels Graph-Matching

Claudia Gönner, Sebastian Schork, Karl-Friedrich Kraiss

Lehrstuhl für Technische Informatik
Rheinisch-Westfälische Technische Hochschule Aachen (RWTH)
Ahornstr. 55, 52074 Aachen

{Goenner,Schork,Kraiss}@techinfo.rwth-aachen.de
www.techinfo.rwth-aachen.de

Zusammenfassung. Korrespondierende Punkte, Linien oder Flächen lassen sich, isoliert betrachtet, oft nicht eindeutig zuordnen. Daher bietet es sich an, räumliche Zusammenhänge zu berücksichtigen. Diese Arbeit erweitert die probabilistische Relaxation um die Bestimmung der optimalen Zuordnung aller direkten Nachbarn mittels Hidden Markov Modellen. Das neue Relaxations-Schemata wird auf vier Bildpaaren ausgewertet und mit der probablistischen Relaxation verglichen.

1 Einleitung

Eines der größten Probleme bei der Registrierung von Innenraumaufnahmen, in denen unifarbene, häufig planare Flächen dominieren, ist die Bestimmung von Korrespondenzen. Isolierte Punkte, Linien oder Flächen lassen sich aufgrund von Farb- oder Texturinformation ihrer Nachbarschaft allein nicht eindeutig zuordnen. Diese Arbeit untersucht daher, wie sich die Zuordnung von Korrespondenzen mit Hilfe räumlicher Zusammenhänge verbessern lässt.

Betrachtet werden korrespondierende Linien, da sich diese in Innenraumaufnahmen gut detektieren lassen. Diese Linien bilden die Knoten $i \in V$ eines Graphen $G = \{V, E\}$. Farb- oder Texturmerkmale entlang einer Linie sind als Attribute a_i an die Knoten geheftet. Zu ausgewählten, nahe gelegenen Linien bestehen Relationen, modelliert als Graphkanten $(i,j) \in E$ und um Attribute A_{ij} ergänzt.

Gesucht ist eine maximale Menge von Korrespondenzen zwischen zwei Aufnahmen, beziehungsweise deren Graphen G und $\breve{G}$, so dass für ein korrespondierendes Linienpaar auch alle benachbarten Linien korrespondieren. Dabei dürfen in beiden Bildern Linien sichtbar bzw. detektiert sein, die in dem anderem Bild keinen Partner haben. Folglich ist die maximale, gemeinsame Clique beider Graphen gesucht. Unter Wahl einer fehlertoleranten Editierdistanz, also einem Distanzmaß inkl. Einfügen, Löschen und Substitution, lässt sich dies in die Bestimmung eines fehlertoleranten Subgraph Isomorphismus überführen [1].

Relaxation Labelling [3,5,6], eingeführt von Rosenfeld et al. [6], ist der bekannteste, approximative Graph-Matching Ansatz. Jedem Knoten i des ersten Graphen G wird jeweils ein Etikett zugeordnet, das dem Knoten α des zweitem Graphen $\breve{G}$ entspricht.

Mit Hilfe der Zuordnungen der Nachbarschaften $N_i = \{j;(i,j) \in E\}$ werden die Etiketten iterativ neu verteilt und bewertet.

Diese Arbeit erweitert die probabilistische Relaxation nach Christmas, Kittler et al. [3] um die Bestimmung optimaler Isomorphismen lokaler Supercliquen $C_i = \{i\} \cup N_i$ des ersten und $\breve{C}_\alpha = \{\alpha\} \cup \breve{N}_\alpha$ des zweiten Graphen (Abb. 1), wobei $\breve{N}_\alpha = \{\beta;(\alpha,\beta) \in \breve{E}\}$. Die Relaxations-Schemata [3,6] betrachten jeden Nachbarknoten isoliert, ohne eine optimale Zuordnung aller Nachbarn zu berechnen. Statt dessen bewerten sie eine paarweise Kombination aller Nachbarn.

Ähnlich zu Myers, Wilson et al. [5] wird hier der optimale Isomorphismus aller Nachbarknoten mittels Dynamischer Programmierung gefunden. Allerdings beschreiben [5] eine diskrete Relaxation. Hier werden, wie bei einer kontinuierlichen Relaxation üblich, die Bewertungen rekursiv definiert. Die Relaxation und die lokale Zuordnung der Nachbarschaft verwenden dieselbe Bewertung. Eine Bewertungsfunktion, die die Ähnlichkeit zweier Supercliquen beschreibt und sowohl Knoten- als auch Kantenattribute berücksichtigt, wird mit Hilfe von Hidden Markov Modellen hergeleitet.

Abb. 1. Die Zuordnung sollte optimal für die gesamten Supercliquen C_i und C_α sein. Die Reihenfolge der Nachbarknoten muss erhalten bleiben. Einige Knoten haben keine Korrespondenz (gekennzeichnet mit ×).

2 Graph Aufbau

Zunächst werden in den Aufnahmen Kanten detektiert, z.B. nach Canny [2], und in gerade Segmenten gruppiert [4]. Segmente ähnlicher Orientierung, die sich überlappen oder aneinander anschließen, werden zu Vektorzügen zusammengefasst. Dies ist notwendig, weil aufgrund von Segmentierungsfehlern und radialer Verzerrungen eine extrahierte Raumkante, unter Umständen in mehrere Segmente unterteilt ist. Der Anschaulichkeit halber wurde in der Einleitung von Linien gesprochen, obwohl Vektorzüge gemeint sind.

Die Vektorzüge entsprechen den Knoten des aufzubauenden Graphen. In Abb. 2 sind diese Knoten an den Mittelpunkten der Vektorzüge eingezeichnet. Zur Aufstellung der Nachbarschaftsbeziehungen, dargestellt als gestrichelte Linien, bieten sich verschiedene Strategien an:

– *Intersektionsgraph:* Wenn sich zwei Vektorzüge schneiden, wird eine Graphkante eingefügt. Aufeinander stoßende Vektorzüge treten auch in Relation.
– *Delaunaygraph:* Auf den Mittelpunkten der Vektorzüge wird eine Delaunay-Triangulation durchgeführt. Die Delaunay-Kanten entsprechen den Kanten des Graphen.
– *Delaunaygraph mit Randbedingungen:* Alle Kanten des Intersektionsgraph müssen in dieser Triangulierung vorkommen. Die Delaunay-Kanten dürfen keine Intersektionskanten schneiden.

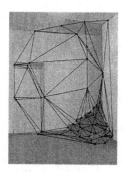 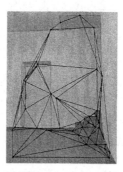

Abb. 2. In Abb. 1 detektierte Linienzüge (rot beziehungsweise grau) und nach dem Delaunaykriterium inklusive Randbedingungen aufgebaute Graphen (schwarz).

Unäre Attribute
Entlang den Vektorzügen werden RGB-Mediane bestimmt, jeweils getrennt für die rechte und linke Seite. Die Mediane werden nach allen drei Kanälen gemeinsam sortiert. Bei identischem Rotanteil erfolgt die Einordung entsprechend dem Blauton als zweiten und dem Grünkanal als dritten Sortierschlüssel.

Binäre Attribute
Monokulare Bildsequenzen weisen mitunter eine deutliche perspektivische Verzerrung zwischen den einzelnen Aufnahmen auf. Hinzu kommen Verdeckungen und eine fehlerhafte Extraktion der Linienzüge. Daher scheiden die gängigen binären Merkmale wie Winkel oder Abstände aus.

Als Alternative bieten sich wiederum RGB-Mediane an, dieses Mal entlang der Vebindungslinie zwischen den Mittelpunkten zweier Vektorzüge, aber ohne Unterscheidung von linker und rechter Seite. Um Verdeckungen Rechnung zu tragen, wird eine multimodale Verteilung der Farbwerte angenommen. In den paarweisen Vergleich dieser Medianvektoren gehen nur die beiden RGB-Mediane mit der kleinsten Distanz ein.

3 Probabilistische Relaxation und lokale Isomorphismen

Die probabilistische Relaxation [3] definiert eine $m{:}1$ Abbildung von V auf $\breve{v}$ und bestimmt in jeder Iteration n eine Wahrscheinlichkeit $P_\alpha^{(n)}(i)$, mit der die Knoten i und α einander zugeordnet sind.

$$P_\alpha^{(0)}(i) = \frac{p(a_i, \breve{a}_\alpha | i, \alpha)\hat{P}_\alpha(i)}{\sum_{\beta \in \breve{v} \cup \epsilon} p(a_i, \breve{a}_\beta | i, \beta)\hat{P}_\beta(i)} \tag{1}$$

$$P_\alpha^{(n+1)}(i) = \frac{P_\alpha^{(n)}(i)Q_\alpha^{(n)}(i)}{\sum_{\beta \in \breve{v} \cup \epsilon} P_\beta^{(n)}(i)Q_\beta^{(n)}(i)} \tag{2}$$

$$\hat{P}_\alpha(i) = \begin{cases} \zeta & \alpha = \epsilon \\ \frac{1-\zeta}{|\breve{v}|} & \text{sonst} \end{cases} \tag{3}$$

$$Q_\alpha^{(n)}(i) = \begin{cases} \prod_{j \in N_i} \zeta \max_{\beta \in \breve{N}_\alpha} \{P_\beta^{(n)}(j)\} & \alpha = \epsilon \\ \prod_{j \in N_i} \sum_{\beta \in \breve{N}_\alpha} P_\beta^{(n)}(j)p(A_{ij}, \breve{A}_{\alpha\beta} | i, j, \alpha, \beta) & \text{sonst} \end{cases} \tag{4}$$

Die RGB-Mediane, gespeichert in den unären und binären Attributen, werden als multivariate Gaußverteilungen $N_x(\mu, \sigma)$ angenommen, $p(a_i, \breve{a}_\alpha | i, \alpha)$ und $p(A_{ij}, \breve{A}_{\alpha\beta} | i, j, \alpha, \beta)$ ergeben sich aus der Differenzverteilung. Die Konstante ζ gibt die Wahrscheinlichkeit an, mit der ein Knoten i keine Korrespondenz hat und dem Null-Label ϵ zugeordnet ist. Alle anderen Zuordnungen sind ohne Kenntnis der Attribute gleichverteilt. Die Hilfsfunktion $Q_\alpha^{(n)}(i)$ berechnet die Wahrscheinlichkeit, die Nachbarschaft N_i zuzuordnen, wenn die Korrespondenz (α, i) vorgegeben ist.

In Graphen, generiert aus Innenraumansichten, existieren viele Knoten mit ähnlichen Attributen, die als sinvolle Kandidaten zur Verfügung stehen. Der Term $Q_\alpha^{(n)}(i)$ ordnet jedem Nachbar j nacheinander alle Nachbarn β zu. Die in Abb. 3(a) skizzierten Supercliquen erzielen dieselbe Bewertung $Q_\alpha^{(n)}(i)$ wie die jeweils korrekten Isomorphismen.

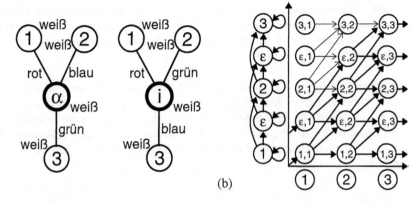

(a)   (b)

Abb. 3. (a) Diese Zuordnung erzielt dieselbe Bewertung $Q_\alpha^{(n)}(i)$ wie der korrekte Isomorphismus. (b) Die Markov Kette für N_α und der aufgespannte Zustandsraum.

Mit Hilfe von Hidden Markov Modellen lässt sich eine Bewertungsfunktion $Q_\alpha^{'(n)}(i)$ definieren, die die Reihenfolge berücksichtigt, leere Zuordnungen für übersprungene Knoten β einfügt und, falls vorteilhaft, einen Knoten β mehrfach zuordnet. Hierzu wird die Nachbarschaft N_i als Folge $[N_i]$ interpretiert. Für die Nachbarschaft N_α wird eine Markov Kette (Abb. 3(b)) generiert, die 0-, 1- und 2-Übergänge erlaubt und in der nach jedem Element Null-Label eingefügt sind. Da der Einsprungknoten unbekannt ist, werden alle zyklischen Permutationen betrachtet. Alle Transitionen sind als gleich wahrscheinlich modelliert. Die Emissionswahrscheinlichkeit ergibt sich aus der aktuellen Bewertung der Zuordnung (j,β) und der Ähnlichkeit der binären Attribute.

$$Q_\alpha^{'(n)}(i) = \sum_{[s_1^N]\in\mathcal{S}} \prod_{t=1}^{N} p(t, s_t | s_{t-1}, N_\alpha^0) \quad \text{mit } N = \|[N_i]\| \tag{5}$$

$$= \sum_{[s_1^N]\in\mathcal{S}} \prod_{j\in[N_i],\,\beta\in[s_1^N]} P_\beta^{(n)}(j) p(A_{ij}, \mathring{A}_{\alpha\beta} | i, j, \alpha, \beta) \tag{6}$$

$$\approx \max_{[s_1^N]\in\mathcal{S}} \{ \prod_{j\in[N_i],\,\beta\in[s_1^N]} P_\beta^{(n)}(j) p(A_{ij}, \mathring{A}_{\alpha\beta} | i, j, \alpha, \beta) \} \tag{7}$$

Bei Betrachtung von $-\ln Q_\alpha^{'(n)}(i)$ lässt sich die beste Zuordnung mittels Dynamischer Programmierung bestimmen. Damit die Bewertung einer Superclique unabhängig von ihrer Größe ist, wird über die Anzahl der Nachbarknoten normalisiert.

4 Diskussion der Ergebnisse

Die Auswertung erfolgt auf vier Bildpaaren exemplarischer Innenraumszenen (Abb. 2 und 4). Zur Analyse des Einflusses von Segmentierungsfehlern sind die Linien sowohl automatisch detektiert, als auch von Hand eingezeichnet. Die automatisch gefundenen Linien sind teilweise in mehrere Segmente unterteilt oder fehlen. In Abb. 4(c) wurden doppelt soviele Linien detektiert, wie manuell vorgegeben.

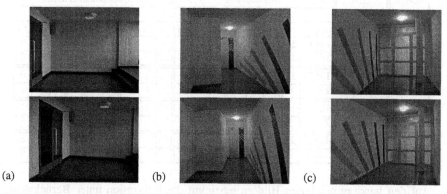

(a)   (b)   (c)

Abb. 4. Die zusätzlich zu Abb. 2 getesteten Szenen. Für die jeweils obere Aufnahme wurden korrespondierende Linienzüge in der unteren Aufnahme gesucht.

Die einem Bildpaar zugeordneten Graphen haben jeweils eine ähnliche Anzahl von Knoten und durchschnittliche Konnektivität. Daher beziehen sich die Angaben in Tabelle 1 nur auf den ersten Graphen. Je nach Graphaufbau unterscheiden sich die Eigenschaften deutlich. Für alle Bilder, Segmentierungsarten, Graphtypen und Relaxationsvarianten wurden fünf Iterationen durchgeführt. Die Null-Label Wahrscheinlichkeit ist auf $\xi = 0{,}1$ eingestellt.

Tabelle 1. Die korrekten Zuordnungen der Knoten V des Datengraphen auf die Knoten $\bar{V}$ des Modellgraphen in Prozent. Es bezeichnen: A automatisch detektierte Linien-, H handgezeichnete Linienzüge, $S \subseteq V$ die Teilmenge der Knoten, für die eine Korrespondenz existiert, $|N_i|$ die mittlere Konnektivität eines Knotens $i \in V$, I den Intersektions-, D den Delaunay- und R den Delaunaygraphen mit Randbedingungen.

Szene		$|N_i|$ I	$|N_i|$ D	$|N_i|$ R	$|V|$	$|S|$	Init	Relaxation I	Relaxation D	Relaxation R	Relax/HMM I	Relax/HMM D	Relax/HMM R
2	A	2.6	5.5	5.5	49	34	18%	44%	42%	38%	18%	30%	30%
	H	3.4	5.5	5.7	42	42	31%	50%	55%	70%	60%	55%	69%
4(a)	A	3.7	5.7	6.0	108	73	18%	23%	26%	31%	21%	24%	25%
	H	3.1	5.5	5.7	42	40	29%	55%	38%	40%	50%	60%	60%
4(b)	A	1.6	5.7	5.7	58	39	36%	28%	38%	40%	31%	40%	36%
	H	3.3	5.5	5.8	42	42	50%	64%	62%	74%	48%	71%	74%
4(c)	A	3.2	5.9	6.3	186	138	31%	30%	33%	33%	30%	38%	36%
	H	3.3	5.8	5.8	91	83	35%	41%	68%	60%	50%	73%	71%

Tabelle 2. Zuordnung der Knoten V des Datengraphen auf die Knoten $\bar{V}$ des Modellgraphen aufgeschlüsselt nach existierenden Korrespondenzen und leeren Zuordnungen. Es bezeichnen: A automatisch detektierte Linien-, H handgezeichnete Linienzüge, I den Intersektions-, D den Delaunay-, R den Delaunaygraphen mit Randbedingungen, c_S die Anzahl der korrekten Korrespondenzen, c_N die Anzahl der korrekt zugeordneten Null-Label und f_N die Anzahl der irrtümlich zugeordneten Null-Label.

Szene		Init c_S	Init c_N	Init f_N	Relax I c_S	Relax I c_N	Relax I f_N	Relax D c_S	Relax D c_N	Relax D f_N	Relax R c_S	Relax R c_N	Relax R f_N	HMM I c_S	HMM I c_N	HMM I f_N	HMM D c_S	HMM D c_N	HMM D f_N	HMM R c_S	HMM R c_N	HMM R f_N
2	A	9	0	0	11	11	14	16	5	4	17	2	7	8	1	1	14	1	0	14	1	0
	H	13	0	0	25	0	11	23	0	8	30	0	4	21	0	0	23	0	0	29	0	0
4(a)	A	18	1	0	14	11	14	15	13	15	21	13	3	20	3	1	24	2	0	25	2	0
	H	12	0	0	21	2	8	15	1	16	17	0	11	20	1	2	24	1	4	24	1	1
4(b)	A	21	0	0	10	6	16	20	2	5	21	2	4	15	3	3	23	0	1	21	0	1
	H	21	0	1	27	0	5	26	0	4	31	0	3	20	0	2	30	0	1	31	0	0
4(c)	A	53	4	3	41	14	26	53	8	10	51	11	15	54	2	4	66	4	3	64	3	3
	H	32	0	0	34	3	9	60	2	6	52	3	7	45	0	1	66	0	0	64	1	0

Auf den handsegmentierten Bildern bestimmt die Relaxation unter Berücksichtigung lokaler Isomorphismen mehr Korrespondenzen als die probabilistische Relaxation. Auch auf der automatisch segmentierten Aufname 4(c) erzielt die Hidden Markov Variante bessere Ergebnisse. Beide Relaxations-Schemata verbessern in der

Regel die initialen Bewertungen deutlich. Einzige Ausnahme ist Aufnahme 4(b), in der viele Linien überhaupt nicht oder nur stückweise detektiert wurden.

Bei fehlerbehafteten Graphen haben beide Verfahren eine niedrige Trefferquote (Tab. 1). Die probabilistische Relaxation schneidet in dieser Situation oft besser ab. Dies ist auf mehr korrekt zugeordnete Null-Label zurückzuführen, allerdings auf Kosten vieler irrtümlicher Null-Zuordnungen (Tab. 2).

Weiterführende Arbeiten könnten ein Graph-Matching in beiden Richtungen kombinieren, um eine $n : m$ Abbildung der Vektorzüge und symmetrische Korrespondenzen zu bestimmen. Ferner ist es denkbar, benachbarte Vektorzüge zu fusionieren, die auf denselben Vektorzug der anderen Aufname abbilden. Dazu ist die neue, erweiterte Relaxation vorteilhaft, weil sie mehr vorhandene Korrespondenzen korrekt zuordnet als die probabilistische Relaxation.

Literaturverzeichnis

1. H. Bunke. Error correcting graph matching: On the influence of the underlying cost function. *IEEE Transactions on Pattern Analysis and Machine Intelligence*, 21(9):917–921, September 1999.
2. J. Canny. A computational approach to edge detection. *IEEE Transactions on Pattern Analysis and Machine Intelligence*, 8(6):679–698, 1986.
3. William J. Christmas, Josef Kittler, and Maria Petrou. Structural matching in computer vision using probablilistic relaxation. *IEEE Transactions on Pattern Analysis and Machine Intelligence*, 17(8):749–764, August 1995.
4. Euijin Kim, Miki Haseyama, and Hideo Kitajima. Fast line extraction from digital images using line segments. *Systems and Computers in Japan*, 34(10):76–88, 2003.
5. Richard Myers, Richard C. Wilson, and Edwin R. Hancock. Bayesian graph edit distance. *IEEE Transactions on Pattern Analysis and Machine Intelligence*, 22(6):628–635, June 2000.
6. Azriel Rosenfeld, Robert A. Hummel, and Steven W. Zucker. Scene labeling by relaxation operations. *IEEE Transactions on Systems, Man and Cybernetics*, SMC-6(6):420–433, June 1976.

Information Integration in a Multi-Stage Object Classifier

Gerd Mayer, Hans Utz, Günther Palm

Department of Neural Information Processing, University of Ulm, Germany

Abstract. Visual sensing systems are one of the most important information sources for autonomous mobile robots. By using temporal information, the object detection can be stabilized over time. In this paper we show, that a tight coupling between the visual system and such a tracking instance allows to integrate a lot more different information sources than it would be possible with two separate modules. This linkage between classification and temporal integration allows to stabilize and accelerate the detection task at the same time as shown in detail in the results section.

1 Introduction

Visual sensing systems are one of the most important information sources for autonomous mobile robots. The robots have to recognize the objects they want to manipulate or other robots or humans they have to interact with. On the other hand, image processing is one of the computational most expensive task within the anyway very computational demanding software system.

In a reliable visual sensing system of a robot, each single detection pass on one image is usually designed to be as specific and robust as possible. But the reliability of the system can also be increased by confirming and refining the classification on images over time. As it is important to be economical with the computational resources available on an autonomous mobile robot, this additional information can be used to lower the amount of processing used on each individual images. The robot has to have enough resources left to reasonably react on the observed situation or to properly take actions based on the concluded conditions.

During the successive steps from grabbing an image up to the final object recognition results a lot of different intermediate results accrue. The challenge is to properly accumulate all this information and to provide a flexible way of knowledge fusion within the visual system itself. This way, the system can profit from the increasing reliability of the resulting information and may even omit further processing steps if the extracted information is already reliable and/or accurate enough for the given task.

In this paper we present an approach that is based on a tight coupling of the visual object detection and the data integration. Each intermediate result is combined as early as possible to allow to leave out further processing steps and to allow to incorporate knowledge on objects not seen in the actual image.

As can be seen in the given example, the system is able to increase the reliability of the found objects up to a chosen level of confidence yet able to reduce the effort needed to retrieve this information. This applies for all objects, no matter if they can be seen actually in the current image or not. Potentially even more important, it is able to take advantage of this for repetitively occurring false detection.

The remainder of this paper is organized as follows. The next section 2 describes the used multi-stage object classification system step by step. In Section 3 the fusion of the different information sources is explained. A detailed evaluation is given in Section 4. Finally section 5 discussed this work in the context of other existing solutions and Section 6 draws conclusion.

2 Multi-Stage Object Classifier

In this section the tightly integrated object classifier system is briefly sketched, before we discuss the system's ability to flexibly fuse the information in section 3. The object classification system is best understood by starting with the single-image multi-stage classifier alone. The temporal integration is then introduced in the second paragraph.

Neural Classifier: The multi-stage object recognition process itself can be divided into the following individual steps:

1. Detect regions of interest
2. Extract features from these regions
3. Classify them using artificial neural networks

In the first step (1), potential object positions are searched within the recorded images to direct the robots attention to possibly interesting places. Subsequent image processing is restricted to these regions of interest (ROI). In order not to miss important information in the image, the ROI detection must not be overly restrictive, so false positives will be upon the ROIs. In the next step (2) additional features are calculated for each of the detected ROIs. They describe generalized attributes of the object specific enough to allow to reliably reject false positives. In step (3), all these features are passed to a neural object classifier. Here, artificial neural networks vote for each feature vector, calculating the certainty of whether it belongs to the expected object type or not, as discussed in more detail in [1] or [2].

Temporal Integration: The above sketched algorithm is extended with a Kalman filter based object tracking and prediction module. The filter module is placed between the attention control and the subsequent classification step. The first objective of providing this additional information is to improve the results of the multistage classifier. The second one is to save computational resources in the evaluation of the single images by providing a more flexible evaluation schema. The idea is, that we can skip parts of the multistage object classification, if their outcome will probably not provide significantly more reliability. This is the case for ROIs for which previous classifications already provide a very certain recognition. The whole processing loop then looks as follows:

1. Detect regions of interest
2. Predict the position of previously classified objects (tracks)
3. Match new ROIs to existing tracks if possible

4. For all tracks with certainty values within thresholds:
5.   Extract and classify features
6.   Integrate classification with previous results if available
7. For all others:
8.   Decrease certainty
9. Start/end tracks for new/unobservable objects
10. Correct the predicted positions using observations

(1) is processed as before. Then (2) all trajectories of already observed objects (called tracks) are updated to their new, predicted positions at this point in time. Next (3) these forecasted positions are matched with the real observations from the attention control. At this point, the procedure is divided into two cords. If the certainty value for a specific object of an already existing track is above or below a specific threshold, we are already sure enough of the ROI's classification (no matter if it is for sure the expected object or if it is for sure a falsely detected region of interest). So only the other regions need to be verified again by the neural classifier (5–6). In (8) the reliability value of all not reclassified tracks are decrease to account for the omitted confirmation. If no matching track is found for an observation, a new track is started, alternatively a track that could not be associated with an observation for a long time is deleted (9). Finally, the predicted positions are adapted and corrected using the observed positions to provide a precise prediction in the next iteration (10). For details and a more exhaustive discussion, please refer to [3].

3 Evidence Fusion

The tight coupling of the ROI tracking with the object classification process allows for a fine grained updating of the tracked classification results. If the neural classifier detects (or negates) an object, this classification result is used directly as new probability value c_i for certainty integration. If the classification step is skipped because of a high (resp. low) enough certainty value, the a-priori detected average correctness of the attention control for this specific object type is used. Both times, the certainty integration is done using the following formula:

$$c_{t+1} = c_t + \tau(-c_t + c_i),$$

where c_t is the old certainty value, c_i the newly retrieved classification result (or a-priori average value), τ a weight factor, denoting how much the new value is taken into account and influences the overall result c_{t+1}.

For tracks which do not result from an actual observation, the certainty c_i is calculated differently: If the object's position is predicted to be outside of the field of view, the a-priori probability c_i of the object being there weighted with the predicted track accuracy error is integrated. If the predicted object is in the field of view, additional verification of the ROI determines, if there is another object which potentially occludes the predicted one. In this case, the same certainty value as above is used. If this additional module however detects a low chance for an occlusion (e.g. there is a high probability of seeing the floor where no object can hide behind), the (most time smaller, see above) probability of a miss during the search for regions of interest again weighted with the forecasted prediction accuracy is used. The weighted certainty value c_i is then calculated using the following formula:

$$c_i = \begin{cases} c_i \dfrac{1}{e_m}(e_m - e_f) & \text{if } e_m > e_f \\ 0 & \text{otherwise} \end{cases}$$

with e_f being the actual forecasted prediction error and e_m used as upper boundary for the maximal tolerable prediction error. All a-priori probabilities as well as the estimated classification and attention control correctness can e.g. be retrieved from a large set of representative images. In our case, these images were collected during several ROBOCUPgames, where this system is used in.

4 Evaluation

To illustrate and measure the improvements of the above described methodology, we performed an illustrative, yet realistic experiment. We let the robot drive past a corner flag; the flag enter the field of view after a couple of frames and then the systems starts recognizing and tracking it. Figure 1 shows some of the recorded images to give an impression on how the experiment look like.

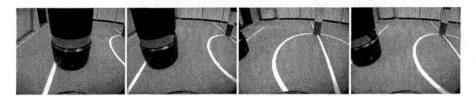

Fig. 1. Example images from the robot's perspective used for evaluation.

Figure 2 shows the different levels of certainty depending on the used integration method. The solid line denotes the pure classification result as retrieved directly from the neural network. The horizontal line with a constant value of 0.85 is the chosen upper threshold above which the system assumes a positive recognition result. Next, the graph labeled "integration" shows the integrated certainty values as described in section 3. It can be seen nicely, that the system needs a number of repetitive validations to affirm the object (i.e. the certainty value is above the dotted threshold line the first time). If classification fails for a number of consecutive frames, the integration plot still stays above the certainty threshold for several images before it falls below the line too. Please note that this is the only possible fusion result if the object recognition and the temporal integration instances are two unconnected and consecutive modules.

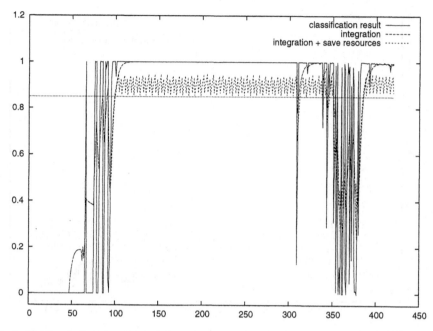

Fig. 2. Certainty values as retrieved from the neural classifier alone, the temporal integration and the integration in combination with partial evaluation skipping.

The last line, labeled "integration + save resources", finally demonstrates the full potential of the presented approach. As soon as the plot is sufficiently above the chosen threshold the first time, the characteristic zigzag-line proves, that the classification stage is only done, if the line is on the verge of falling below the certainty limit. So in this example, in the middle of the test run the corner flag is validated by the neural classifier in every second image only. This can only be done, if the tracking instance is directly interwoven with the visual object detection and is therefore able to influence the degree of image interpretation necessary for a desired level of certainty.

In the above example the individual classification results are 317 times above the recognition threshold. Integrated over time as plotted in the second graph, the fused evidence is 286 times above the threshold. Although this look like a performance loss, the fused result is considerably more trustworthy than the individual results. When using the integration method proposed in the paper the final result is still 283 times above the certainty threshold. At the same time, a considerable amount of computational resources is saved.

Depending on the desired application field, it may not be wanted to reduce the upper certainty limit even to a such high value as the used 0.85. On the other hand, rejected object hypotheses (i.e. found ROIs which are proven to be false assumptions in later classification steps) seldom need to be confirmed again and again. Most time, there is no essential cognition benefit between unsureness levels of e.g. 0.0 or 0.3. When using the proposed solution, the system benefits from this also as can be seen in Fig. 3. This plot is made with the same image sequence as used, above but with a modified neural network so that the classifier rejects the found ROIs repetitively.

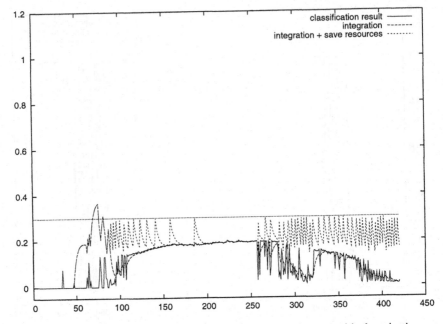

Fig. 3. Classification results and integrated certainty values when repetitively rejecting unwanted objects. The horizontal line with a constant value of 0.3 is the chosen lower threshold below which the system assumes a negative recognition result.

5 Related Work

A lot of research is done in the field of object recognition and tracking to either keep track of object positions in the environment or the robot's position itself.

A lot of this work is also done within the ROBOCUP environment. For example Schmitt et. al. [4][5] presented an approach for the simultaneous tracking of multiple robots. After visually recognizing the robots, a multiple hypothesis tracker then estimates and predict their movement. Another quite similar application is described by Kwok and Fox in [6]. They're using a Rao-Blackwellised particle filter system to track a moving ball in a rather comparable scenario. They even model different motion modules for the ball depending on the different states the ball is in (e.g. bounced, grabbed or kicked). This way they can cope with unsteadiness in the ball's movement. Another system is described by Weiss and Hildebrand in [7]. In contrast to the above described workings which operate on the robot's local field of view, this work is done using a global camera. Again, the (multi-stage) vision system is followed by a object tracking instance. However, this tracking module is not described in detail here.

The common denominator of all these approaches is, that the object recognition and the tracking part are two separated modules. This way, the only available information which can be integrated over time is the final detection result of the object recognition module. Beside that, the whole processing task has to be performed completely every single image frame. This in turn hinders vision systems on mobile robots in highly dynamic environments to evolve as necessary to be applicable in "real-world" scenarios.

6 Conclusion

This paper discussed the advanced possibilities of evidence fusion, if tracking of objects over time is recurrently coupled with object classification. Using temporal integration in an object classification system provides two structural advantages that seem to be contradictory at first sight. On the one hand, the classification results are stabilized by the temporal integration, on the other hand, the effort needed for the object retrieval can be reduced.

Furthermore this approach provides a natural basis for extending the modeling capabilities for a detailed, flexible verification of previous classification results. The presented methodical improvements in object classification and tracking are of great advantage especially in highly dynamic environments like e.g. the ROBOCUP, as they allow for a highly reliable, precise and fast visual object classification system.

7 Acknowledgment

The work described in this paper was partially funded by the DFG SPP-1125 in the project *Adaptivity and Learning in Teams of Cooperating Mobile Robots*

References

1. Ulrich Kaufmann, Gerd Mayer, Gerhard K. Kraetzschmar, and Günther Palm. Visual robot detection in robocup using neural networks. In RoboCup 2004: *Robot Soccer World Cup VIII*, Lecture Notes in Arti_cial Intelligence, pages 262{273, Berlin, Heidelberg, Germany, 2004. Springer.

2. Gerd Mayer, Ulrich Kaufmann, Gerhard Kraetzschmar, and Günther Palm. Neural robot detection in robocup. In Günther Palm and StefanWermter, editors, *27th German Conference on Arti_cial Intelligence (KI2004), NeuroBotics Workshop*, pages 22–31, University of Ulm, Germany, September 2004.

3. Gerd Mayer, Jonas Melchert, Hans Utz, Gerhard Kraetzschmar, and Günther Palm. Neural object classi_cation and tracking. In *Proceedings of the IEEE SMC UKRI Chapter Conference 2005 on Applied Cybernetics*, London, United Kingdom, September 2005. IEEE Systems, Man and Cybernetics Society, United Kingdom and Republic of Ireland Chapter.

4. T. Schmitt, M. Beetz, R. Hanek, and S. Buck. Watch their moves: Applying probabilistic multiple object tracking to autonomous robot soccer. In *The Eighteenth National Conference on Artificial Intelligence*, Edmonton, Alberta, Canada, 2002.

5. Thorsten Schmitt, Robert Hanek, Michael Beetz, Sebastian Buck, and Bernd Radig. Cooperative probabilistic state estimation for vision-based autonomous mobile robots. *IEEE Transactions on Robotics and Automation*, 18(5), October 2002.

6. C. Kwok and D. Fox. Map-based multiple model tracking of a moving object. In *Proc. of the RoboCup 2004 International Symposium*, 2004.

7. Norman Weiss and Lars Hildebrand. An examplary robot soccer vision system. In *CLAWAR/EURON Workshop on Robots in Entertainment*, Leisure and Hobby, Vienna, Austria, December 2004.

Finding Rooms on Probabilistic Quadtrees

Guillem Pagès Gassull, Gerhard K. Kraetzschmar, Günther Palm

University of Ulm Department of Neuroinformatics

E-mail: {guillem.pages,gerhard.kraetzschmar,guenther.palm}@uni-ulm.de

Abstract. Probabilistic quadtrees are a recently developed concept for compact storage of probabilistic occupancy data. In this paper, we introduce a new method for finding high-level regions ("rooms") on a probabilistic quadtree, exploiting features like its underlying neighborhood graph and binary flags.

1 Introduction

Innovative service robots need a multi-level spatial representation in order to fulfill a wide spectrum of tasks involving spatial knowledge about the environment. Two levels of particular interest are the occupancy map level[4,1] and the region map level [5]. After an occupancy map has been created, one can try to discover regions, such as obstacles, rooms, and hallways, and add them to the region map.

The available resolution of nowadays most widely used sensors allows to increase the resolution of occupancy maps, which increases memory requirements for map storage. In addition, several realistic robotic applications require the mapping of larger areas (e.g. hospitals, universities,...) than the traditional single floor office or laboratory, which also increases memory use. Both developments together can easily result in prohibitive memory use.

To solve this problem, we introduced *probabilistic quadtrees* [2], a compact representation for probabilistic occupancy grids that also provides other improvements to the original quadtrees [7].

If a service robot wants to fully integrate with a human environment, it has to be able to understand concepts like "John's office" or "Peter's room", so that the user does not have to point at a map on a screen. Thus, the robot needs to be able to automatically detect region information, such as rooms, corridors, doors, elevator shafts or staircases, while exploring the environment.

Algorithms to extract high level region information from occupancy gridmaps are not yet fully reliable or automated, although some of them already exist [3], [6], [5]. But the regions on most office environments share a common property: they are rectangular and aligned or can be approximated quite accurately by rectangles. Quadtrees benefit from this structure, by grouping free (and occupied respectively) space in square regions. Then if an algorithm could exploit this quadtree property, it could both benefit from the memory saves inherent to PQTs and reduce the number of calculations by not running through every cell on the occupancy gridmap (All cells on a PQT leaf node can be regarded as one big cell with the same properties).

In this paper, we present a method for extracting this region information from an existing probabilistic quadtree, using the underlying neighborhood graph and the flag markers it provides.

2 Probabilistic Quadtrees

Quadtrees are a storage method used mainly in monochrome computer images, that divides the image in four square regions and subdivides them further until a region is either completely white or completely black. Probabilistic quadtrees extend this concept further, allowing k different gray levels (*classes*) to be stored. The case $k = 2$ is equivalent to a classical quadtree, while $k = 256$ would allow lossless compression of a *8bit / cell* gridmap.

Furthermore, each node of a PQT also contains statistical data (minimum, maximum, mean and variance) about its offspring nodes (or itself, if it is a leaf node), pointers to its geographical neighbors and a vector of binary markers. This structure is shown in Fig. 1.

Fig. 1. Structure of a PQT node.

3 Finding Regions on PQTs

Probabilistic quadtrees are not meant to be used alone, but integrated in an existing multi-level representation. In our environment, we use a four level architecture. The robot has very simple stop and avoid behaviors that work on an egocentric map. This egocentric map is also used to update the allocentric map, which contains the probabilistic representation of the explored environment, and is the base on which higher levels are built and many algorithms work (SLAM, low-level motion planning,...). On top of the allocentric map, a region map is built, which can be used to direct the navigation commands. And from that region map, a topological graph is inferred, where the high-level motion planning works.

PQTs are meant to substitute the occupancy gridmaps on the allocentric level, and thus, need to be able to provide the upper levels with the kind of information they need, as well as work with the algorithms on its level. Region finding is not a trivial task, even on standard occupancy gridmaps. Therefore, a new algorithm is needed that both works on PQTs and is able to find regions which better approach rooms.

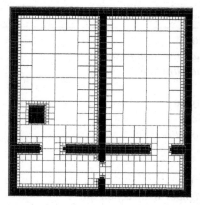

Fig. 2. Original PQT used to demonstrate the algorithm.

One such algorithm should treat isolated obstacles inside a big free region as part of this region (just as furniture is considered part of a room), but recognize walls as region separators and not furniture.

Taking those considerations into account, on the example used to demonstrate our method (Fig. 2), four regions should be found, corresponding to the four rooms that a person would immediately see. The number of four regions could grow up to seven if the doorways were considered separate regions.

The first step towards recognizing rooms, corridors and other geometrical structures is to find their common properties. On a rectangular environment, those common properties are: having a rectangular form and being mostly composed of free space; allowing the possibility of small occupied regions inside (noise, small objects,...).

Then, finding those regions is equivalent to finding the biggest rectangles that fit inside the walls, ignoring closed small occupied regions (obstacles).

4 Region Finding Method

The main idea behind this method is the flood fill algorithm, used in computer graphics to fill closed forms with a single color. But instead of growing the seed in the four directions, it is only propagated in a single direction.

The algorithm (Alg. 1) needs a quadtree node marked free to start. The result region is initialized with this node. From this seed it goes south until it finds an occupied node. While doing this, all visited nodes will be marked and the result region will be grown by each of the visited nodes (the southern boundary will be extended further south).

Once the seed cannot go further south, it turns east and repeats the same sequence, growing the result region according to the visited nodes (extending the border in the new direction). Successively, this will be done following a counter-clockwise scheme, until a rectangle is finally closed. The resulting region will then be a rectangular area, and contain any object whose boundaries are completely inside this region. That allows disregarding tables, chairs and other objects that are in a room (commonly regarded as obstacles), but should not lead to undesired regions representing arbitrary parts of rooms.

Algorithm 1 FindRegion(seed)

---

initialDirection ⇐ South
stack.push(seed, initialDirection.next)
stack.push(seed, initialDirection)
result ⇐ *seed*
while not *stack.empty* **do**
  node ⇐ *stack.top.node*
  direction ⇐ *stack.top.direction*
  stack.pop
  node.markVisited
  result.growBy(node)
  for all *n* = *node.neighbor* **in** *direction* **do**
    if *n.free* **and not** *n.visited* **then**
      stack.push(n, direction.next)
      stack.push(n, direction)
  return *result*

Figure 2 shows the original PQT, Fig. 3 the result of applying the described algorithm to it with a random seed, and Fig. 4 the results with a methodically chosen seed.

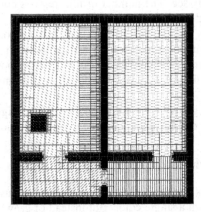

Fig. 3. PQT showing the different fornd regions.

Due to the strictness of the traverse method, there are places where two regions share a border and are distinctly part of the same room yet are classified as two separate regions (this can be clearly seen on the large upper left room, Fig. 3). If the seed is poorly chosen, once the traversal has taken the northwards direction, it will not find any part of the region that lies eastwards, even though they share more than 99% of their borders.

Thus, instead of choosing the seed at random or with a simple tree traversal schema, a new method had to be developed. The most straightforward way to go is to assume that bigger leaves are most likely to produce "good" regions and use those as seeds. Testing some basic cases produced promising results, which were equivalent or better to those previously produced by a post-processing join algorithm. It also

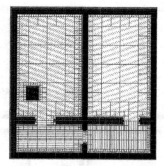

Fig. 4. PQTshowing the different fornd regions using the breadth-first approach to look for a seed.

avoided some problems that the latter algorithm had; mainly the time overhead (each region had to be tested against all others to see whether it could be joined) and the joining of adjacent regions separated by a wall (specifically the join algorithm joined the two regions on Fig. 6 B).

Due to the properties of the quadtrees, a breadth-first traversal method will return the leaves in the desired order, as in a quadtree, nodes are classified on levels according to size (the root node being the biggest one, i.e. the whole map).

To find all the regions on a PQT, the *FindRegion* algorithm (Alg. 1) must be applied iteratively with new unmarked seeds, chosen on a size basis, until there are no more unvisited free nodes. This yields the *FindRooms* algorithm (Alg. 2).

Algorithm 2 FindRooms()
regionList ⟸ { }
while ∈ *node* | *node.unvisited* **and** *node.free* **do**
region ⟸ FindRegion(*node*)
for all *node* **in** *region* **do**
node.markVisited
node ⟸ GetNextSeed(*node*)
returm *regionList*

5 Experimental Evaluation

The proposed method was tested on eight different gridmaps (Fig. 5) with promising although not perfect results. Figure 6 shows the results using a simple recursive traversal method to get the seeds, and Fig. 7 shows the results using a breadth-first traversal method, which will give seeds in order of size; (free) bigger nodes first. It is clear from the examples that the use of methodically chosen seeds is better than a simple random choice. On maps C, D and E, the random seed split regions that clearly are a single room, whereas using a breadth-first search for the new seed, correctly assigns a single region to a single room.

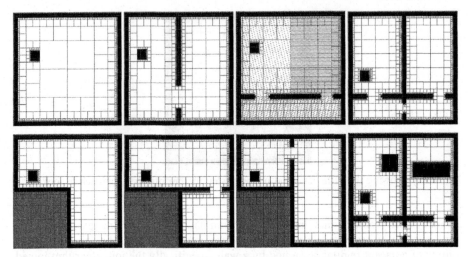

Fig. 5. The different maps used for testing.

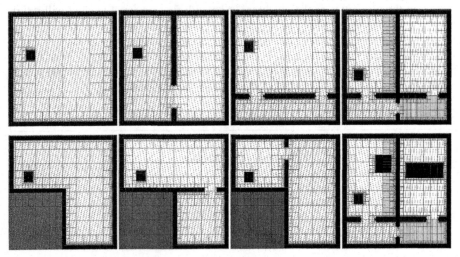

Fig. 6. The test maps after finding the regions, using a random seed node.

However in both cases we can see a problem on map E. The algorithm is not capable of detecting the L-shaped room as such, and returns instead the square region that contains it. The problem is not on the algorithm itself but the region representation it uses. Regions are represented by two sets of coordinates describing a rectangle. This region is not a rectangle though, and needs additional points to define it.

Although the L-shaped room could be separated into two regions with a carefully chosen seed, there is no simple way to fully automate this, and another solution must be found. Possible solutions are a new region representation (e.g. polygonal regions) or a post-processing split-and-resize algorithm.

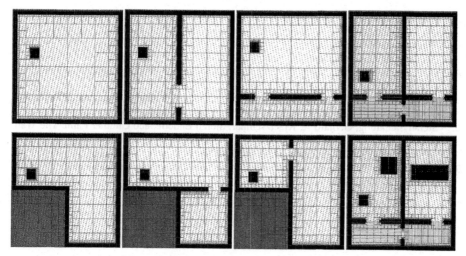

Fig. 7. The test maps after finding the regions, using the breadth-first approach.

On the other hand, if we allow quadtree nodes to be part of more than one region simultaneously, the L-shaped room could still be classified as a single region (if we classified it manually, it would actually be a single room), whilst allowing the bottom left part to be part of another region if it were free space. Depending on the application, that might be desirable or not.

6 Conclusions

Probabilistic quadtrees turn out to be a viable and practical solution for variable-resolution mapping of large environments. They are comparatively easy to build and maintain, and they achieve compact representations while allowing for increased map resolution. A new representation structure requires new algorithms working on it. Some of the algorithms developed for occupancy grid maps can be ported to PQTs, but it is interesting to develop new algorithms which exploit the PQT characteristics.

One example for such an algorithm is the room finding algorithm introduced here, that uses the aggregation properties of quadtrees to minimize the cells traversed, the underlying neighborhood graph to avoid converting back to gridmap, and the binary flags so that no additional memory structure is needed. Our method is able to find rectangular regions even in the case that they contain obstacles inside.

References

1. Alberto Elfes. Using occupancy grids for mobile robot perception and navigation. *IEEE Computer*, July 1989.
2. Gerhard K. Kraetzschmar, Guillem Pagès Gassull, and Klaus Uhl. Probabilistic quadtrees for variable-resolution mapping of large environments. In M. I. Ribeiro and J. Santos Victor, editors, *Proceedings of the 5th IFAC/EURON Symposium on Intelligent Autonomous Vehicles*, Lisbon, Portugal, July 2004. Elsevier Science.

3. Gerhard K. Kraetzschmar, Stefan Sablatnög, Stefan Enderle, and Günther Palm. Application of Neurosymbolic Integration for Environment Modelling in Mobile Robots. In Stefan Wermter and Ron Sun, editors, *Hybrid Neural Systems*, number 1778 in Lecture Notes in Artificial Intelligence, Berlin, Germany, March 2000. Springer. ISBN 3-540-67305-9.

4. H. P. Moravec. Sensor fusion in certainty grids for mobile robots. *AI Magazine*, 9(2):61–74, 1988.

5. Stefan Sablatnög. *Region-Based Representation of Spatiotemporal Concepts*. Dissertation, University of Ulm, Neuroinformatics, Ulm, Germany, October 2001.

6. Stefan Sablatnög, Gerhard K. Kraetzschmar, Stefan Enderle, and Günther Palm. The Wall Histogram Method. *In Proceedings of the International Conference on Artificial Neural Networks (ICANN-99)*, Edinburgh, Scotland, 1999.

7. Hanan Samet. *The Design and Analysis of Spatial Data Structures*. Computer Science. Addison-Wesley, 1989.

Active Autonomous Object Modeling for Recognition and Manipulation
Towards a Unified Object Model and Learning Cycle

Jens Kubacki, Björn Giesler, Christopher Parlitz

Fraunhofer IPA, University of Karlsruhe (TH)

Abstract. In this paper the aim is combine the principle of active autonomous object modeling with results from the field of computer vision and 3D geometrical modeling for recognition purposes focusing on modern robotics. The goal is to make a first step towards a unified object model and learning cycle that allow for integration of inputs from different research activities related to recognition and modeling in order to enable a robot to actively develop models over operation time.

1 Introduction

In the European project COGNIRON [1] we study the perceptual, representational, reasoning and learning capabilities of embodied robots in human centered environments. The goal is that such robots will be able to learn new skills and tasks in an active and open-ended manner.

In order to recognize artifacts in their environment robots need internal representations or *models* that describe certain properties of objects so that they can be detected in the environment or so that a robot can manipulate them.

There are difficult problems involved in designing such powerful recognition systems. One is to find the right *type* of models and the other is the requirement for open-endedness. If virtually *all* objects are to be recognized then the robot should be equipped with capabilities that allow it to *learn* models either by interacting with the human or autonomously by interacting with the object.

In this paper the aim is combine the principle of active autonomous object modeling with results from the field of computer vision and 3D geometrical modeling. The goal is to make a first step towards a unified object model and learning cycle that will lead to continuous learning of object properties.

2 Related Research

Firstly, the work in this paper is influenced by the field of 3D object modeling of objects (see e.g. the related chapters in [2]). For complex physical manipulation it is e.g. required to grasp an object or to plan manipulator paths around it. 3D structure provides a bidirectional mapping from texture space to object coordinates, yielding accurate view synthesis and object pose reconstruction. Grasp planning can be done using shape-primitive object approximation [3], which can be produced from 3D structure information.

Secondly, the work presented in this paper is influenced by results from appearance-based computer vision. There are approaches that rely on training an object recognition system with specific views from the object. Promising results have been achieved with scale-invariant key-points [4]. Also *pure* learning approaches seem to be suited for robust object labeling [5] on the basis of example views.

Since the robot's actuators are used for the training process the work can also be related to [6].

In contrast to other papers we take a rather pragmatic view on object modeling. The goal is to combine different methods that are already used in robotics with a focus on typical manipulation tasks that are investigated in the context of modern robotics. This view also includes the use of modern sensors (e.g. [7]) and typical three-layer control architectures.

3 Unified Object Model and Learning Cycle

In order to structure research work and to visualize basic problems and concepts involved in continuous object learning the unified object model and learning cycle shown in Fig. 1 is presented. The aim of these concepts is to serve as a basic guide to structure inputs from different research activities related to object modeling.

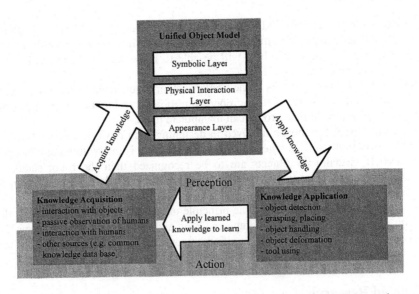

Fig. 1. The unified object model and object learning cycle as general concepts to enhance object models over operation time.

The unified object model (Fig. 1, top) consists of three main layers. The lowest layer is the *appearance layer*. The work and results from classical computer vision related to object recognition with a strong focus on learning from example views fits into this level. However, also shape-based matching would be located here if 3D information is available.

The next layer is the *physical interaction layer*. This incorporates e.g. the association of grasping points to a geometric 3D model of the object. However, if the object has degrees of freedom (e.g. a pair of scissors or other tools) then descriptions of the object's mechanics would also belong into this layer.

The *symbolic layer* associates symbolic information to the object. Entries in this layer may contain class information such as part-of and kind-of relationships as well as object names or ownerships. If the object has 'functions' or actions associated with it then this information would also be included in this layer.

The object learning cycle is divided into two main groups of competences that include perception and action (Fig. 1, bottom). The first group relates to the acquisition of knowledge i.e. to the *filling* of the of the object model's content. The second group deals with the application of the acquired knowledge to recognition tasks. There is a link between both groups that describes the possibility that improved recognition and manipulation competences may also allow for more sophisticated learning strategies.

4 Implementations and Results

For the unified object model and learning cycle are rather superficial concepts in this chapter a specific learning cycle related to a pick and place task is introduced to investigate object learning in more detail. Assumed are two sensors available: a range imaging sensor [7] and a conventional color imaging sensor mounted next to each other and looking into the same direction.

The specific object learning cycle assumes that the human supervisor places an object onto a *learning* table (Step 1). Then (Step 2) the robot fits a first but incomplete geometric model to the object that relates to the physical interaction layer of the unified object model. The object is separated from the background incorporation knowledge of the learning table and the background. The first model is used to grasp the object (Step 3) and to feed an appearance-based classifier model with example views by rotation the object in front of the sensors (Step 4). Here figure-background separation can be achieved using the measurements of the range imaging sensor and the known position of the object. The classifier model is related to the appearance layer of the unified object model. Then Step 2 to Step 4 are repeated until the geometric model and the classifier models are complete. In the following sub-chapters partial implementations of the two models are described.

4.1 Learning a Geometric Model for Grasping

Two methods are proposed for geometric object modeling: model completion and basic-body extraction.

Model Completion. A single view does not contain enough information to construct a complete surface model of the object. Therefore, several views are taken from different angles, until the robot has exhausted the sphere segment centered around the object over which it can move its sensor. The resulting scans are registered using a fast variant of the ICP algorithm. Even if the sphere segment of possible sensor perspectives is exhausted, the resulting surface model will still not be complete in the general case. For instance the object might be located on a table in front of a wall. In this case the object's back and bottom are not observable. For real autonomous learning the object must be picked up by the robot. A partial model cannot supply manipulation information in the general case, however. This chicken-and-egg problem is solved using basic-body extraction.

Basic-Body Extraction. If only partial 3D information is available, for everyday objects it is often possible to approximate those using basic bodies sufficient for careful manipulation. This has been proved in literature using different shapes such as spheres, cylinders and cones [8], but it is more convenient to use a single versatile model such as the generalized cylinder [9] or, more recently, superquadric ellipsoids [10,11]. We choose superquadrics since they are capable of describing complex shapes ranging from cuboids over ellipses to spheres.

To adapt one or more superquadrics to a set of object points, a robust segmentation process first extracts those points from the background. In our initial experiments, we use an object standing on a table. The table plane is automatically extracted and defines background (everything outside the bounds of the table, projected upward and downward) and object (everything inside the bounds of the table but not part of the table plane itself). After this initial segmentation, clustering is performed to remove outliers and segmentation errors.

Fig: 2. Left: Object on table, middle and right: segmented range image from front and left (red: background, blue: table plane, green: object including segmentation error)

A superquadric is parameterized by the 11-vector $\mathbf{a}=(\mathbf{a}_r|\mathbf{a}_t|\mathbf{a}_s)$, where $\mathbf{a}_r=(\varphi, \vartheta, \theta)$ is the vector of Euler rotation angles (below represented as rotation matrix $\mathbf{R}$), $\mathbf{a}_t=(x,y,z)$ is the translation and $\mathbf{a}_s=(a_1,a_2,a_3,\varepsilon_1,\varepsilon_2)$ is the 5-vector of shape parameters. With this parameterization, the superquadric is given by the equation

$$\mathbf{x}(\eta, \omega) = \mathbf{R} \begin{pmatrix} a_1 \cos^{\varepsilon_1}(\eta) \cos^{\varepsilon_2}(\omega) \\ a_2 \cos^{\varepsilon_1}(\eta) \sin^{\varepsilon_2}(\omega) \\ a_3 \sin^{\varepsilon_1}(\eta) \end{pmatrix} + \mathbf{a}_t, \text{ for } -\frac{\pi}{2} \leq \eta < \frac{\pi}{2} \text{ and } -\pi \leq \omega < \pi. \quad (1)$$

Fitting this parameterized equation into a set of N points $\mathbf{x}=(\mathbf{x}_1,...,\mathbf{x}_N)$ is a matter of minimizing the error function

$$E(\mathbf{a}) = \frac{1}{N} \sum_{i=1}^{N} \chi^2(\mathbf{a}, \mathbf{x}_i) \quad (2)$$

where χ is a distance function relating the measurement $\mathbf{x}_i$ to the superquadric's surface near it. The obvious choice for χ would be the Euclidean distance of $\mathbf{x}_i$ to its orthogonal projection $\mathbf{x}'_i$ onto the surface defined by $\mathbf{a}$. Since there exists no closed-form solution for determining $\mathbf{x}'_i$, we use the approximation suggested in [9], which uses the inside-outside function $F(\mathbf{a,x})$ (<1 if $\mathbf{x}$ is inside $\mathbf{a}$, =1 if $\mathbf{x}$ is on $\mathbf{a}$, >1 otherwise):

$$\chi(\mathbf{a,x}) = \sqrt{a_1 a_2 a_3}(F(\mathbf{a,x}) - 1). \tag{3}$$

Minimization of the error function is needed using least-squares approximation and refined using the Levenberg-Marquardt method. Since an object will not generally be well approximated by a single superquadric, a split-and-merge approach is used to first split the object point set into as few subsets as possible that can be approximated with superquadrics while keeping the error lower than a certain threshold T_s. A second threshold T_m is used as an upper error threshold for merging.

Fig: 3. Left: Object points after clustering and outlier removal, middle: Superquadric split, right: superquadric merge

After split and merge, grasp points are determined from the resulting superquadric cluster. Automatic grasp point detection for arbitrary grippers has only been solved for objects made up of spheres, cuboids, cylinders and cones [12]. But since we use a simple two-finger parallel jaw gripper, we use a heuristic: A line is defined running parallel to the floor plane and at 45° angle to the robot's front and side axes and passing through the cluster's barycentre. This line's outermost intersections with the cluster are used as initial grasping point estimates. They are then refined by gradient descent to minimize object rotation due to grasping.

After basic-body adaptation and grasping point analysis, the object can now be grasped and moved in front of the sensors to complete the surface model and to feed the appearance-based classifier part.

Fig: 4. Resulting suggested grasping points.

4.2 Learning an Appearance Model for Invariant Object Detection

The goal of this part is to design an appearance-based object detection method that is trained when the object is grasped by the robot. The idea is to combine the advantages of scale stable key-points [4] with the advantages of learning algorithms [13].

The results of a first experiment are shown in Table 1. The SIFT filter was applied to all objects appearances of the first five objects of the COIL data base [14]. Each color layer (RGB) was processed separately. Then only the key point descriptors of the objects were extracted. Three numbers were appended to describe the color at the key-point's location. The result is a large list of vectors with 131 dimensions (128 numbers for the key-point descriptors and 3 for color). The data was then reduced by a threshold applied to distances between learning samples (set to 160.0). The resulting numbers of key-point descriptors per object are shown in the second row of Table 1. Then one-class RBF-SVMs [15] were trained using the software from [16]. The parameter $\gamma=0.00019$ was used for the radial basis function. The parameter $v=0.001$ was set to allow for errors on the training set. See [15] for an explanation of these parameters.

Table 1. Results of the SVM experiment. Key-point descriptors of an object can be learned by a single one-class RBF-SVM

COIL Object Number	1	2	3	4	5
No of overall key-points	9624	913	3889	1477	4359
No of support vectors	2250	572	1716	805	1777
Hit rate on object 1	0.89	0.12	0.52	0.16	0.57
Hit rate on object 2	0.58	0.68	0.48	0.24	0.51
Hit rate on object 3	0.47	0.16	0.80	0.15	0.43
Hit rate on object 4	0.60	0.24	0.58	0.74	0.43
Hit rate on object 5	0.12	0.08	0.34	0.57	0.81

This experiment shows that a single key-point descriptor can be associated correctly with high probability to the target object (see the gray-shaded recognition rates). The new and interesting point is that this is done with one-class SVMs. If n-class classifiers are used then the original views of all objects must be stored over operation time and the time for training one new object increases with the number of objects already known. Current work focuses on the reduction of the false hits on other objects then the one being searched for.

5 Summary and Conclusions

The aim of this paper is to combine different research activities related to object recognition in modern robots. For this purpose a unified object model with different layers related to recognition tasks is presented. A specific learning cycle for a pick and place task is introduced to investigate active modeling in more detail by describing current results in 3D modeling and appearance-based recognition. Further work is will be conducted to fully integrate of the two modeling parts into a real robot. Further conceptual work will focus on the unified object model and learning cycle.

6 Acknowledgements

The work described in this paper was conducted within the EU Integrated Project COGNIRON ("The Cognitive Companion") and was funded by the European Commission Division FP6-IST Future and Emerging Technologies under Contract FP6-002020.

References

1. Chatila R: The Cognitive Robot Companion and the European Beyond Robotics Initiative. 6th EAJ International Symposium "Living with Robots", 2004
2. Forsyth DA and Ponce J: Computer Vision: A Modern Approach. Prentice Hall 2002.
3. Miller AT, Knoop S, Allen PK, Christensen HI: Automatic grasp planning using shape primitives. In Proc. IEEE International Conference on Robotics and Automation, Taipei, Taiwan, pp. 1824-1829, 2003.
4. Lowe DG: Distinctive image features from scale-invariant keypoints. International Journal of Computer Vision, Vol. 60, pp. 91-110, 2004.
5. Roobaert D, Zillich M, Eklundh JO: A Pure Learning Approach to Background-Invariant Object Recognition using Pedagogical Support Vector Learning. CVPR, Vol. 2, No. 2, pp. 351, 2001.
6. Fitzpatrick P, Metta G, Natale L, Rao S and Sandini G: Learning About Objects Through Action – Initial Steps Towards Artificial Cognition.
Web page: citeseer.ist.psu.edu/fitzpatrick03learning.html
7. Oggier T, et al.: An all-solid-state optical range camera for 3D real-time imaging with sub-centimeter depth resolution (SwissRangerTM). In: Proceedings of the SPIE, Vol. 5249, No. 65, 2003.
8. Lukacs G, Marshall AD, and Martin RR: Geometric least-squares fitting of spheres, cylinders, cones and tori. Report GML 1997/5, 1997.
9. Nevatia R and Binford TO: Description and recognition of curved objects. Artificial Intelligence, 8(1):77–98, 1977.
10. Leonardis A, Jaklic A, and Solina F: Superquadrics for segmenting and modeling range data. IEEE Transactions on Pattern Analysis and Machine Intelligence, 19(11):1289–1295, 1997.
11. Jaklic A, Leonardis A, and Solina F: Segmentation and Recovery of Superquadrics. Kluwer, 2000.
12. F. Solina and R.K. Bajcsy: Recovery of parametric models from range images: The case for superquadrics with global deformations. IEEE Transactions on Pattern Analysis and Machine Intelligence, 12(2):131–147, 1990.
13. Vapnik, VN: The Nature of Statistical Learning Theory. New York: Springer-Verlag, 1998.
14. Nene SA, Nayar SK and Murase H: Columbia Object Image Library (COIL-100). Technical Report CUCS-006-96, Columbia University, 1996.
15. Schölkopf B, Platt JC, Shawe-Taylor J, Smola AJ, and Williamson RC: Estimating the support of a high-dimensional distribution. Neural Computation, 13(7):1443–1471, 2001.
16. Chang CC and Lin CJ: LIBSVM: a Library for Support Vector Machines (Version 2.31) Wep page: citeseer.ist.psu.edu/chang01libsvm.html

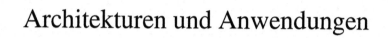

Architekturen und Anwendungen

Die Softwarearchitektur eines Laufroboters für RoboCup Rescue
AIMEE

Robert Borchers, Larbi Abdenebaoui, Malte Römmermann, Dirk Spenneberg

AG Robotik, Universität Bremen, Fachbereich 3, Bibliotheksstr.1, 28359 Bremen

Zusammenfassung. Dieser Artikel beschreibt den verhaltensbasierten Kontrollansatz des Laufroboters AIMEE, des ersten vierbeinigen Laufroboters, der am RoboCup Rescue teilnahm. Zur Bewegungssteuerung des Roboters kommen bio-inspirierten Ansätze zum Einsatz. Zur Erzeugung der rhythmischen Bewegung werden Modelle für Zentrale Mustergeneratoren (CPG) auf Basis von Bezier-Splines verwendet. Zur Nachahmung biologischer Reflexe werden trainierte sensorgetriggerte neuronale Netze verwendet.

1 Einleitung

Im folgenden Artikel wird die bio-inspirierte Softwarekontrolle des vierbeinigen Laufroboters AIMEE (siehe Abb. 1) vorgestellt. AIMEE wurde im Rahmen eines 2-jährigen studentischen Projekts der AG Robotik an der Universität Bremen für den RoboCup Rescue Wettbewerb entwickelt. Dieser Roboter war der bisher komplexeste vierbeinige Laufroboter, der an dem RoboCup Rescue Wettbewerb teilgenommen hat.

Dabei zeigte sich, dass laufende Systeme bereits jetzt in der Lage sind, mit der Konkurrenz, welche beim Robocup Rescue vornehmlich aus ketten-, raupen- und radgetriebenen Systemen besteht, mitzuhalten. Dabei ist das Potential laufender Systeme bei weitem noch nicht ausgeschöpft. Insbesondere die simultane Kontrolle der

Abb. 1. Der Vierbeiner AIMEE im Random Stapping Field

vielen aktiven Freiheitsgrade stellt hohe Anforderungen an die Kontrollansätze. Erschwert wird die Entwicklung geeigneter Ansätze noch dadurch, dass gerade bei Laufrobotern im Vergleich zu üblichen rad- und kettengetriebenen Systemen die zur Verfügung stehenden Ressourcen eingeschränkter sind, denn das Verhältnis von Gesamtgewicht zur Aktuatorleistungsfähigkeit ist ein wesentliches Kriterium für die Mobilität eines Laufroboters. Beim Robotergewicht stellen derzeit insbesondere die Batterien einen wesentlichen Faktor dar. Aus diesem Grund muss zwingend auf Strom sparende und leichte Kontroll-Hardware geachtet werden. Daher kommen für kleine Laufrobotersysteme wie den hier vorgestellten AIMEE keine Hochleistungsprozessoren in Frage, wie man sie für die Berechnung aufwendiger Kinematikmodelle benötigt. Auch das Gewicht der dafür notwendigen Hardware ist nicht unerheblich. Somit lassen sich die in der Industrie verbreiteten Ansätze der Kinematikkontrole in kleinen Laufrobotersystemen nicht einsetzen. Hier bieten bio-inspirierte Ansätze [1,2,3,7] eine mögliche Lösung, da diese aufgrund einfacher Elementarberechnungen/-mechanismen einen wesentlich geringeren Rechenaufwand haben. Diese Ansätze werden im folgenden aufgegriffen und in ein verhaltensbasiertes Programmierkonzept integriert.

Die Steuerung des Laufverhaltens erfolgt in unserem Roboter nach Vorbild des SCORPION Roboters [9], bei welchem die rhythmische Bewegung der Gelenke durch Überlagerung von rhythmischen Kurven, die mittels Beziersplines beschrieben sind, erzeugt wird. Diese Bewegungsmuster ermöglichen das Laufen selbst bei geringen Störungen. Für die Behandlung von Störungen werden Reflexkonzepte benötigt.

2 Software-Architektur

In der verhaltensbasierten Robotik wird häufig davon ausgegangen, dass die einem Behavior zugrunde liegenden Mechanismen/Prozesse innerhalb eines Agenten parallel ausgeführt werden. Da dies auf einem sequentiellen Prozessor nicht umsetzbar ist, verwenden viele Systeme ein Single-Loop-Verfahren, bei dem alle Prozesse in einer konstanter Zeit durchlaufen werden (z.B. in [10]). Während die Prozesse durchlaufen werden, werden deren Effekte auf interne oder Aktuatorwerte akkumuliert und erst nach Ablauf der konstanten Zeit werden diese Werte gesetzt, sodass eine quasi-parallele und getaktete Ausführung ermöglicht wird. Dieses Verfahren stösst jedoch an seine Grenzen, sobald der Durchlauf durch alle Prozesse mehr Zeit braucht, als die dafür vorgesehene konstante Zeit und somit die Prozesse nicht mehr auf dieser festen Frequenz laufen.

Z.B. unterbrechen Reflexe auf Interruptbasis die Prozessschleife und verlangsamen deren Ablauf. Bei Schleifenzeitüberschreitung kann der Roboter somit ein ungewolltes Verhalten zeigen, da die Zeitbasis nicht mehr korrekt ist. Das Reflexverhalten lässt sich aber nur schwer im Voraus vorhersehen, genauso wenig, wie man zu Beginn der Bottom-Up Programmierung eines verhaltenbasierten Roboters schon genau weiß, wie viel Zeit der Durchlauf aller später implementierten Prozesse benötigen wird. Der Ansatz, dieses, sobald das Problem auftritt, durch Verringerung des Zeittaktes zu lösen, führt nur zur Aufschiebung des Problems und bedeutet zusätzlich jeweils eine notwendige Adaption/Umprogrammierung der Prozesse auf diese neue Zeitbasis. Zudem bleibt bei geringer Auslastung viel Prozessorzeit ungenutzt. Eingehend sind diese Probleme in [8] beschrieben.

Komplette RTOS kamen zur Lösung nicht in Frage, da sie für unseren MPC565 Mikrocontroller schon einen sehr hohen Overhead bedeuten. Des weiteren sind viele RTOS-Systeme auch sehr teuer.

Daher verwenden wir einen neu entwickelten Mikrokernel namens M.O.N.S.T.E.R. [8], der speziell auf unsere Anforderungen einer verhaltensbasierten Programmierung zugeschnitten ist und nur die nötigsten Erweiterungen zu einem Single-Loop-System beinhaltet.

Das Single-Loop-System wird hier durch ein 2-Level-Background/Foreground System ersetzt. Bei diesem Ansatz werden Prozesse in harte Echtzeit-Prozesse, z.B. Hardwaretreiber oder Low-Level Motoransteuerungsprozesse, und zeitunkritischere Prozesse (Verhaltens-Prozesse) unterschieden und zusätzlich ein Hintergrundprozess eingeführt. Hierbei wird auch in einer Situation in der die Schleife über ihre ge-wünschte Periode hinaus ausgeführt wird, gewährleistet, dass alle Prozesse korrekt arbeiten. Den Behaviorprozessen wird Ihre jeweilige Ausführungsfrequenz mitgeteilt, sodass sie sich daran adaptieren können.

Des weiteren gilt:

Sei q_i ein interner oder ein Aktuatorwert auf den Einfluss genommen werden kann und sei weiter $inf_{b,i}$ der Einfluss des Behaviorprozesses b auf diesen Wert und $w_{b,i}$ dessen Gewichtung, dann wird der neue Wert $q_i(t)$ im Zeitschritt t wie folgt berechnet:

$$q_i(t) = \frac{\sum_{b=0}^{n} w_{b,i}(t) \cdot inf_{b,i}(t)}{\sum_{b=0}^{n} w_{b,i}(t)}$$

$$w_{b,i}(t) = \{ \begin{array}{l} w_{b,i}^{set}(t), \text{if } b \text{ active in step } t \\ w_{b,i}(t-1) \cdot dec(b,i), \text{else} \end{array}$$

Wenn ein Prozess b nun im Zeitschritt t nicht ausgeführt wurde, wird sein Einfluss auf q_i mit der Dekrementerfkt. dec(b,i), die auf das Gewicht wirkt, abgeschwächt, ansonsten wird der Einfluß mit dem aktuellen Gewicht berechnet.

Eine eingehende Beschreibung von M.O.N.S.T.E.R und seiner Arbeitsweise findet sich in [8].

3 Zentrale Mustergeneratoren

Über Modelle von Zentralen Mustergeneratoren (CPGs) (siehe [4]) die Fortbewegung zu kontrollieren ist eine mit geringem Aufwand und Rechenleistung verbundene Möglichkeit. Diese rhythmischen Kurven können, vergleichbar zu ihrem biologi-schem Vorbild, in ihrer Amplitude, Frequenz und Phase moduliert werden und sind zusätzlich auch überlagerbar. Zum Beispiel lässt sich ein Kurvenlaufen durch den gleichzeitigen Einfluss der Behavior-Prozesse „Vorwärtslaufen" und „Drehung im Stand" auf die Systemressourcen erzeugen. Zur Beschreibung der CPG-Muster bieten sich Beziersplines an. Diese ermöglichen durch die Beschreibung mit wenigen Punk-ten alle möglichen Arten von Kurven für die Beinkoordination und implizieren auf-grund ihrer Glattheit weiche Bewegungen der Gelenke.

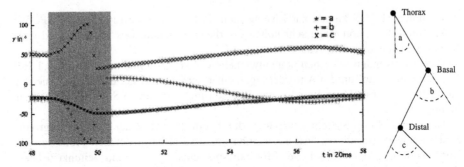

Abb. 2. Der Graph zeigt eine Swing- und nachfolgende Stancephase der drei Gelenke Thorax, Basal und Distal (siehe schematische Darstellung des Beines). Dabei gibt die y-Achse die Gradzahlen der Gelenke an. 0 Grad entspricht der senkrechten Haltung des Beines zum Körper. Negative Werte ergeben eine Haltung des Beines nach vorne. Die x-Achse des Graphen entspricht einem Zeitfenster, das in 20 ms Schritten skaliert ist.

Abbildung 2 zeigt ein rhythmisches CPG-Muster für die Vorwärtsbewegung an einem Bein. Die Beziersplines werden als Ring durchlaufen, wozu der erste Punkt der Kurve den gleichen Winkelwert und die gleiche Steigung haben muss wie der letzte Punkt der Kurve.

Zu jedem Punkt der Beziersplines werden vier Werte angegeben: x, y, m und *scales*. Der Wert x steht für die Zeit und y ist die Sollposition des jeweiligen Servos. Die Steigung an dem Punkt beschreibt der Wert m. Der Wert *scales* kann 1 oder 0 sein und nimmt Einfluss auf die Skalierung der Kurve bezüglich der Zeit. Zur Verwendung der Beziersplines zur Kontrolle der Gelenke werden zunächst Teilstücke aus den festen Punkten errechnet und in einer Liste verwaltet. Ein Teilstück beschreibt den Abschnitt zwischen zwei festen Punkten mit einer Bezierkurve 3ten Grades. Ein Teilstück ist eine Struktur und beinhaltet die Länge des Abschnittes (l_n, Länge des n'ten Teilstück), den Phasenversatz (*displace*) und die Polynomkoeffizienten (k_0, k_1, k_2, k_3)). Über das Polynom werden die y Werte berechnet.

Eine Funktion $BEZIER_{ij}(t)$ ermittelt den Winkelwert des Gelenks j am Bein i zu dem Zeitpunkt t. Dabei wird zunächst anhand des aktuellen x der richtige Part in der Liste ermittelt und dann das jeweilige y aus der zugehörigen Bezierkurve errechnet. r auf die aktuelle Position in der Kurve verweist und mit der Zeit inkrementiert wird. Die CPGs können über drei Parameter beeinflusst werden, Phase (Phasenversatz), Amplitude (Amplitudenmodulation) und Frequenz (Periodenmodulation). Zur Laufzeit des Roboters können die Parameter verändert werden, um zum Beispiel über die Frequenz die Geschwindigkeit des Laufens zu regulieren. Eine weitere Anwendung für die Adaptation der Parameter wäre ein Behavior-Prozess, welcher aktiv auf dem Roboter das Laufmuster an die Umgebung (rutschiger Untergrund, Sand, etc.) anpasst.

Der Phasenversatz wird in den Teilstücken direkt abgespeichert. Zum Anpassen der Amplitude werden einmalig die Koeffizienten der Polynome in den jeweiligen Teilstücken neu skaliert. echnung der aktuellen Gelenkposition: Beim Verändern der Frequenz müssen die Längen der Teilstücke neu skaliert werden und die Koeffizienten erneut berechnet werden. Dabei gibt es Abschnitte der Beziersplines, die nicht skaliert werden sollen. Typischerweise ändern Tiere mit 4 oder mehr Beinen nur sehr begrenzt ihre Swingphase(Flugphase). Diese wird typischerweise so schnell als möglich ausgeführt und nur die Stancephase(Stemmphase) wird tatsächlich verlängert oder verkürzt. Welche Abschnitte der Beziersplines skaliert werden, kann mittels des Parameter *scales* der Bezierpunkte angegeben werden. Nach der Neuberechnung der Teilstücke können die jeweiligen Servowerte wieder über die Funktion $BEZIER_{ij}(t)$ abgerufen werden.

Zum Finden geeigneter Punkte für die Beziersplines können diese in dem Kontrollinterface des Roboters angelegt und auch online auf den Roboter gespielt werden, was das Austesten neuer Bewegungsmuster sehr komfortabel gestaltet.

4 Reflexe

Insgesamt haben wir drei Reflexe in Aimee implementiert: den Stolperreflex zur Hindernisüberwindung, den Gleichgewichtsreflex und den Bodenkontaktreflex, um das Fehlen eines Bodens zu erkennen. Eine gute Kombination von CPGs und Reflexen bietet ein sehr robustes Verhalten.

Die Reflexe sind in unserer Architektur als harte periodische Prozesse implementiert. Das Hauptproblem bei der Implementierung von Reflexen ist der Umgang mit unsicheren Daten. Für eine flexible und robuste Lösung gegenüber unterschiedlichen Parametern, die das Material und das Verhalten von unserem System beeinflussen, haben wir die Stimulation des jeweiligen Reflexes hierarchisch eingebaut.

Erst werden die Sensordaten und andere Informationen über das System innerhalb eines Zeitraumes T bewertet. Das Ergebnis wird in ein trainiertes 3-schichtiges neuronales Netz weitergeleitet, welches die Eingabe in zwei Klassen einordnet: Eine richtige Stimulation oder ein falscher Alarm. Diese Architektur ist in Abb. 3 dargestellt (basierend auf die Architektur von Brotherton und Pollard, 1992 [6]).

Der Gleichgewichtsreflex soll eine ungewollte Schräglage des Roboters ausgleichen. Dafür wird der Wert eines eingebauten Neigungssensors ausgelesen und je nach Schräglage des Roboters bei einzelnen Beinen deren Höhe angepasst. Durch das verhaltensbasierte Konzept und der dadurch möglichen Überlagerung der Servoansteuerung muss dafür kein CPG beim Laufen verändert werden und der Roboter kann so unter anderem ohne spezielle Feinanpassung Schrägen bewältigen.

Der Bodenkontaktreflex soll die Fortbewegung stoppen, wenn der Roboter beim Eintreten in die Stancephase keinen Untergrund findet, und damit einen eventuellen Sturz verhindern. Die Stimulation erfolgt, wenn ein Bein am Anfang der Stancephase den Boden nicht berührt. Als Reaktion erfolgt hierbei das sofortige Ausstrecken des Beines, um einen eventuell tiefer als geplant befindlichen Boden zu erfassen. Sollte dies fehlschlagen, wird das aktuelle Laufverhalten des Roboters gestoppt, um ihn nicht zu beschädigen.

Der wichtigste und komplizierteste Reflex ist momentan der Stolperreflex, der bei Kontakt eines Beines mit einem Hindernis in der Luft-Phase (Swing-Phase) dieses Bein höher anzieht, um das vorhandene Hindernis überschreiten zu können. Dieser Reflex ist abgeleitet von Untersuchungen an Katzen in [5].

Dem Stolperreflex dient als Eingabe, in welcher Bewegungsphase sich die Beine befinden, wie weit sie darin fortgeschritten sind sowie der Soll- und Istwert der Schultergelenke. Für die erste Bewertung muss sich ein Bein in einem sinnvollen Zeitraum der Swingphase befinden, da ein Reflex z.B. am Ende der Swingphase nicht mehr zu einer Überwindung des Hindernisses führen kann und daher nur unnötige Störungen verursacht. Wenn dann die Differenzen der Soll- und Istwerte der zeitlich letzten fünf

Abfragen alle über einem Schwellwert liegen, werden die Werte und der Zeitraum an ein entsprechendes neuronales Netz weitergegeben, welches entscheiden soll, ob es sich um eine Stimulation oder eine Reihe von Ausreissern der Sensorwerte handelt. Abbildung 4 zeigt den Soll- und Istwert eines Schulterservos. In der Abbildung sind die Swingphasen grau unterlegt. Für den Stolperreflex ist die erste Hälfte der Swing-phase von Interesse, da hier die Trajektorie des Beines noch so verändert werden kann, dass eine Hindernisübergehung möglich ist. Eine Störung macht sich wie in der zwei-ten Swingphase aus Abb. 4 bemerkbar.

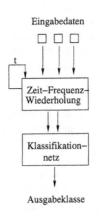

Abb. 3. Stimulation-Architektur

Abb. 4. Soll- und Istwert eines Schultergelenkes bei unge-störter und gestörter Swingphase

5 Fazit

Der erste Einsatz unseres Laufroboters bei den German Open 2005 in Paderborn hat uns gezeigt, dass unser Ansatz sehr viel versprechend ist. Das „Bremen Rescue Wal-kers" Studententeam erreichte den dritten Platz.

Durch geringes Gewicht und Größe des Roboters erreichten wir Orte, an die die anderen Teams nicht herankamen und konnten zudem den Roboter sehr schnell mit nur einer Person in Betrieb nehmen.

Die behaviorbasierte Softwarearchitektur ist so modular aufgebaut, dass wir sie während des Wettbewerbes an vielen Stellen leicht umbauen und der aktuellen Situa-tion anpassen konnten. Dies ist speziell dann von Vorteil, wenn man die Sensoren und Aktuatoren auf andere Umstände wie neuem Untergrund anpassen muss. Durch die behaviorbasierte Programmierung konnten verschiedene Elemente der Software de-zentral geändert und angepasst werden.

Tests an dem Random Stepping Field (siehe Abb. 1) im RoboCup Rescue haben gezeigt, dass unser System auch während eines permanent reflexgetriebenen Laufens, was verstärkt dem Walknetansatz [3] entspricht, stabil bleibt und keine Fehlfunktio-nen zeigt.

Durch die unkomplizierte Weise, mit Bezierkurven neue CPGs zu basteln, war es uns möglich, nicht nur viele verschiedene Bewegungen wie Kriechen oder das Drehen auf der Stelle, sondern auch innerhalb von 3 Tagen ein Treppenlaufen zu implementieren.

Während die anderen Systeme, wie rad- und kettengetriebene Fahrzeuge schon relativ ausgereizt in ihren Möglichkeiten sind, haben wir gezeigt, dass ein laufendes System großes Potential in sich birgt.

Literaturverzeichnis

1. Joseph Ayers. A conservative biomimetic control architecture for autonomous underwater robots. In Ayers, Davis, and Rudolph, editors, *Neurotechnology for Biomimetic Robots*, pages 241–260. MIT Press, 2002.
2. R.D. Beer, R.D. Quinn, H.J. Chiel, and R.E. Ritzmann. Biologically-inspired approaches to robotics. *Communications of the ACM*, 40(3):30–38, 1997.
3. H. Cruse, J. Dean, T. Kindermann, J. Schmitz, and M. Schumm. Walknet -a decentralized architecture for the control of walking behavior based on insect studies. In G. Palm, editor, *Hybrid Information Processing in Adaptive Autonomous Vehicles*. Springer, 1999.
4. F. Delcomyn. Walking robots and the central and pheripheral control of locomotion in insects. *Autonomous Robots*, 7:259–270, 1999.
5. H. Forssberg. Stumbling corrective reaction: A phase-dependant compensatory reaction during locomotion. *Journal of Neurophysiology*, 42(4), July 1979.
6. Dan W. Patterson. *Künstliche neuronale Netze, Das Lehrbuch Ch.8.* 1 edition, 1996.
7. Dirk Spenneberg. A hybrid locomotion control approach. In *Proceeding of CLAWAR 2005*, 2005.
8. Dirk Spenneberg, Martin Albrecht, and Till Backhaus. M.o.n.s.t.e.r.: A new behavior-based microkernel for mobile robots. In *accepted for ECMR 2005*, 2005.
9. Dirk Spenneberg and Frank Kirchner. Scorpion: A biomimetic walking robot. *Robotik*, 1679:677–682, 2002.
10. Luc Steels, Peter Stuer, and Danny Vereertbrugghen. Issues in the physical realisation of autonomous robotic agents. In *From Animals To Animats 4: Proceedings of the Forth International Conference on Simulation of Adaptive Behavior, SAB'96*, 1996.

Flexible Combination of Vision, Control and Drive in Autonomous Mobile Robots

Wolfgang Ertel, Joachim Fessler, Nico Hochgeschwender

University of Applied Sciences Weingarten, Department of Computer Science, Doggenriedstraße, 88250 Weingarten

Abstract. We present a universal modular robot architecture. A robot consists of the following intelligent modules: central control unit (CCU), drive, actuators, a vision unit and sensor input unit. Software and hardware of the robot fit into this structure. We define generic interface protocols between these units. If the robot has to solve a new application and is equipped with a different drive, new actuators and different sensors, only the program for the new application has to be loaded into the CCU. The interfaces to the drive, the vision unit and the other sensors are plug-and-play interfaces. The only constraint for the CCU-program is the set of commands for the actuators.

1 Introduction

In the literature many approaches for software-modularity [2], hardware-modularity [4] and framework projects exist.

Two particularly interesting framework projects are Player/Stage [7], which comes with a nice simulation environment, and the component based system Orca [1] with special modularity features. While these systems are mainly focused in software modularity in our approach we provide methods for a smooth combination of software- and hardware-modularity.

1.1 Overview of the Architecture

As shown in Fig. 1 the hardware of the robot consists of five flexibly exchangeable parts or groups of parts. The mechanical structure provides an easy way of exchanging these hardware modules. The architecture comprises a central control unit, a drive, a vision unit and a sensor input unit and an actuator unit, which work as follows.

Central control unit (CCU): This unit contains a powerful embedded PC-board running linux with all standard interfaces such as USB, AGP, firewire, ethernet, wlan, PC-card-slots and its own power supply built in a rugged cage with many slots and mountings. In order to exchange information with all the other components in a simple and uniform way, a CAN bus [3] acting as the information backbone completes this module. The high level intelligence utilizing machine learning techniques on the behaviour level of course is application dependent and at least parts of it have to be exchanged for new applications.

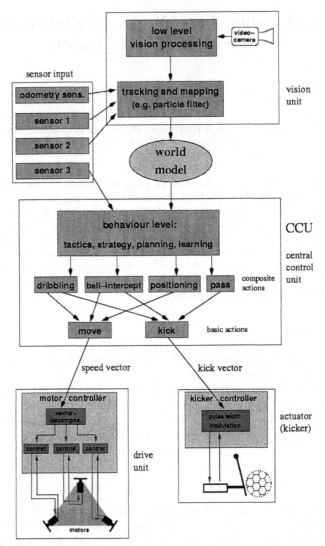

Fig. 1. A robot soccer player as an example for the functional structure of the modular robot architecture.

Vision unit: A VGA color CCD-camera is connected via firewire to a graphic processing unit which does the low level image processing. The high level image processing includes detection of objects, lines, structures as well as mapping, tracking and positioning of the robot. For this resource comsuming task we spend a second embedded PC-board. The software interface between the vision unit and the CCU is the world model consisting of coordinates, shape, size and speed of all detected objects, which will be described in section 2. In subsection 2.1 we describe a concept which allows for an easy replacement of the vision unit by a different one, for example a stereo vision unit.

Drive: The drive of the robot drive must be easily exchangeable. Drive replacement is based on a fixed high level protocol for motion commands which have to be interpreted by the intelligent microcontroller of the drive unit.

Sensor input: Due to the flexible mounts at the cage of the CCU and the drive, various secondary sensors can easily be attached on any side or on top of the robot. Depending on the particular task each sensor may communicate via the CAN bus either with the CCU, vision unit or even directly with the drive.

Actuator(s): Actuators have to be connected via the CAN-bus. A well defined and fixed set of (low-level) action commands to be sent from the CCU to the actuators is the basis for easy replacement of actuators.

2 Software Architecture

A reconfigurable robot hardware needs on the software side a flexible and modular architecture. To achieve this goal we need a middleware which bridges the gap between operating system and applications. Our object oriented software architecture is based on linux and the ICE middleware [5]. The Internet Communication Engine (ICE) gives us the possibility to develop autonomous software modules with different programming languages e.g Java, C++ and Python. As shown in Fig. 1 the software architecture is divided into a vision unit with low- and high level processing, a world model and a central control unit (CCU) for planning the actions of the robot.

2.1 Software Modularity

Suppose we have configured the five units of the robot to solve a particular task. If we now want to reconfigure the system for a new, completely different task, we may for example need a different drive and a different actuator. Of course for a new task the software of the CCU has to be replaced by a new one. For exchanging the drive, due to the application independent fixed interface between CCU and drive, we just attach any new drive (with its built in motor controller) to the robot with no software changes at all in the CCU or in the motor controller. Thus exchanging drives works in a plug-and-play manner. The interface between CCU and the actuator(s) is more complex, because the commands to the actuators are application specific. This means that the CCU must send only commands which the actuator(s) can interpret. The CCU programmer of course has to know the actuator commands before programming any actions. The interface between CCU and the vision unit seems to be even more complex because both units have to use the same structure of the world model. Thus, at least the mapping and tracking part of the vision unit has to be reprogrammed for every new CCU software. A solution for this problem is based on a **generic vision unit** which is able to detect objects with a wide range of different shapes and colours. When the CCU and the vision unit are connected, they start an initial dialogue in which the CCU sends a specification of the required world model structure to the vision unit. For example in a soccer application the CCU may send an object-description-list with items like

```
object("ball", moving,
       geometry(circle(20,25)),
       color([0.9,1],[0.7,0.8],[0,0.5]),
object("goal1", fixed,
       geometry(rectangle(200,100)),
       color([0,0.3],[0,0.3],[0.9,1]),
```

describing an object "ball" as a moving object with the shape of a circle, a diameter between 20 and 25 cm and orange colour. "goal1" is a fixed blue 200 × 100 cm rectangle. After this initialization, when the robot starts working, in each elementary perception-action-cycle the vision unit tries to detect objects of the types specified by the object-description-list and returns them together with their current position. For example in the simplest case the vision unit sends a list like

```
detected("ball", pos(2.47,11.93)), detected("goal1",
pos(5.12,3.70)),
```

to the CCU. This description may be more complex, including for example the size and color of the detected objects. This protocol allows us to work with the same vision unit for a large class of different control units and the interface between CCU and vision unit becomes a plug-and-play interface. With other sensors like infrared detectors which are simpler than a CCD-camera this interface may become simpler, but nevertheless it may use the same object description language.

2.2 The Vision Unit

The vision unit is an autonomous software module responsible for low level and high level image processing.

Low Level Processing: In this unit we use the power of a common Graphic Processing Unit (GPU) to extract information about the relevant objects from a raw picture. We developed a library which has implementations for standard image processing tasks like normalization of a picture which is useful for many mobile robot applications.

High Level Processing: Based on the data from the low level processing unit and the odometry the high level processing unit is responsible for more intelligent and complex tasks such as self localization and mapping using a particle filter [6].

2.3 The World Model

The derived information from the high level vision processing unit is saved in a data structure called world model. The world model contains information about position, velocity and acceleration of the robot and detected obstacles. It forms the basis for decision-making in the central conrol unit.

2.4 The Central Control Unit

The central control unit (CCU) as shown in Fig. 1 provides the "intelligence" of the robot. It is structured in three levels of functions. The lowest level of **basic actions** outputs the control-commands for the drive or the actuators. The inputs for these basic actions are very simple. For example the call of the "move"-action of a robot as described in section 3.3 may be *move(2.4,35,0)*, meaning, that the robot has to accelerate with 2.4m/s in the relative direction 35° with no rotation of its body, i.e. the alignment does not change. The next level of **composite actions** aggregates a sequence of basic actions, which may form quite complex motion pattern. For example dribbling from position A to position B (while guiding the ball) is a composite action. In complex cases like this, programming an optimal motion sequence may become very hard, especially because we have no formal model of the motion sequence. Therefore machine learning techniques, such as reinforcement learning have to be used in order to optimize these composite actions. At the top level, the **behaviour** of the agent has to be implemented. A wide variety of techniques from artificial intelligence can be applied, depending on the particular application.

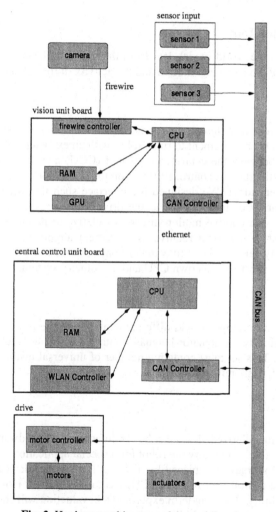

Fig. 2. Hardware architecture of the mobile robot.

3 Hardware Architecture

To increase CPU performance and support modularity of the system the vision unit and the central control unit are distributed on two PC-boards 2.

3.1 Vision Unit

We use a color CCD camera to capture the pictures of the environment. The CPU sends the picture to the graphic processing unit (GPU) which executes the low level vision processing with a high degree of parallelism resulting in high performance. The high level vision processing and mapping as described in section 2.2 is performed on the CPU of the vision board.

3.2 Central Control Unit

The CCU receives the world model data from the vision unit via ethernet and computes the actions of the robot e.g. motion and the activity of the actuators.

3.3 Drive

If a robot should work in different environments the possibility to easily exchange the drive is essential. One requirement to run a robot with an exchangeable drive is a well defined interface between the central control unit (CCU) and the drive. The second goal is to have all the motion control software on the drive unit to make the CCU and the drive unit independent. We designed the interface such that the CCU asynchronously sends an array of three values to the drive. This array contains the desired acceleration value, the relative motion direction (relative to its current heading) and the new alignment of the robot relative to its current alignment. To guarantee the required modularity (all low level movement control loops must be implemented in the drive) the drive must have its own CPU and a odometry system.

3.4 Actuators

The modular structure allows to add different actuators to the robot. To guarantee a maximum of flexibility an actuator-language with a fixed low level instruction set must be designed. This set must contain a number of universal instructions like grab, release, drop, hold, push, pull or kick.

4 Conclusion

We introduced a modularized flexible robot architecture. All hardware and software parts are exchangeable. To modify the robot for a special application drive, vision unit, sensors and actuators may be replaced by different (intelligent) devices. On the software side only the central control unit (CCU) has to be adjusted. However, there are limits to this flexibility. The long term experimental evaluation will show where these limits are. For the future we would like to develop hardware components that adapt to each other by means of machine learning techniques.

References

1. A. Brooks, T. Kaupp, A. Makarenko, S. Williams, and A. Orebäck. Towards component-based robotics. In *Proceedings of the IEEE/RSJ International Conference on Intelligent Robots and Systems*, 2005.
2. J. Corder, O. Hsu, A. Stout, and B. A. Maxwell. A modular software architecture for heterogeneous robot tasks. In *AAAI Mobile Robot Competition & Exhibition Workshop*. AAAI Press, 2002.
3. Konrad Etschberger. *Controller-Area-Network*. Fachbuchverlag Leipzig, 2002.
4. Yang Guilin and I.-M. Chen. Task-based optimization of modular robot configurations - mdof approach. *Mechanism and Machine Theory*, 35(4):517–540, 2000.
5. Michi Henning and Mark Spruiell. *Distributed Programming with Ice*. http://www.zeroc.com, 2004.
6. W. Burgard S. Thrun, D. Fox and F. Dellaert. Robust monte carlo localization for mobile robots. *Artificial Intelligence*, 128(1–2):99–141, 2001.
7. T. Vaughan, B. Gerkey, and A. Howard. On device abstractions for portable, reusable robot code. In *Proceedings of the IEEE/RSJ International Conference on Intel ligent Robots and Systems*, 2003.

Zentrale Aufgabenverteilung in einem fahrerlosen Transportsystem

Kay-Ulrich Scholl, Markus Klein, Bernd Gaßmann

Forschungszentrum Informatik, Interaktive Diagnose- und Servicesysteme, Haid-und-Neu-Straße 10-14, 76131 Karslruhe

1 Einleitung

Fahrerlose Transportsysteme (FTS) bestehen aus einem Fuhrpark von unterschiedlichen Fahrzeugen und einem zentralen Leitstand, welchem die Aufgabe der Betriebsmittelverwaltung zukommt: Zuweisung der Transportaufgaben auf Fahrzeuge, Freigabe von Wegstrecken und Ansteuerung sekundärer Komponenten wie Aufzüge, Ladestationen, Automatiktüren etc. Ein FTS gewinnt gerade dann an Attraktivität, wenn es mit nur geringem Aufwand installiert werden kann und eine hohe Flexibilität bezüglich der Fahrspuren und Einsatzmöglichkeiten bietet. Aus diesem Grund wurden am FZI Softwarekomponenten entwickelt, die es ermöglichen, Fahrzeuge und auch eine Leitstelle mit intelligenten Strategien auszustatten. Diese Komponenten haben sich zum Teil bereits im Einsatz bewährt. Auf Basis der hierbei gemachten Erfahrungen konnten neue Erkenntnisse gewonnen werden, die in die Entwicklung einer neuen innovativen Repräsentation von FTS-Anlagen und eines neuen zentralen Leitstands eingeflossen sind. Zur Repräsentation einer Anlage wurde ein topologischer Graph mit einem Zustandsautomaten verknüpft, wodurch das Verhalten und die Fahrspuren der Fahrzeuge definiert sind. Die Verknüpfung zwischen konservativer Fahrspurdefinition und dem durch den Zustandsautomaten definierten Verhaltensrepertoir der Fahrzeuge stellt ein Novum in der Realisierung von Steuerungen Fahrerloser Transportfahrzeuge dar und wurde seitens verschiedener FTS-Hersteller mit großem Interesse wahrgenommen. Logisch oder topologisch zusammenhängende Strukturen können darüber hinaus in hierarchischen Gruppen angeordnet und zusätzlich mit sekundären Komponenten (Aufzüge etc.) verknüpft werden. Hierdurch bleibt die Übersichtlichkeit der gesamten Anlage gewahrt. Es entsteht ein hierarchischer, topologischer Zustandsautomat, den wir HTFM (Hierarchical Topological Finite State Machine) nennen.

Sowohl eine Fahrzeugsteuerung als auch ein von uns entwickelter zentraler Leitstand basieren auf diesem Ansatz der Anlagenrepräsentation. Letzterer berücksichtigt insbesondere in äußerst flexibler Weise die Ansteuerung zusätzlicher Peripherie, die sich in jeder Anlage anders darstellt. Das TaPAC (Transports and Peripherals Administrative Centre) führt die Kommunikation zu solchen externen Einheiten mit Hilfe von Plugins aus, die jeweils individuell auf die Anlage vor Ort angepasst werden.

Fahrzeugsteuerung und Zentrale führen gemeinsam zu einem hochflexiblen FTS, das auf einfachste Weise installiert, modifiziert und simuliert werden kann.

2 Aufgaben eines zentralen Leitstands

Eine zentrale Steuerung für ein FTS muss verschiedene Aufgaben erledigen. Dies sind im Folgenden:

Wegewahl Die Wegewahl ist das Finden eines Pfades vom momentanen Standort zum Ziel. Sie kann sowohl im Fahrzeug selbst, als auch in einer zentralen Steuerung stattfinden. Eine zentrale Steuerung kann die Gesamtsituation aller Fahrzeuge betrachten, und daher die Wegewahl flexibler und effektiver gestalten. In der Regel wird hier die A-Suche im gegebenen Graphen der Fahrstrecken angewendet, um die optimalen Wege unter Berücksichtigung von Nebenbedingungen zu finden.

Ressourcenverwaltung Ressourcen des Systems, die von den Fahrzeugen gemeinsam benutzt werden, müssen in geeigneter Weise verwaltet werden, um Kollisionen und Verklemmungen der Fahrzeuge zu vermeiden. Zu solchen Ressourcen gehören z.b. Aufzüge, Türen, aber auch einfache Wegkreuzungen.

Aufgabenverteilung Die Transportaufträge, die das System erledigen soll, müssen erfasst, verwaltet und den Fahrzeugen geeignet zugewiesen werden. Hierbei ist der zentrale Ansatz einem verteilten Ansatz (Agentensysteme, Taxibetrieb) überlegen, da die Aufträge effektiver und vorausschauender auf die Fahrzeuge verteilt werden können.

Insbesondere zur Aufgabenverteilung auf die Fahrzeuge und möglicher Kriterien, die hierbei zu beachten sind, finden sich zahlreiche Ansätze in der Literatur. Die folgenden Arbeiten stellen einen kleine, relevante Auswahl dar. Klein und Kim [1] untersuchten die Auswirkungen verschiedener Algorithmen zur Steuerung von automatischen Fahrzeugen. Sie stellten fest, dass kombinierten Verfahren eine signifikant bessere Leistung erbringen. In [2] wird Active Rescheduling untersucht. Angestoßen von so genannten "cues" (Ereignissen im System) wird ein Netz von intelligenten Agenten angestoßen, die für jedes Subsystem überprüfen, ob eine Änderung des Ablaufplans erforderlich ist. Umgeplant werden kann hier auch die Wegewahl, indem FTF umgeleitet werden. Reveliotis [3] stellt ein Verfahren zur konfliktfreien Wegewahl von FTF vor, das keine starken Restriktionen stellt und trotzdem Robustheit erreicht. Eine sehr gute Übersicht über bisherige Forschungsergebnisse ist in [4] zu finden.

Es lässt sich zusammenfassen, dass aufwändige Optimierungsmaßnahmen nur bei komplexeren Anlagen mit einer großen Anzahl Fahrzeuge rentabel sind. Neben großen Distributionszentren ist dies bei der Anwendung von FTS in Krankenhäusern der Fall. Hier finden sich Systeme mit bis zu 100 Fahrzeugen. Bei dieser Anzahl Fahrzeuge und entsprechend zu erwartender hoher Anzahl Transporte ist eine Berechnung der optimalen Lösung bei einem Aufwand von $O(n^m)$ nicht möglich. (Wobei n für die Anzahl der Fahrzeuge steht und m für die der Aufträge.) Der gängigste Weg ist die ereignisgesteuerte Verteilung der Aufträge mittels Heuristiken. Die sinnvollen Stationsinitiierten (neuer Auftrag vorhanden) bzw. Fahrzeuginitiierten (Fahrzeug wieder verfügbar) Bewertungsfunktionen beachten hierbei Wartezeiten der Aufträge, Länge des Weges zum Transportgut, Betriebsdauer oder die Zeit, wie lange das Fahrzeug zuletzt oder insgesamt nichts zu tun hatte. Es werden auch andere Kriterien genannt, die jedoch lediglich von akademischen Wert sind.

Nachteile der ereignisgesteuerten Verfahren ist, dass zumeist mögliche ideale Verkettungen von Aufträgen nicht beachtet werden.

3 Repräsentation der Anlagendefinition

Für die Wegplanung ist es Stand der Technik, einen Graphen mit möglichen Abzweigungen und Zielknoten zu definieren, um mittels Graphensuchalgorithmen (A^*) den besten Weg bezüglich verschiedener Nebenbedingungen zu finden. Im HTFM werden in einem solchen Graphen weitere Informationen gespeichert, die auch das Verhalten der Fahrzeuge definieren.

Im einfachsten Fall fahren die FTF die Kanten des Graphen entlang, so dass diese quasi auch die Fahrspuren selbst definieren. Es wurden aber auch andere Fahrstrategien implementiert, die ein freies Fahren ermöglichen, wodurch die topologische Information im Graphen in den Hintergrund rückt. Als ein Beispiel sei hier ein verhaltensbasierter Ansatz genannt, der die Umfahrung von Hindernissen und die frei Navigation in sich stark verändernder Umgebung ermöglicht.

Neben den Bedingungen, ob ein Fahrzeug in bestimmten Konfigurationen und Modi eine Kante befahren darf, enhalten auch die Knoten Informationen darüber, wann diese logisch als erreicht gelten. Neben der bereits erreichten Position des Fahrzeugs sind auch Abhängigkeiten von weiteren Ereignissen, die mittels verschiedener Sensoren erfasst werden, definierbar. Auf diese Weise lassen sich im Graphen logische Abfolgen von Aktionen und Ereignissen definieren (wie z.B. die Aufnahme eines Transportguts), die mit der Transportfahrt selbst nur indirekt zusammenhängen. Dies schließt auch die automatische Fehlerbehandlung mit ein.

In großen Anlagen gerät ein solcher Graph recht schnell unübersichtlich. Desweiteren ist zu bemerken, dass viele Teilgraphen – abgesehen von ihrer örtlichen Anordnung – vollkommen identisch sind (z.B. Ladestationen oder Lastübergabestationen). Aus diesem Grund ist der HTFM hierarchisch ausgeprägt. Bereiche können sowohl nach ihrer Funktion als auch nach anderen sinnvollen Kriterien gruppiert werden. Die Möglichkeit, Gruppen von parametriesierten Schablonen abzuleiten, die eine einmal definierte Handlungsabfolge wiederverwendbar machen, führt zu einer Vereinfachung des in XML-Notation repräsentierten Graphen und zu einer Aufwandsminderung bei der Erstellung.

Gruppen dienen gleichzeitig auch der Betriebsmittelverwaltung. Zum einen können hiermit Kreuzungsbereiche und Engstellen mittels festzulegender maximaler Nutzeranzahl kontrolliert werden. Zum anderen ist eine Verknüpfung mit Peripheriegeräten möglich. Die Ansteuerung der Peripherie wie Lastübergabestationen, Ladestationen, Automatiktüren, Aufzüge, Warngeräte (Lautsprecher,Lichter) ist im TaPAC mittels Plugins realisiert worden. Diese werden anlagenabhängig erstellt und lesen die notwendigen Parameter direkt aus dem HTFM aus, so dass bereits in dieser Datei die örtlichen und logischen Zusammenhänge zwischen Fahrzeugen und Peripherie definiert sind. Durch die globale Sicht auf alle Betriebsmittel ist eine vorausschauende Belegung und Bereitstellung der Resourcen möglich, wodurch sich die Fahrzeiten nochmals reduzieren. Hierfür wurde das Verfahren ORANGE (optimized resource allocating dynamic routing engine) entwickelt.

4 Auftragsvergabe

4.1 Anforderungen

Die in einem FTS auftretenden Transportaufträge sind definiert durch einen Startpunkt und einen Endpunkt. Am Startpunkt befindet sich eine Nutzlast, die zum Endpunkt transportiert werden soll. Trifft ein Transportauftrag im System ein, hat das zwei Fahrten eines Fahrzeugs zur Folge. Zuerst führt ein Fahrzeug eine **Holfahrt** aus, bei der das Fahrzeug zum Startpunkt des Auftrags fährt, um die Nutzlast abzuholen. Nach Aufnahme der Nutzlast führt das Fahrzeug eine **Transportfahrt** aus, indem es die Nutzlast zum Ziel fährt, und dort absetzt. Nach Absetzen der Nutzlast ist der Transportauftrag beendet und das Fahrzeug ist wieder verfügbar.

Um maximale Flexibilität bei der Aufgabenverteilung auf die Fahrzeuge zu erhalten, muss der Algorithmus zur Aufgabenverteilung vier Voraussetzungen erfüllen.

global Einfache Steuerungen für automatische Transportsysteme, wie sie z.B. in [5] vorgestellt wurden, vergeben Aufträge an Fahrzeuge, nachdem eine lokale Auswertung der Situation vorgenommen wurde. Dabei kommt es zu Zuweisungen zwischen Fahrzeugen und Aufträgen, die aus der lokalen Sicht heraus sehr gut sind. Für das Gesamtsystem kann es jedoch bessere Lösungen geben, die auch Verkettungen von Aufträgen berücksichtigen. Hierzu ist es allerdings notwendig zu jedem zeitpunkt sämtliche Aufträge und alle Fahrzeuge zu berücksichtigen.

dynamisch Es gibt auch in bisherigen FTS verschiedene Ausführungen von dynamischen Aufgabenverteilungen. Die in [2] vorgestellten „cues" stoßen beispielsweise eine Neuberechnung der Aufgabenverteilung an. Letztendlich können aber viele Ursachen zu einer Veränderung im System führen, so dass die Definition solcher Ereignisse sehr aufwändig werden kann. Veränderungen der Gesamtsituation können z.B. auftreten, wenn ein Fahrzeug auf seinem Weg zu einem Ziel aufgehalten wird. Dies können kurzfristige Verzögerungen sein, verursacht durch von anderen Fahrzeugen belegte Ressourcen, oder länger andauernde Verzögerungen durch Verklemmungen oder technische Ausfälle. Aus diesem Grund wird vorgeschlagen, die Aufgabenverteilung auszuführen, wenn ein neuer Auftrag hinzukommt, ein Fahrzeug frei wird und zusätzlich periodisch, um weitere Veränderungen der Gesamtsituation zu beachten.

verdrängend Ist eine Aufgabenverteilung nicht verdrängend, so kann sie einen einmal zugewiesenen Auftrag dem Fahrzeug nicht mehr wegnehmen. Durch neu eintreffende Aufträge oder andere Veränderungen im System kann es für das Gesamtsystem jedoch vorteilhaft sein, eine Holfahrt oder sogar eine Transportfahrt abzubrechen, und die Zuordnungen von Aufträgen auf die Fahrzeuge neu zu verteilen.

multi-attributive Bewertung Einige FTS legen sich nur auf ein Kriterium fest, die Zuordnung zwischen Fahrzeug und Auftrag zu bestimmen. Sie werden dadurch starr und unflexibel. Mit solchen Verfahren lässt sich meist nur eine bestimmte Situation im System positiv oder negativ bewerten. Oft gibt es aber mehrere Kriterien, die zu beachten sind. Mit multi-attributiven Bewertungen können mehrere Bewertungskriterien in die Gesamtbewertung einer Zuordnung eingehen und dadurch verschiedene Situationen differenziert bewerten. Hierdurch sind Anpassungen an verschiedene Situationen und Erweiterungen sehr viel einfacher umzusetzen.

4.2 Das Verfahren Magycs (multi-attributed, global, dynamic, supplanting)

Eine mulit-attributive Bewertung kann nach der **simple additive weighting Methode** erfolgen. Eine Zuordnung A_i (Zuordnung eines Auftrags i an ein Fahrzeug) wird mit jeder einzelnen Bewertungsfunktion j bewertet. Die Ergebnisse x_{ij} ergeben dann zusammen die Bewertung für eine Zuordnung. Die beste Bewertung aller Zuordnungen wird ausgewählt: $A_{max} = \{A_i \mid max_i \sum_{j=1}^{J} w_j x_{ij}\}$. Die Interpretationen der Bewertungsfunktionen müssen gleichwertig sein, d.h. für eine hohe Bewertung auch einen hohen Wert liefern. Desweiteren sollten sie streng monoton wachsen, um eindeutige Ergebnisse zu liefern und ein ständiges Wechseln der (gleichwertigen) Zuordnungen zu verhindern.

Verdrängungen und die dynamische Anpassung an relevante Änderungen im System können erreicht werden, indem zu jedem Planungszeitpunkt zunächst sämtliche Zuordnungen zwischen Fahrzeugen und Aufträgen gelöscht werden, um dann die Zuordnungen wieder vollständig neu zu berechnen.

Die wichtige globale Sicht auf alle Fahrzeuge und Aufträge sieht vor, dass nicht jedem Fahrzeug genau ein Auftrag zugeordnet wird, sondern vielmehr eine Kette von Aufträgen. Die Verteilung der Aufträge auf die Fahrzeuge wird nach folgenden Regeln durchgeführt:

Zunächst sind sämtliche Zuordnungenbewertungen zwischen Fahrzeugen und Aufträgen zu berechnen. Die beste Zuordnung wird durchgeführt. Danach müssen alle Zuordnungen zwischen dem betroffen Fahrzeug und alle anderen Aufträgen neu berechnet werden, da die Berechnungsgrundlagen für dieses Fahrzeug durch die Erstzuordnung verändert wurde. Die Berechungsgrundlagen für die anderen Fahrzeuge ändert sich nicht, so dass hier eine Neuberechnung nicht notwendig ist. Aus den Bewertungen wird wieder die beste Zuordnung ausgewählt und übernommen. Das Verfahren wird so lange wiederholt, bis allen Fahrzeuge mindestens ein Auftrag zugewiesen wurde oder kein Auftrag mehr übrig ist. Der Aufwand für diese Berechnung kann mit $O(n \cdot m + m^2)$ oder vereinfacht mit $O(m^2)$ angegeben werden, da die Aufgabe erst mit $n < m$ interessant wird.

5 Evaluierung des Verfahrens

Für die Bewertung des entwickelten Ansatzes wurden Daten einer realen Krankenhausanlage verwendet. Die Anlage wird derzeit mit einer bereits bestehenden zentralen Leitsteuerung betrieben, so dass sowohl betriebsrelevante Datensätze (Transportaufgaben über die Zeit) und Ergebnisstatistiken zur Verfügung stehen [6].

Die Verifikation des entwickelten Systems musste in einer simulierten Anlage erfolgen, da eine interessante reale Anlage mit mehr als einem Fahrzeug nicht zur Verfügung stand.

Die betrachtete Anlage hat eine Ausdehnung von ca. 120 auf 60 Meter. Sechs Fahrzeuge transportieren darin Container zwischen ca. 50 Abhol- und Ablieferstationen. Sie bewegen sich in insgesamt 6 Stockwerken, die durch 5 Aufzüge verbunden sind. Im System sind sechs Ladestationen vorhanden. Das FTS führt täglich ca. 300 Transporte durch.

5.1 Bewertungsfunktion

Als Bewertungskriterien wurden die geschätzte Zeit bis Auftragsende, die Wartezeit des Auftrags und eine Strafe für das Verdrängen berücksichtigt. Letzteres Kriterium sollte lediglich ein ständiges Springen von Aufträgen zwischen zwei gleichwertigen Fahrzeugen verhindern. Als Gesamtbewertungsfunktion ergab sich daher:

$$A^* = w_1 \frac{q}{t_i} + w_2 e^{pa_i} + w_3 \begin{cases} 0 & \text{falls keine Holfahrt verdrängt wird} \\ -1 & \text{falls eine Holfahrt verdrängt wird} \\ -2 & \text{falls zwei Holfahrten verdrängt werden} \end{cases}$$

mit t_i als Zeit bis Auftragsende und ai der Wartezeit des Auftrags, wobei als Paramter die Werte w_1, w_2, w_3 als Gewichtungsfaktoren der Kriterien und q und p als Skalierungsfaktoren mit in die Gleichung eingehen.

5.2 Experimente

Es lagen Datensätze einer durchschnittlichen Woche vor. Diese wurden um Transportfahrten bereinigt, die durch Systemfehler verfälscht wurden. Die Übereinstimmung zwischen Simulation und realer Anlage konnte durch den Vergleich von Einzelfahrten ohne Einfluss anderer Fahrzeuge und Transportaufträge verifiziert werden.

Die sieben Tage wurden mehrfach simuliert, wobei die in den Fahrzeugsimulationen eingebauten Sensorrauschwerte zur Folge hatten, dass die Ergebnisse der verschiedenen Durchläufe nicht exakt übereinstimmten.

Das neue Verfahren schnitt hinsichtlich der Auftragszeit um 25% besser ab, als das bestehende System. Als Auftragszeit wurde die Zeit zwischen dem Auftreten eines Transportbedarfs und der Ablieferung am Zielort berechnet. Das bestehende System betrachtet bei der Zuordnung der Fahrzeuge leidglich eine beste Lösung (lokale Sicht). Eine Verdrängung von Aufträgen ist nicht möglich, so dass auch eine dynamische Anpassung nicht erfolgt.

Um den Einfluss der fünf Parameter (w1,w2,w3,q,p) zu erfassen, wurde ein typischer, aber stark verkürter Datensatz erzeugt und mit verschiedenen Parametereinstellungen ausführlicher getestet. Es stellte sich heraus, dass der Einfluss der Parameter als sehr gering zu bezeichnen ist. Die besten Parametereinstellungen wichen im Vergleich zu den anfangs durch Überlegungen gewählten lediglich um 1%-Punkt im Vergleich zum Originalsystem ab.

6 Zusammenfassung

Mit dem Ziel eine sehr flexible Grundlage für die Definition und Verwaltung von Fahrerlosen Transportsystemen zu entwickeln und mit dem Blick auf die konkrete Anwendung eines FTS im Krankenhaus, entstand das in XML kodierte, erweiterbare Schema des HTFM. Darauf basierend wurde bereits eine Fahrzeugsteuerung und auch eine zentrale Leitsteuerung realisiert. Die Leitsteuerung verwendet ein neues Verfahren zur Zuweisung von Transportaufträgen an Fahrzeuge, welches durch die Eigenschaften global, dynamisch, verdrängend und multi-attributiv zumeist zu einer optimalen Betriebsmittelnutzung führt.

Experimente, die auf Basis von realen Daten in einer Simulation durchgeführt wurden, haben gezeigt, dass diese Methoden sowohl bei der Definition einer Anlage als auch im Betrieb signifikante Vorteile im Vergleich zu einem herkömmlichen FTS zeigen.

Literaturverzeichnis

1. C.M. Klein and J. Kim. Agv dispatching. *The International Journal of Production Research*, 34(1):95–110, March 1996.
2. Leslie D. Interrante and Daniel M. Rochowiak. Active rescheduling for automated guided vehicle systems, 1994.
3. S. Reveliotis. Conflict resolution in agv systems. *IEEE Transactions*, 32(7):647–659, 2000.
4. Tharma Ganesharajah, Nicholas G. Hall, and Chelliah Shriskandarajah. Design and operational issues in agv-served manufacturing systems. 76:109–154, September 1998.
5. Pius J. Egbelu and Jose M.A. Tanchoco. Characterization of automatic guided vehicle dispatching rules. *The International Journal of Production Research*, 22(3):359–374, August 1984.
6. Johannes Fottner. FTS-Einsatz in der Krankenhauslogostik. *Fahrerlose Transportsystem (FTS) und mobile Roboter*, Franhofer IPA Technologieforum F115:59–68, April 2005.

Autonom navigierende Fahrerlose Transportsysteme in der Produktion

Klaus Feldmann, Wolfgang Wolf

Lehrstuhl für Fertigungsautomatisierung und Produktionssystematik (FAPS)
Universität Erlangen-Nürnberg, Egerlandstr. 7–9, 91058 Erlangen

Zusammenfassung. Die zunehmende Forderung nach höherer Flexibilität an die Produktion zwingt die Unternehmen innovative Systemlösungen einzuführen. Die neuen Anforderungen sind sowohl in der Herstellung variantenreicher Produkte mit niedrigen Stückzahlen, als auch in der schnellen Lieferung und Bereitstellung von Gütern (Just-in-Time) erkennbar. Die erforderliche Dynamik ist insbesondere im Materialflusssystem zu berücksichtigen, das hierbei den hohen Anforderungen hinsichtlich einer optimierten Auslastung der Stationen genügen und eine hohe Gesamtverfügbarkeit gewährleisten muss, um nachhaltige Lieferverzögerungen zu vermeiden. Diese Forderungen fließen bei diesem Beitrag in ein neuartiges Steuerungskonzept für Fahrerlose Transportsysteme (FTF) ein, das zudem eine schnelle Inbetriebnahme, hohe Flexibilität und die unmittelbare Anpassung an sich ständig verändernde Produktionsbedingungen gewährleistet. Die Konzepte werden an einem eigens konstruierten und entwickelten Fahrerlosen Transportfahrzeugs umgesetzt und im Labor validiert.

1 Einleitung

Die unmittelbare Anpassung der Fahrerlosen Transportfahrzeuge mit kurzen Inbetriebnahmezeigen an die veränderten Produktionsbedingungen bzw. das veränderte Produktionsumfeld ist mit bisherigen Steuerungslösungen nicht zu erzielen. Analog zu allgemeinen Steuerungsstrukturen für Anlagen werden nach heutigem Stand der Technik auch Fahrerlose Transportfahrzeuge von einer zentralen übergeordneten Koordinierungseinheit überwacht [1,2]. Die Fahrzeuge verfügen nur über eine eingeschränkte Entscheidungsgewalt, die Fahrwege sind fest definiert und die Routen werden von zentraler Stelle vorgegeben. Befindet sich im Fahrweg ein Hindernis, so können Staus entstehen, die nur dann umgangen werden können, wenn Ausweichrouten existieren. Service-Fahrzeuge in Kliniken, in öffentlichen Einrichtungen oder im privaten Bereich sind mit Mechanismen zur Abtastung der Umgebung und zum automatischen Ausweichen von Hindernissen ausgestattet [1]. Allerdings sind in Produktionen solche Mechanismen noch nicht umgesetzt worden. Darüber hinaus stellen zentrale Steuerungssysteme einen Single-Point-of-Failure dar, deren Ausfall einen Totalausfall des gesamten Transportsystems bedeuten würde. Fahrerlose Transportsysteme nach dem heutigen Stand der Technik bieten nicht die notwendige Dynamik, um in Produktionen mit stark variierenden Bedingungen und hochfrequenten Umbaumaßnahmen wirtschaftlich eingesetzt werden zu können.

2 Neues Konzept für Fahrerlose Transportfahrzeuge

Um den oben beschriebenen Nachteilen zentralistischer Steuerungsansätze entgegen zu wirken und maximalen Durchsatz beim Materialtransport – selbst im Fehlerfall – zu gewährleisten, wird im vorliegenden Ansatz eine verteilte, dezentrale Steuerungsstruktur umgesetzt. Um der Forderung nach mehr Dynamik in Produktionssystemen nachzukommen sollen die Fahrerlosen Transportfahrzeuge autark Entscheidungen treffen und Bahnen selbständig planen und berechnen können. Dadurch entfällt die manuelle Festlegung der Bahnen im Raum und die Validierung dieser mittels Testfahrten. Somit wird zum einen die Akzeptanz Fahrerloser Transportsysteme in der Fertigung durch eine vereinfachte Handhabung gestärkt und zum anderen der wirtschaftliche Anreiz geschaffen, diese selbst in hochflexiblen Fertigungen einzusetzen.

Abb. 1. Abbildung FTF mit Transportmodul

3 Mechanischer Aufbau

Am Lehrstuhl wurde ein Fahrerloses Transportfahrzeug (FTF) entwickelt, das auf Grund seines modularen Aufbaus unterschiedliche Funktionsmodule aufnehmen kann und somit sowohl Transportaufgaben wahrnehmen, sowie als modulare Bearbeitungsstation mit aufgesetztem Gerät agieren kann. Die derzeitige Aufgabe des FTF ist der Transport von Werkstückträgern zwischen Anlagen und Zellen (Abb. 1). Unter Abb. 2 ist der Grundriss des Fahrzeuges dargestellt. Mit den gelenkten Antriebsmotoren ist das Fahrzeug sogar in der Lage seitlich zu fahren. Zur Bestimmung der Position im Raum verfügt das Fahrzeug über einen Laserscanner, der mit Hilfe von bekannten Reflektormarkierungen an den Wänden im Raum die Position und Drehlage des Fahrzeuges ermitteln kann. Da das Fahrzeug gleichförmig sowohl nach vorne, als auch nach hinten fahren kann, sind in beiden Richtungen Laserscanner angebracht, die in der Lage sind Hindernisse zu erkennen. Infrarot-Sensoren an den Seiten gewährleisten die Erkennung von Hindernissen beim Seitwärtsfahren.

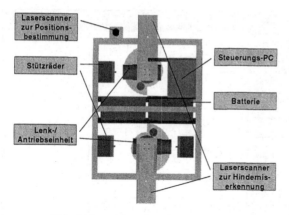

Abb. 2. Grundriss mechanischer Aufbau

4 Konzeption eines verteilten Steuerungsansatzes

Neben der Bahnberechnung und Fahrzeugführung übernimmt die Steuerung des FTF alle Entscheidungs- und Koordinierungsaufgaben. Dadurch existiert keine zentrale Koordinierungsinstanz, die alle Fahrzeuge überwacht, Kollisionen erkennt und Aufträge verteilt. In diesem Fall liegt ein echt verteiltes System autonomer Fahrzeuge vor, die sich untereinander hinsichtlich der Auftragsvergabe und Kollisionserkennung abstimmen müssen. Der Vorteil autonomer Systeme liegt in der einfachen Handhabbarkeit und zusammen mit der autonomen Navigation in der schnellen Inbetriebnahme neuer Fahrzeuge innerhalb der Produktion. Neue Fahrzeuge können sich selbständig im Netzwerk registrieren und erkennen welche weiteren FTF sich in der Produktion befinden. Außerdem können sie auch unmittelbar von diesen alle relevanten Informationen zur Navigation beziehen. Da jedes FTF seinen Aufbau hinsichtlich der Maße und der Aufnahme von Werkstückträgern und sein eigenes spezifisches Verhalten hinsichtlich des Lenkverhaltens und der Geschwindigkeit kennt, ist es ohne weitere Konfiguration in der Lage, Transportaufträge selbständig durchzuführen.

4.1 Kartographie, Hinderniserkennung und Kollisionsvermeidung

Das Fahrzeug ist in der Lage sich selbständig die Bahn durch den Fertigungsbereich anhand einer Karte zu berechnen. Diese Karte wird dem FTF vorgegeben, indem mit Hilfe eines Handwagens, der ebenfalls einen Laserscanner zur Positionsbestimmung besitzt, die Halle vermessen wird. In dieser Karte werden alle Zellen und ihre Bahnhöfe registriert (Abb. 3). Die Karte ist dabei in verschiedene Zonen unterteilt:

- In grüne Zonen darf sich das FTF völlig frei bewegen.
- Gelbe Zonen sollten nur dann vom FTF befahren werden, wenn sich z.B. im grünen Bereich ein Hindernis befindet.
- Rote Zonen markieren Hindernisse oder sonstige verbotene Bereiche, die das FTF niemals befahren darf.

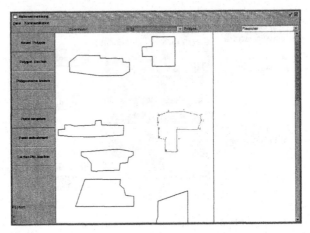

Abb. 3. Erster Prototyp zur Vermessung der Produktionshalle

Die spezielle Anforderung beim Entwurf der Datenstruktur zur Ablage der Karteninformationen ist das schnelle Auffinden von Hindernissen innerhalb dieser. Dies ist notwendig beim Abgleich der virtuellen Karte mit den Messdaten der Laserscanner aus der realen Welt. Somit ist das Fahrzeug in der Lage selbst neue bzw. temporäre Hindernisse zu erkennen und diese in die Karte für eine neue Planung der aktuellen Route und für spätere Bahnplanungen einzutragen. Die temporären Hindernisse erhalten in der Karte eine Lebensdauer, damit nach einer bestimmten Zeit der ggf. blockierte Fahrweg nochmals überprüft wird, ob das Hindernis noch immer den Weg versperrt.

Zur Kollisionsvermeidung zwischen Fahrzeugen ist die Erkennung auf Kollisionskurs beweglichen FTF mittels der Laserscanner zur Hinderniserkennung geplant. Dabei muss zunächst zwischen statischen und beweglichen Hindernissen unterschieden werden. Die Fahrzeuge werden hierbei aus der Menge beweglicher Hindernisse anhand ihrer Maße herausgefiltert. Befindet sich ein solches auf Kollisionskurs, wird eine Mitteilung an alle Fahrzeuge über WLAN versendet, um das betreffende FTF zu identifizieren. Mit Hilfe von Verhandlungsmechanismen sind diese dann in der Lage anhand der Lieferfrist und der Priorität des aktuellen Transportauftrages zu entscheiden, welches FTF zuerst passieren darf. Alternativ können Ausweichrouten bestimmt werden, wobei hier von den Ergebnissen aus dem Bereich der Service-Robotik profitiert werden kann [4]. Dabei entstehende Konflikte müssen im Modell berücksichtigt und anhand eingehender Tests validiert werden.

4.2 Bahnplanung

Die Aufgabe der Bahnplanung besteht darin, den kürzesten Weg zwischen zwei Punkten in der Fertigung anhand der Karteninformationen zu bestimmen. Das Ergebnis dieser Berechnung ist eine Sequenz von aufeinanderfolgenden Vektoren, die der Bahnverfolgung zur Verfügung gestellt werden. Zur Berechnung dieser Bahn gibt es verschiedene Ansätze, die derzeit noch untersucht werden. Erste Algorithmen sind in einer Simulation umgesetzt worden (Abb. 4). Eine erste geeignete Auswahl ist im Folgenden skizziert:

- Zellenmethode
  Die Produktionshalle wird in gleich große Zellen unterteilt, deren Größe maximal der des Fahrzeuges entspricht. Die Zellen werden ausgehend vom Zielpunkt mit Distanzinformationen gefüllt, die zur Suche des kürzesten Weges von Startpunkt ausgehend herangezogen werden. Zudem ist eine Glättung der Bahn erforderlich.
- Bahnplanung mittels topologischer Karten
  Ein Nachteil von Gitterkarten ist der enorme Speicherbedarf und die hohe Rechenleistung, die zur Pfadsuche erforderlich ist. Dieser Nachteil kann durch den Einsatz von topologischen Karten, wie Voronoi-Diagrammen [5], behoben werden. Diese bestehen aus Punktmengen, bei denen jeder Punkt von mindestens zwei unterschiedlichen Hindernissen gleich weit entfernt ist. Mit deren Hilfe kann die globale Bahnplanung erheblich beschleunigt werden. [6]
- Potentialfeld
  Bei der Potentialfeldmethode besteht ein virtuelles Potentialgefälle vom Start- zum Zielpunkt, wobei Hindernisse mit nach außen zeigenden virtuellen Kräften versehen sind und somit das Fahrzeug davon abstoßen. [7]

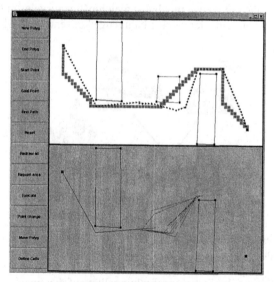

Abb. 4. Simulation der Bahnplanung am PC mit Karte der „realen" Welt (oben) und der Karte im FTF (unten)

In der Simulation werden die Bahnplanungsalgorithmen exemplarisch implementiert. Dabei wird eine Karte gemäß der realen Welt mit permanenten und temporären Hindernissen durch den Benutzer definiert (Abb. 4). Während der Simulation fährt ein FTF entlang der Bahn und tastet die Umgebung mit einem virtuellen Sensor ab. Wird ein Hindernis erkannt, prüft eine Routine, ob es sich um ein bekanntes oder neues Hindernis handelt, wobei im letzteren Fall die Bahnplanung eine Alternativroute mit den neuen Karteninformationen berechnet.

4.3 Agentenmechanismen

Zur Vergabe von Transportaufträgen mit bestmöglicher Auslastung des Fahrerlosen Transportsystems werden Agentenmechanismen [8,9] eingesetzt, die zur Laufzeit Verhandlungen zwischen den Fahrzeugen und den Stationen vornehmen (Abb. 5). Liegen Transportaufgaben bei der Station vor, teilt die dortige lokale Steuerung den Auftrag allen Fahrzeugen mit. Die Fahrzeuge geben daraufhin entsprechend den bereits eingelasteten Aufträgen in ihrer Warteschlange, der Entfernung von der Station und der benötigten Zeit für den Transport ein Angebot in Zeit und Kosten ab. Die Berechnung der benötigten Zeit erfolgt anhand von Tabellenwerten, die den Stationen initial zur Verfügung stehen, bzw. die während der Produktion mit reellen Messdaten aktualisiert werden. Parallel dazu werden die benötigten Betriebskosten errechnet, damit der Agent entscheiden kann, ob ein hoch priorer Auftrag ggf. mit wenig Gewinn produziert wird, um die Frist einzuhalten, oder üblicherweise versucht den Gewinn zu maximieren. Der Entscheidungsalgorithmus ist im Werkstückagenten hinterlegt, der das Werkstück begleitet und von Station zu Station migriert. Aus einer Menge von Geboten wird der Zuschlag für das günstigste erteilt. Mit diesem Mechanismus wird gewährleistet, dass alle Fahrzeuge möglichst gleichmäßig ausgelastet sind und sich die Beanspruchung nicht auf ein oder wenige Fahrzeuge konzentriert [10].

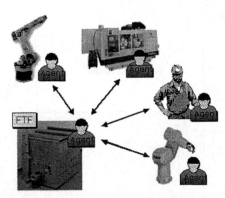

Abb. 5. Direkte Interaktion zwischen dem FTF und den Bearbeitungsstationen über eine gemeinsame Agentenplattform

Durch die Verteilung der Steuerungsfunktionalität auf die einzelnen Fahrzeuge und dem Fehlen eines Single-Point-of-Failures weist das Fahrerlose Transportsystem eine sehr niedrige Gesamtausfallwahrscheinlichkeit auf. Somit können Ausfälle einzelner Fahrzeuge automatisch kompensiert werden, da sich diese an der Ausschreibung nicht beteiligen.

5 Zusammenfassung

Mit dem vorliegenden Fahrzeug können geeignete Mechanismen zur autonomen Navigation, zur selbständigen Entscheidungsfindung und zur Verhandlung mit stationären und mobilen Agenten zur Lösung des gemeinsamen Ziels untersucht und entwickelt werden. Das Fahrzeug kennt seinen Aufbau, seine Eigenschaften und ebenso seine Umgebung, die es mittels Sensoren abtasten kann. Ein dezentrales, hoch entwickeltes Steuerungssystem kann die Akzeptanz von Fahrerlosen Transportfahrzeugen stärken und unterstützt den Einsatz in Fertigungsbereichen mit sich ständig verändernden Bedingungen.

Durch den Einsatz von Agentenmechanismen wird ein optimierter Materialfluss mit gleichzeitig hoher Flexibilität gewährleistet, die selbst im Falle von Teilausfällen eine optimale Auslastung erzielen. Mit diesem Vorhaben wird ein Beitrag dem allgemeinen Vorhaben zur Entwicklung von steuerungstechnisch hoch entwickelten Geräten geleistet, die ein entscheidendes Maß an Flexibilität und Ausfallsicherheit aufweisen und gleichzeitig durch die vernetzte Zusammenarbeit einen hohen Durchsatz zum Ziel haben.

Literatur

1. Schmitt M: Ein Leitstand zur Einsatzplanung und Überwachung mobiler Roboter. In: Schmidt G (Hrsg.), et al.: Autonome Mobile Systeme 1999. Springer, Berlin, 1999.
2. Open Transportation Control System (OpenTCS): www.opentcs.de. (12.09.2005).
3. Graf B: Zuverlässige Zielführung mobiler Serviceroboter durch dynamische Hindernisumfahrung in Echtzeit. In: Schraft (Hrsg.), et al.: Fraunhofer-Institut für Produktionstechnik und Automatisierung IPA – Fahrerlose Transportsysteme (FTS) und mobile Roboter. Fraunhofer IPA Technologieforum, Stuttgart, 29. April 2005.
4. Jäger M: Collision Avoidance for Cooperating Cleaning Robots. In: Schmidt G (Hrsg.), et al.: Autonome Mobile Systeme 2001. Springer, Berlin, 2001.
5. Aurenhammer F, Klein R, Diagrams V: Handbook of Computational Geometry. Elsevier Science Publishers B. V. North-Holland, Amsterdam, 2000.
6. Mojaev A, Zell A: Aufbau topologischer Karten und schnelle globale Bahnplanung für mobile Roboter. In: Schmidt G (Hrsg.), et al: Autonome Mobile Systeme 2000. Springer, Berlin, 2001.
7. Latombe, J-C.: Robot Motion Planning. Kluwer Academic Publishers, Massachusetts, 2004.
8. Beetz M: Plan-Based Control of Robotic Agents. Springer, Berlin, 2002.
9. The Foundation for Intelligent Physical Agents: www.fipa.org. (12.09.2005).
10. Feldmann, K, Wolf W, Weber M: Performance Analysis of Distributed Strategies for Material Flow Control of Modular Production Systems Based on Discrete Event Simulation. In: Schoop R (Hrsg.), et al.: 2nd IEEE International Conference on Industrial Informatics INDIN 2004, Fraunhofer IRB Verlag, Berlin, 2004.

Sensorgestützte Bewegungssynchronisation von Operationsinstrumenten am schlagenden Herzen

Kathrin Roberts, Gábor Szabó, Uwe D. Hanebeck

Universität Karlsruhe (TH), Institut für Technische Informatik, Lehrstuhl für Intelligente Sensor-Aktor-Systeme, 76128 Karlsruhe
Universitätsklinik Heidelberg, Abteilung Herzchirurgie, 69120 Heidelberg

Zusammenfassung. Offene oder minimal invasive Operationen am schlagenden Herzen erfordern von dem Chirurgen eine hohe Konzentrationsfähigkeit über einen längeren Zeitraum. Daher ist es für den Chirurg sehr hilfreich durch ein robotergestütztes Chirurgiesystem unterstützt zu werden, das die Instrumente im Interventionsareal mit der Herzbewegung synchronisiert. Um eine Bewegungskompensation durchzuführen, muss ein Mechanismus gefunden werden, der aufgrund einer Prädiktion der Herzbewegung die Instrumente nachführt. Für die Prädiktion der Herzbewegung wird in diesem Artikel ein Verfahren zum Entwurf eines stochastischen 3D-Bewegungsmodells für die Herzoberfläche gezeigt. Ein Schätzer nimmt dieses Modell als Grundlage und verwendet die verrauschten Sensormessungen von Landmarken der Herzoberfläche um die Herzoberflächenbewegung zu prädizieren.

1 Einleitung

Die häufigste Herzkrankheit ist die Erkrankung der Herzkranzgefäße, welche durch Bypass-Operationen gelindert werden. In der Herzchirurgie werden Bypass-Operationen mit der konventionell offenen Operation oder mit der minimal invasiven Operation auf zwei Arten durchgeführt: zum einem am stillgelegten Herzen mit dem Einsatz der Herz-Lungen-Maschine, zum anderen am schlagenden Herzen. Für den Patienten ist eine Operation am schlagenden Herzen mit weniger Risiken (wie z.B. Schlaganfälle, Infektionen) verbunden als bei einem stillgelegten Herzen, wo die Herz-Lungen-Maschine den Kreislauf des vorher unterkühlten Patienten übernimmt. Bei den vorzuziehenden Operationen am schlagenden Herzen wird das Herz mechanisch stabilisiert und der Operateur muss der verbleibenden Bewegung konzentriert manuell folgen. Bei längeren Bypass-Operationen wäre es für den Chirurgen daher sehr vorteilhaft mit einem Master-Slave-System unterstützt zu werden, welches die Operationsinstrumente im Interventionsareal mit der Herzbewegung synchronisiert. Bei den am häufig benutzten Master-Slave-Systeme in der minimal invasiven Chirurgie, wie z.B. da Vinci und ZEUS, wurde bist jetzt noch keine autonome Bewegungssynchronisation für Operationen am schlagenden Herzen realisiert. Die Hauptaufgabe für die Umsetzung der Synchronisation ist es ein Mechanismus für die Prädiktion der Herzbewegung zu finden. Dazu gibt es bereits wissenschaftliche Arbeiten, z.B. [1,2,3]. Die Verfahren von [1,2,3] können leider nicht auf Extrasystolen des Herzens reagieren und sind zumindest bei [2,3] auf periodische Bewegungen des Herzens angewiesen. Um die obigen Nachteile zu umgehen, wird in diesem Artikel für die Prädiktion der Herzbewegung ein modellbasierter Ansatz verwendet. In unserem Ansatz entwerfen wir ein stochastisches Bewegungsmodell der Herzoberfläche, welche dabei durch eine oder mehrere verknüpft schwingende Membranen (siehe z.B [4]) modelliert wird. Die Membranen weisen linear oder nichtlinear elastisches Materialverhalten auf und werden als ein dynamisch verteiltes Phänomen gesehen, dessen Verhalten mit partiellen Differentialgleichungen (PDGL) beschrieben wird. Zur Beschreibung des Schwingungsverhaltens der einzelnen Membran kommt die Wellengleichung zum Einsatz. Die Membranen werden je nach Bewegung mit bestimmten Anregungsfunktionen geformt. In diesem Artikel wird eine einzelne rechteckig schwingende Membran mit linearem Materialverhalten betrachtet, mit der es möglich ist einen Teilbereich der Herzoberfläche zu modellieren.

Der Artikel ist in mehrere Abschnitte untergliedert. Im nächsten Abschnitt wird erläutert aus welchen Komponenten ein robotergestütztes Chirurgiesystem zur Bewegungssynchronisation mit dem schlagenden Herzen bestehen kann. In Abschnitt 3 werden folgende Teilaspekte des Herzoberflächemodells erläutert: In Abschnitt 3.1 wird das stochastische Bewegungsmodells für eine rechteckig schwingende Membran hergeleitet, in Abschnitt 3.2 wird das Prädiktionsergebnis des Modells mit einer Finite-Elementen-Simulation verglichen und in Abschnitt 3.3 wird diskutiert, wie Messungen der sich bewegenden Herzoberfläche mittels Sensoren ermittelt werden können. In Abschnitt 4 werden die wichtigsten Punkte noch einmal zusammengefasst und ein Ausblick gegeben.

2 Problemformulierung

Um die Vision, mit einem robotergestützten Chirurgiesystem am schlagenden Herzen operieren zu können, zu realisieren, müssen die folgenden Komponenten zusammenspielen: Eine Master-Slave-Einheit mit Manipulatoren und einer haptischen Schnittstelle, eine Nachführeinheit für die Manipulatoren, ein Kamerasystem angebunden an eine Bildverarbeitungseinheit, eine Visualisierungseinheit und eine Einheit zur Prädiktion der Herzbewegung. Im Operationsgebiet wird das Kamerasystem platziert, um dem Chirurgen eine Sicht auf die Szene zu geben. Die Bildaufnahmen von dem schlagenden Herzen werden in der Bildverarbeitungseinheit so bearbeitet, dass auf der Visualisierungseinheit für den Chirurgen ein virtuell stillstehendes Herz angezeigt wird. Dies hat den Vorteil, dass der Chirurg wie bei einer Operation am stillstehenden Herzen operieren kann. Zur Visualisierung kann ein Head Mounted Display oder ein Monitor dienen. Das Kernstück des Gesamtaufbaus ist die Prädiktionseinheit, in der ein Modell für die Herzbewegung simuliert wird, um eine Prädiktion der Bewegung durchzuführen. Für die Anpassung des Modells an die reale Herzbewegung muss das Modell ständig Messdaten vom schlagenden Herzen erhalten. Die Bereitstellung der Messdaten wird durch eine Bildverarbeitungseinheit übernommen, welche das Kamerasystem ansteuert und die aufgenommenen Bildsequenzen ständig auswertet. Die Nachführeinheit für die Positionierung der Manipulatoren muss sowohl die Steuerbefehle des Chirurgen als auch die Prädiktionsergebnisse für die Herzbewegung berücksichtigen. Nun kann der Chirurg anhand der Visualisierung des Operationsgebietes die sich synchronisierenden Instrumente über die haptische Schnittstelle steuern. In Abb. 1 findet man eine Gesamtübersicht.

In diesem Artikel konzentrieren wir uns nur auf die Erfassung von Messdaten und auf die Einheit für die Prädiktion der Herzbewegung. Um die verteilte Bewegung des Herzens zu prädizieren, entwerfen wir für einen Teilbereich der Herzoberfläche ein 3D-Bewegungsmodell. Die notwendigen Messdaten für das Modell bestehen z.B. aus verrauschten Messungen für einzelne Landmarken der Herzoberfläche. Eine Anforderung an das Modell ist es daher, dass es die verteilte Oberflächenbewegung mit Hilfe der diskreten Messungen rekonstruieren muss, d.h. auch an Nicht-Messstellen. Wegen der auftretenden Rauschprozesse bei Messungen und auch möglichen Unsicherheiten im Modell wird ein Schätzer verwendet. Der Schätzer kann auf Basis des entworfenen Bewegungsmodells und der erfassten Messdaten aus der Bildverarbeitung die Position für alle Landmarken im betrachteten Teilbereich prädizieren.

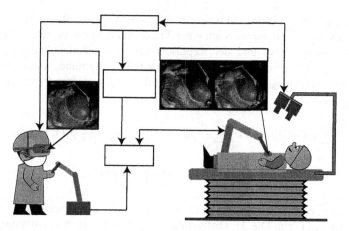

Abb. 1. Operation am schlagenden Herzen mit Hilfe eines Master-Slave-Systems.

3 Herzoberflächenmodell

Für das Bewegungsmodell wird in diesem Artikel eine rechteckig linear schwingende Membran behandelt, mit der ein Teilbereich der Herzoberfläche geschätzt wird. Als Ausgangsbasis für die Beschreibung des physikalischen Verhaltens der Membran dient dabei die zwei-dimensionale Wellengleichung

$$\frac{\partial^2 u}{\partial t^2} - c\left(\frac{\partial^2 u}{\partial x^2} + \frac{\partial^2 u}{\partial y^2}\right) + d\frac{\partial u}{\partial t} = f(x,y,t); \quad 0 \le x \le a, \quad 0 \le y \le b, \tag{1}$$

deren gesuchte Lösungsfunktion u die Bewegung in z-Richtung an der Position (x,y) zum Zeitpunkt t angibt. In (1) ist c ein Materialparameter, du_t der geschwindigkeitsabhängige Dämpfungsterm und die Funktion $f(x,y,t)$ beschreibt die äußere Anregung der Membran. Für (1) sollen die Randbedingungen

$$u(0,y,t) = 0, \quad u(a,y,t) = 0, \quad u_y(x,0,t) = 0, \quad u_y(x,b,t) = 0, \tag{2}$$

gelten. Diese Bedingungen stehen für eine an zwei gegenüberliegenden Rändern fest gelagerten Membran (siehe Abb.3(a)).

3.1 Herleitung der System-, Mess- und Rekonstruktionsgleichung

Um die erforderlichen Gleichungen für das Bewegungsmodell herzuleiten wird das in [6] beschriebene Verfahren auf die lineare PDGL (1) mit den Randbedingungen (2) angewendet. Im ersten Schritt wird mit Hilfe des Separationsansatzes die partikuläre Lösung der homogenen PDGL von (1) ermittelt. Danach wird die partikuläre Lösung an die Randbedingungen (2) angepasst. Man erhält unterschiedliche Lösungen $u_i(x, y, t) = \psi_i(x, y)\alpha_i(t)$ mit den orts- und zeitabhängigen Eigenfunktionen ψ_i und α_i. Die Superposition aller Lösungen mit

$$u(x,y,t) = \sum_{i=1}^{\infty} u_i(x,y,t) \approx \sum_{i=1}^{N} \psi_i(x,y)\alpha_i(t) = \underline{\psi}(x,y)^T \underline{\alpha}(t) \tag{3}$$

ergibt ebenfalls eine Lösung. In unserem Fall wollen wir die Reihe in (3) mit N Eigenfunktionen approximieren. Nach einer Transformation der partikulären Lösung in eine normalisierte Form und der Repräsentation der äußeren Anregung $f(x, y, t)$ durch eine mit N Eigenfunktionen approximierten Fourierreihe, werden beide Kompontenten in Gleichung (1) eingesetzt. Man erhält das in jedem Punkt zu erfüllende Gleichungssystem

$$\sum_{i=1}^{N} \psi_i(x,y)\ddot{\alpha}_i(t) = c\left(\sum_{i=1}^{N} \frac{\partial^2}{\partial x^2}\psi_i(x,y)\alpha_i(t) + \sum_{i=1}^{N} \frac{\partial^2}{\partial y^2}\psi_i(x,y)\alpha_i(t) \right)$$
$$-d\sum_{i=1}^{N} \psi_i(x,y)\dot{\alpha}_i(t) + \sum_{i=1}^{N} \psi_i(p)f_i(t). \tag{4}$$

mit den Fourierkoeffizienten f_i wie in [6]. Man wählt M verschiedene Stützpunkte und setzt sie in (4) ein. Die M Differentialgleichungen werden zusammengefasst und reduziert auf ein Gleichungssystem erster Ordnung der Form

$$\underbrace{\begin{pmatrix} \mathbf{I}_{M\times N} & 0_{M\times N}(5) \\ 0_{M\times N} & \psi \end{pmatrix}}_{E} \underline{\dot{\beta}}(t) = \underbrace{\begin{pmatrix} 0_{M\times N} & \mathbf{I}_{M\times N}(6) \\ c\psi^c & d\psi \end{pmatrix}}_{A} \underline{\beta}(t) + \underbrace{\left(0_{M\times N} \mid \psi\right)}_{B} \underbrace{\begin{pmatrix} 0_{N\times1} \\ \underline{f}(t) \end{pmatrix}}_{u(t)}, \tag{5}$$

$$\text{mit } \underline{\beta}(t) = \begin{pmatrix} \underline{\alpha}(t) \\ \underline{\dot{\alpha}}(t) \end{pmatrix}, \quad \psi = \begin{pmatrix} \psi_1(x_1, y_1) & \cdots & \psi_N(x_1, y_1) \\ \vdots & \ddots & \vdots \\ \psi_1(x_M, y_M) & \cdots & \psi_N(x_M, y_M) \end{pmatrix}, \tag{6}$$

$$\psi^c = \begin{pmatrix} \Delta\psi_1(x_1, y_1) & \cdots & \Delta\psi_N(x, y) \\ \vdots & \ddots & \vdots \\ \Delta\psi_1(x_M, y_M) & \cdots & \Delta\psi_N(x_M, y_M) \end{pmatrix}. \tag{7}$$

Wir nehmen an, dass in (8) Prozessrauschen auftritt, welches wir mit dem additiven Term $\mathbf{D}\underline{w}(t)$ beschreiben. Dabei ist $\mathbf{D}$ eine Matrix und $\underline{w}(t)$ ein mittelwertfreies

Rauschen mit der Kovarianzmatrix $\mathbf{Q}$. Ist $\mathbf{E}$ invertierbar, dann erhält man die lineare Systemgleichung mit dem Zustand $\underline{\beta}(t)$

$$\underline{\dot{\beta}}(t) = \mathbf{E}^{-1}\mathbf{A}\underline{\beta}(t) + \mathbf{E}^{-1}\mathbf{B}\underline{f}(t) + \mathbf{E}^{-1}\mathbf{D}\underline{w}(t). \tag{8}$$

Ist $\mathbf{E}$ aber nicht regulär, dann liegt ein Deskriptorsystem vor. In [6] wurden einige Literaturangaben zur Lösung dieses Problems angegeben. Bei $M > N$ wenden wir unter der Annahme, dass die Spalten von $\mathbf{E}$ und $\mathbf{Q}$ linear unabhängig sind, das Least-Squares-Verfahren an und erhalten die Systemgleichung

$$\underline{\dot{\beta}}(t) = (\mathbf{E}^{T}\mathbf{Q}^{-1}\mathbf{E})^{-1}\mathbf{E}^{T}\mathbf{Q}^{-1}(\mathbf{A}\underline{\beta}(t) + \mathbf{B}\underline{u}(t) + \mathbf{D}\underline{w}(t)). \tag{9}$$

Die lineare Messgleichung setzt sich mit L Messpunkten und dem additiven weissen Rauschwert $\underline{v}_k$ zusammen aus

$$\underline{y}_k = \begin{pmatrix} \psi^{H} & 0_{L \times N} \\ 0_{L \times N} & \psi^{H} \end{pmatrix} \begin{pmatrix} \underline{\alpha}(t_k) \\ \underline{\dot{\alpha}}(t_k) \end{pmatrix} + \underline{v}_k. \tag{10}$$

ψ^{H} hat die Form wie in (9), dabei wird aber M durch L ersetzt. Um den Zustand $\underline{\beta}(t)$ der konzentriert parametrischen Systeme abzuschätzen, wird aufgrund der linearen System- und Messgleichung das lineare Kalman Filter verwendet. Im Falle nichtlinearer System- und Messgleichungen können nichtlineare Filter benutzt werden, z.B. das Erweiterte Kalman Filter oder das Progressive Bayes-Verfahren [4]. Die verwendeten Gleichungen für das lineare Kalman Filter sind in [6] beschrieben. Bei einem überbestimmten System ändert sich lediglich die Berechnung der Kovarianzmatrix im Prädiktionschritt. Sie ist nun

$$\mathbf{C}_{k+1} = (\mathbf{E}^{T}\mathbf{Q}^{-1}\mathbf{E})^{-1}\mathbf{E}^{T}\mathbf{Q}^{-1}(\mathbf{A}\mathbf{C}_{k}\mathbf{A}^{T} + \mathbf{D}\mathbf{Q}\mathbf{D}^{T})\mathbf{Q}^{-1}\mathbf{E}(\mathbf{E}^{T}\mathbf{Q}^{-1}\mathbf{E})^{-1} \tag{11}$$

Mit dem aktuell abgeschätzten $\hat{\underline{\alpha}}_k$ aus dem Zustand, der gleichzeitig den zeitabhängigen Teil der Lösung der PDGL darstellt, können wir die Lösung u für einen beliebigen Punkt (x, y) des abzuschätzenden Systems rekonstruieren mit

$$u(x, y, t) \approx \underline{\psi}(x, y)^{T}\hat{\underline{\alpha}}(t_k). \tag{12}$$

Nachdem nun alle wichtigen Komponenten des stochstischen 3D-Be–wegungs–modells angesprochen wurden, wenden wir uns kurz den Simulationsergebnissen zu.

3.2 Simulationsergebnisse

Um die Prädiktions- und Rekonstruktionsgenauigkeit unseres hergeleiteten Bewegungsmodells zu überprüfen wird die Lösung einer Finiten-Elemente-Simulation abgeschätzt. In der Finite-Elemente Simulation wurde das Bewegungsverhalten für ein Membran mit der PDGL (1) der Größe $a = 5$ und $b = 3$ für das Zeitintervall $[0, 10]$ berechnet. Um die Messungen für unser Modell zu generieren wurde auf die FEM-Lösung ein Rauschen addiert. In der Abb. 2 sind die Funktionsverläufe der FEM-Simulation und der abgeschätzten Auslenkung durch unser hergeleitetes Modell für einem Messpunkt und einen Nicht-Messpunkt dargestellt. Dabei waren $N = 16$, $M = 20$, $L = 10$. Die äußere Anregung hatte die Form $f(x, y, t) = \sin(\pi/5x)\sin(\pi/2t)$ und die Messrate lag bei 5,5 Messungen je Zeiteinheit. In beiden Abbildungen ist zu sehen wie sich das Modell ausgehend von einem anderen Startwert an den wahren Funktionsverlauf anpasst.

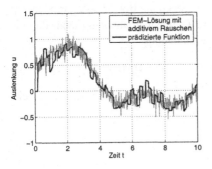
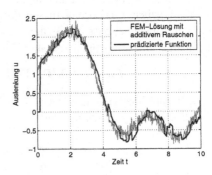

(a) Ergebnis am Nicht-Messpunkt (2.11,2.67)   (b) Ergebnis am Messpunkt (4.36,0.28)

Abb. 2. Prädiktions- und Rekonstruktionsergebnisse für eine schwingende Membran.

3.3 Ermittlung von 3D-Messungen der Herzoberfläche

In der praktischen Anwendung möchten wir später die für das Modell notwendigen 3D-Messungen $\underline{y}_k$ von der Herzoberfläche mit der Hilfe von optischen Sensoren (z.B. Stereo-Kamerasystem) ermitteln. Mit den Kameras verfolgen wir die Lage von natürlichen oder aufgeklebten künstlichen Landmarken (siehe Abb. 3.3) und ermitteln die jeweilige 3D-Position und Geschwindigkeit. Es ist auch denkbar, andere Sensoren zur Messung zur verwenden, z.B. sonometrische Sensoren, Drucksensoren oder Beschleunigungssensoren.

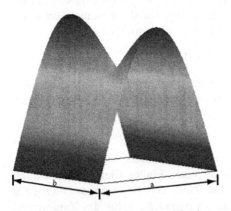
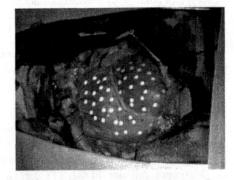

(a) Auslenkung einer rechteckigen Membran   (b) Kameraaufnahme vom Herzen mit aufgeklebten Markern.

Abb. 3. Membran zur Oberflächenmodellierung und Kameraaufnahme des Herzens.

4 Zusammenfassung und Ausblick

Um in der robotergestützten Chirurgie die Instrumente mit dem schlagenden Herzen zu synchronisieren, wurde ein stochastisches Bewegungsmodell für einen Teilbereich der Herzoberfläche hergeleitet. Die Herzoberfläche wird dabei durch eine schwingende rechteckige Membran modelliert. Die gezeigten Prädiktions- und Rekonstruktionsergebnisse der Auslenkung sind sehr vielversprechend und lassen auf eine hohe Praxistauglichkeit hoffen. Die weiteren Forschungsziele sind die Erweiterung der Auslenkung auf die x- und y-Richtung, die Verknüpfung von Membranen und die Betrachtung schwingender Membranen mit nichtlinearem Materialverhalten. Es ist auch denkbar, dass man bei guten Ergebnissen in der Modellverknüpfung auf die mechanische Stabilisierung verzichten kann und somit das Herz in seiner Funktionalität während den Operationen nicht mehr einschränkt.

Literaturverzeichnis

1. Y. Nakamura, K. Kishi, H. Kawakami: Heartbeat synchronization for Robotic Cardiac Surgery. Proceedings of the 2001 IEEE Conference on Robotics and Automation, pp. 2014–2019, 2001.
2. T. Ortmaier: Motion compensation in Minimally Invasive Robotic Surgery. Fortschritt-Berichte VDI, Reihe 17, Nr. 234, 2002.
3. R. Ginhoux, J. Gangloff et al.: Active Filtering of Physiological Motion in Robotized Surgery Using Predictive Control. IEEE Transactions on Robotics, vol. 21, no.1, pp. 67-78, 2005.
4. H. Tao, T. S. Huang: Connected Vibrations: A Modal Analysis Approach for Non-rigid Motion Tracking. Proceedings 1998 IEEE Conference on Computer Vision and Pattern Recognition, pp. 735–740, 1998.
5. U. D. Hanebeck, K. Briechle, A. Rauh: Progressive Bayes: A New Framework for Nonlinear State Estimation. Proceedings of SPIE, Volume 5099, AeroSense Symposium, pp. 256–267, 2003.
6. K. Roberts, U. D. Hanebeck: Prediction and Reconstruction of Distributed Dynamic Phenomena Characterized by Linear Partial Differential Equations. 8th Int. Conference on Information Fusion, 2005.

Steuerung und Navigation

Robot Guidance Navigation with Stereo-Vision and a Limited Field of View

André Treptow, Benjamin Huhle, Andreas Zell

University of Tuebingen, Department of Computer Science WSI-RA, Sand 1,
D-72076 Tuebingen, Germany

E-mail: treptow@informatik.uni-tuebingen.de

Abstract. We describe a method for robot guidance navigation (homing) that is purely based on vision with a stereo camera system which has a limited field of view. Instead of using specific landmarks our algorithm is based on warping of snapshots. Therefore it is meant to work even in environments where it is difficult to extract distinctive visual features. Stereo-vision is used instead of an equal distance assumption to improve the performance especially in outdoor environments.

1 Introduction

In nature, animals and insects have the ability to recognize and remember places so that they are able to return to them. This behavior is called homing and belongs to the group of guidance navigation approaches in robot navigation. In guidance navigation, the robot only knows a constellation of certain environmental features near the target position and the task is to navigate to the target by re-establishing the original feature constellation. Cartwright and Collet [2] had been the first who simulated the homing behavior of bees using image snapshots. In the published work about homing navigation on mobile robots one can find mainly two different approaches: Image and landmark-based homing. For landmark-based homing, distinctive landmarks (e.g. edges, corners) have to be extracted from the image and are used to calculate the home vector ([1], [8]). In image- or snapshot-based homing oneuses the whole image to predict the direction to the home position ([3],[6],[7]). Both approaches are mostly implemented by using a panoramic sensor. To our knowledge, the only approach that uses stereo information by the calculation of a disparity signature from a panoramic stereo sensor is described by Stürzl et al. in [7]. The novelty of our method is twofold: First we deal with a limited field of view and consider camera rotations so that we do not need other sensor modalities. Second, stereo information is used instead of an equal distance assumption to improve the resulting home vector. Therefore, our approach can be seen as an extension of the work proposed by Rizzi et al. [6] who also deal with a limited field of view but do not consider camera rotations and assume equal distances instead of using 3D data.

2 Our approach

Our approach is image-based: The robot stores a color image of the home position V_H which is a local view with a field of view of about 90°. To find the way back to this location, the robot's current view V_P is used to calculate the best driving direction. Calculation is done by the following steps:

1. If the difference between $\tilde{V}_P$ and V_H is too high: Find new camera orientation.
2. Sample over possible home vectors $P'_{ij} = (x_i, y_i, \theta_j)$, for each position do
   a) Calculate expected image $V_{P_{ij}}$ by image warping from V_P.
   b) Match expected image $V_{P_{ij}}$ with image of home position V_H
3. Take best match V_{P_*} and transform corresponding home vector p'_* into driving commando

The sampling over different possible home vectors (see step 2(a)) is done by selecting 144 points that lie in different distances around the current robot position P. The sample positions are depicted in Fig. 1. At each position we simulate a camera rotation θ from -0.4rad to 0.4rad in 9 steps so that altogether 1296 different warped images for all sampled positions have to be calculated.

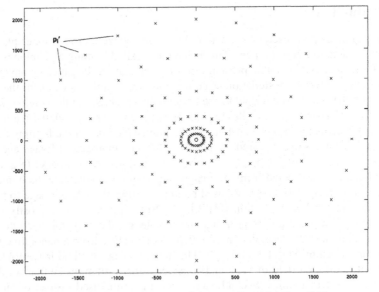

Fig. 1. Sampled points around the actual position P at (0,0).

2.1 Image warping

The central component of our algorithm is the procedure of predicting an expected image $V_{P'}$ at a sample position P' which is derived from the current position P by a translation T and a rotation R. Pixels $(x_{p'}, y_{p'})$ in the image $V_{p'}$ can be calculated from the pixels (x_p, y_p) in the current image V_p as follows:

$$z_{p'} \frac{x_{p'}}{f} = r_{11} \left(z_p \frac{x_p}{f} - t_1 \right) + r_{12} \left(z_p \frac{y_p}{f} - t_2 \right) + r_{13} \left(z_p - t_3 \right) \tag{1}$$

$$z_{p'} \frac{y_{p'}}{f} = r_{21} \left(z_p \frac{x_p}{f} - t_1 \right) + r_{22} \left(z_p \frac{y_p}{f} - t_2 \right) + r_{23} \left(z_p - t_3 \right) \tag{2}$$

$$z_{p'} = r_{31} \left(z_p \frac{x_p}{f} - t_1 \right) + r_{32} \left(z_p \frac{y_p}{f} - t_2 \right) + r_{33} \left(z_p - t_3 \right) \tag{3}$$

where z_p is the depth information from a 3D sensor, f is the focal length of the camera, t_i are the components of T and rij describe the entries of R. Figure 2 shows a current view V_p and a warped view V_p which is located 20cm left and 20cm backward from the current position with a camera view rotated by 10 degrees. The 3D sensor that we use on our robot is a stereo camera (see Fig. 3) where the 3D information is calculated for each pixel using the disparity between images from two aligned cameras. The correlation based method that is used with the camera for stereo computation (small vision system, see [4]) only provides 3D information in image areas that are textured. Therefore, we use a fixed distance assumption for pixel positions where no depth value can be calculated.

Fig. 2. Image warping: V_p left, $V_{p'}$ right.

2.2 Image matching

Every warped image is compared to the image V_H at the home position. This matching is done by the calculation of the mean squared error between the two images:

$$d = \sum_{y=1}^{h} \sum_{x=1}^{w} \frac{\omega_G (G_1(x,y) - g_2(x,y))^2 + \omega_H (H_1(x,y) - H_2(x,y))^2}{w \cdot h} \tag{4}$$

$G_i(x,y)$ defines the gray value in image i at the pixel position (x,y) and $H_i(x,y)$ is the hue value. w and h are width and height of the image and ω_G, ω_H are different weights for the pixel differences in gray and hue values. The constant factor δ is used for normalisation.

To deal with the limited field of view of the camera, one has to find the best camera orientation (see step 1 of the algorithm). This is achieved by a stepwise rotation of the camera to get a number of views $V_1 ... V_n$. For each view a home vector is calculated and the view with the best match with the home view is used. The camera is only completely reoriented if the the difference between the current image and the snapshot at the home position is above a certain threshold. Small orientation changes are covered by the fact that sampling over possible home vectors includes different angles θ_j (see step 2).

3 Experiments

The experimental platform was a mobile ATRV-Jr robot with a stereo camera head mounted on a pan-tilt unit (see Fig. 3). The resolution of the images was 320x240. The robot was equipped with a dual Pentium III 1GHz board. In a first experiment we recorded images at 27 fixed positions on a grid of 4x8 meters in an office environment and at 56 fixed positions within 6x7 meters in an outdoor environment. In the outdoor environment, the robot was placed on a nearly flat lawn near a building and the robot's field of view included two bushes that had been located around 8 meters away from the home position. The light conditions changed during image recording due to the fact that sometimes the sun had been covered by clouds.

The orientation of the camera while recording images was fixed. For every position p_i the next best location toward the home position p_i was calculated by the homing algorithm together with a value Δp which describes the mean position improvement that is achieved over the number of n test positions:

$$\Delta p = \frac{1}{n}\sum_{i=1}^{n} d_i, \quad d_i = \begin{cases} \dfrac{d(p_i,h)-d(p_i,h)}{d(p_i,h)} & : d(p_i,h) < 2000 \\[2ex] \dfrac{d(p_i,h)-d(p_i,h)}{2000} & : d(p_i,h) \ge 2000 \end{cases}$$

where $d(p_i,h)$ is the Euclidean distance (in mm) between the current position p_i and the home position h and $d(p_i,h)$ is the distance between the new position p_i and the home position. The constant factor of 2000 allows for the fact that a maximal distance

Fig. 3. Mobile robot ATRVjr (left) and stereo camera head (right).

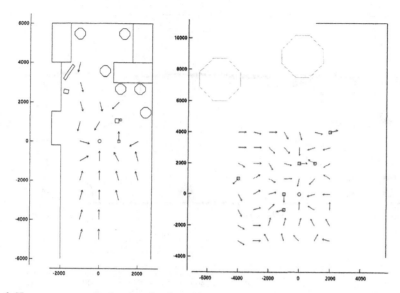

Fig. 4. Home vectors calculated at fixed positions in indoor (left) and outdoor (right) environment. The home position is located at (0,0).

of 2 meters is covered by one single homing step. Figure 4 shows the environments and the resulting home vectors at each position. The accuracy Δp for different image scales is depicted in table 1. As one can see, the use of depth information improves the results in the outdoor environment drastically, from $\Delta p = 0.10$ to $\Delta p = 0.42$. The results in the indoor environment could not be improved by including depth information due to the fact that in this environment one gets only very sparse stereo images. Comparing the different image scale factors one can find a local optimum at 1/14: More sub-sampling provided images with a resolution that is too low while lesser sub-sampling introduced more details and noise so that errors in the matching process increased. Therefore, image sub-sampling provided reduced runtimes for image warping and matching while acting as a noise filter. The runtime for one homing step ranged from 416ms (scale 1/20) up to 612ms (scale 1/10) on a Pentium III 1GHz processor.

Table. 1. Average home vector position improvement.

Image scale	Stereo	Δp indoor	Δp outdoor
1/10	on	0.60	0.26
1/14	on	0.65	**0.42**
1/14	off	**0.69**	0.10
1/20	on	0.42	0.19

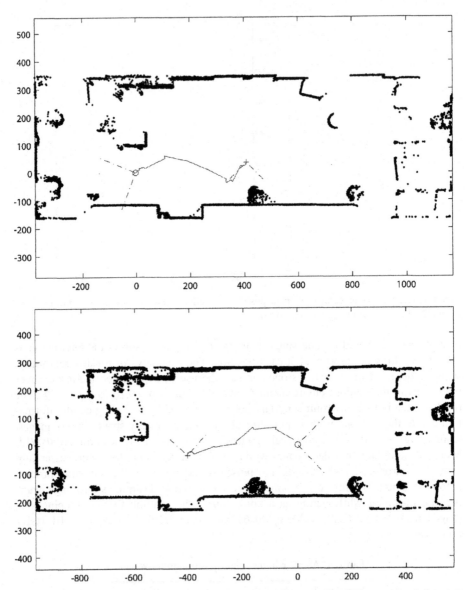

Fig. 5. Two different runs in an unconstrained office environment. The home position is located at (0,0), marked with a circle. The plotted line shows the path of the robot and the camera orientation is shown with a dotted line. The map is calculated based on laserscanner measurements.

To test our approach under real conditions, homing-runs in an office environment had been performed. Figure 5 exemplarily shows 2 different runs. In all experiments, the robot's starting position had been approximately 4 meters away from the home position. The camera orientation had not been the same as at the home location so that the robot had to find the right orientation first. In all runs the home position was reached with an average position error between 10 and 20cm.

4 Conclusion and future work

In this paper, we proposed a new image based robot homing algorithm that is able to cope with a limited camera view and uses depth information provided by a stereo camera. With this algorithm it is possible to perform short range robot guidance navigationwithout the usage of additional sensors. Especially in outdoor environments which provide dense stereo information our approach showed to be much more suitable than using an equal distance assumption as done in previous work about robot homing. The approach is suitable for the future application of a tracking algorithm (e.g. Particle Filtering) to adapt the number and positions of the home vectors that are sampled so that accuracy and runtime could possibly be improved further. The proposed algorithm could also be used for homing tasks in three dimensions, like for example on an autonomous helicopter or blimp and in the future we will implement a global navigation method using so called view graphs [5] to extend the local homing behavior.

References

1. A.A. Argyros, C. Bekris, S.C. Orphanoudakis, and L.E. Kavrak. Robot Homing by Exploiting Panoramic Vision. *Autonomous Robots*, 19(1):7–25, 2005.
2. B.A. Cartwright and T.S. Collet. Landmark Learning in Bees. *Journal of Computational Physiology*, A 151:521–543, 1983.
3. M.O. Franz and H.A. Mallot. Where did I take this snapshot? Scene-based homing by image matching. *Biological Cybernetics*, 79:191–202, 1998.
4. K. Konolige. Small Vision Systems: Hardware and Implementation. In *Proc. of the 8th International Symposium in Robotic Research*, pages 203–212, 1997.
5. H.A. Mallot M.O. Franz, B. Schölkopf and H.H. Bültho_. Learning View Graphs for Robot Navigation. *Autonomous Robots*, 5:111–125, 1998.
6. A. Rizzi, G. Bianco, and R. Cassinis. A Bee-Inspired Visual Homing using Color Images. *Robotics and Autonomous Systems*, 25:159–164, 1998.
7. W. Stürzl and H.A. Mallot. Vision-Based Homing with a Panoramic Stereo Sensor. *In Biologically Motivated Computer Vision*, pages 620–628, 2002.
8. K. Weber, S. Venkatesh, and M. Srinivasan. Insect-Inspired Robotic Homing. *Adaptive Behavior*, 7(1):65–97, 1999.

Einfaches Steuerungskonzept für mobile Roboter in dynamischen unstrukturierten Umgebungen

Thomas Krause, Peter Protzel

Technische Universität Chemnitz, Fakultät für Elektrotechnik und Informationstechnik,
Institut für Automatisierung

E-mail: thomas.krause@etit.tu-chemnitz.de, peter.protzel@etit.tu-chemnitz.de

Zusammenfassung. Es wurde eine Steuerungsarchitektur für mobile Roboter entwickelt, die den Anforderungen einer teilstrukturierten, weitläufigen Umgebung im Außenbereich gerecht wird. Die Architektur ist hierarchisch vernetzt aufgebaut und in verschiedene parallel arbeitende Prozesse aufgeteilt. Dabei wird auch der notwendige Informationsfluss von den Rohdaten bis zu Informationen aus dem Weltmodell betrachtet. Die Architektur wurde auf zwei verschiedenen Roboterplattformen in unterschiedlichen Umgebungen getestet.

1 Einleitung

Eine leistungsfähige Steuerungsarchitektur für mobile Roboter ist die Grundlage für ein komplexes Verhalten und eine geschickte Interaktion mit der Umgebung. Eine geeignete Architektur kann helfen, selbst bei komplexeren Aufgaben und Strukturen den Überblick zu behalten und erleichtert die Implementation neuer Elemente. Für den Innenbereich existieren bereits einige Strukturen, die den Bedürfnissen einer vorhersehbaren Umgebungssituation in einer strukturierten Umgebung gerecht werden [3]. Der Einsatzbereich in dynamischen und teilstrukturierten Außenbereichen stellt neue Anforderungen an die Leistungsfähigkeit solcher Strukturen. In diesem Beitrag wird eine diesen Bedingungen angepasste leistungsfähige Steuerungsarchitektur vorgestellt. Dabei werden Testergebnisse auf zwei verschiedenen Plattformen undin unterschiedlichen Umgebungen präsentiert.

2 Definition der Einsatzumgebung

Der Roboter soll in einem relativ großen Gelände außerhalb von Gebäuden navigieren. Das Gelände kann als teilstrukturiert betrachtet werden. Dort können geometrisch einfache Objekte, wie Häuser, Container oder Straßen, aber auch komplexe Objekte, wie Autos, Bäume, Sträucher oder Parkbänke vorhanden sein. Neben diesen stationären Objektenkönnen aber auch dynamische Objekte, wie Menschen, Tiere oder fahrende Autos auftreten. Es ist derzeit praktisch unmöglich alle Fälle, die in so einer Umgebung auftreten können, mit Sensoren zu erfassen oder gar vorher einzuplanen. Hinzu kommt dass ein Roboter auf den unterschiedlichen Untergründen (Sand, Kies, Asphalt, Wiese...) sehr unterschiedlich fährt, was die Erfassung der Bewegungen anhand von Odometrie stark beeinflusst. Daraus ergeben sich einige Anforderungen an die Leistungsfähigkeit der Steuerungsarchitektur.

1. Der Roboter muss schnell auf plötzliche Ereignisse, wie vor dem Roboter auftauchende Objekte (Hindernisse) reagieren können.
2. Der Roboter muss einfache Bewegungen auf unterer Ebene ausführen können, ohne dass eine übergeordnete Instanz, ständig regeln muss.
3. Es müssen dennoch komplexe Bewegungsformen ausführbar sein.
4. Trotz der Komplexität darf die Übersicht nicht verloren gehen.

Es gibt verschiedene Ansätze für die Steuerung von Roboterverhalten. Das Spektrum reicht dabei von Zustandsgrafen über reaktives Verhalten bis zu planenden Instanzen anhand von komplexen Weltmodellen. Diese Ansätze haben ihre spezifischen Vor- und Nachteile und sind in den entsprechenden Situationen sehr effektiv. So kann man mit dem Zustandsgrafen viele Pläne aufbauen und komplexe Verhalten kreieren. Vorraussetzung dabei ist jedoch, dass man alle möglichen auftretenden Fälle im Zustandsgrafen abdeckt. Bei größeren Strukturen wird der Zustandsgraf sehr schnell unübersichtlich. Das reaktive Verhalten reagiert schnell und zuverlässig auf plötzliche Ereignisse. Komplexere Verhalten sind damit jedoch schwer umzusetzen. Die Planung anhand von komplexen Weltmodellen ermöglicht sehr komplexes und vorrausschauendes Verhalten. Jedoch ist dieses Verfahren oft sehr langsam und kann auf plötzlich eintretende Ereignisse meist nicht schnell genug reagieren [4].

3 Die Steuerungsarchitektur

Angelehnt an die Steuerungsarchitektur eines Reinigungsroboters aus [1], in der eine hierarchische Struktur für eine Reinigungsmaschine für Explorationsaufgaben beschrieben wird, wurde versucht, die Vorteile der verschiedenen Strukturen in einer kombinierten Architektur miteinander zu vereinen. Abbildung 1 zeigt die hierarchisch vernetzte Struktur der Steuerung.

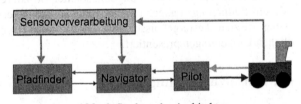

Abb. 1. Struktur der Architektur

Die verschiedenen Anforderungen werden auf verschiedenen Ebenen (Hierarchien) bearbeitet. Dabei nutzt jede Ebene verschiedene Informationen als Eingang. Die Architektur ist von der Aufgabenteilung her mit der einer Besatzung einer Schiffsbrücke vergleichbar. Diese Idee wurde bereits in der Arbeit von [5] aufgegriffen.

3.1 Der Pilot

Auf der untersten Ebene arbeitet der Pilot, der als alleiniger Agent Zugriff auf die Antriebe des Roboters hat. Er bekommt als Eingabe vom Roboter die Rohdaten der

Sensoren, die Hindernisse direkt erkennen lassen, wie zum Beispiel die Entfernungs-sensoren. Aus Modulen, die eine Sensordatenvorverarbeitung vornehmen und dem Weltmodell können ebenfalls Informationen zuHindernissen als Eingabe genutzt werden, insofern sie vorliegen. Der Pilot prüft anhand dieser Informationen und sei-nem aktuellen Bewegungsbefehl, ob sich ein Hindernis in Fahrtrichtung befindet oder nicht. Je nachdem ob und wie weit ein Hindernis entfernt ist, führt er die gewünschte Bewegung aus oder stoppt den Roboter. Vor dem Hindernis verringert er automatisch die Geschwindigkeit so weit, dass er den Roboter noch vor dem Hindernis zum Ste-hen bringen kann. Er ist damit rein reaktiv aufgebaut und kannso schnell auf unvor-hergesehene Situationen reagieren. Damit ist der Roboter in der Lage gezielte Bewe-gungen relativ zum Koordinatensystem des Roboters auszuführen und in einfachen Ansätzen auf seine Umgebung zu reagieren.

3.2 Der Navigator

Der Navigator in der zweiten Ebene berechnet aus globalen Koordinaten die Bewe-gungsbefehle für den Piloten, die ausgeführt werden müssen, um den vorgegebenen Zielpunkt zu erreichen. Zusätzlich beinhaltet er eine Regelung, die den Roboter auf einer geradenLinie zum Zielpunkt steuert. Dazu wird bei Empfang eines neuen Fahr-befehls in globalen Koordinaten eine Linie zum Zielpunkt berechnet. Während der Roboter zum Ziel fährt, wird der Abstand zur virtuellen Linie zyklisch berechnet. Resultierend aus dieser Abweichung wird zu dem Richtungsvektor der virtuellen Linie ein Korrekturvektor addiert. Daraus ergibt sich ein neuer Richtungsvektor, der Richtung Linie und Zielpunkt zeigt und den Roboter so wieder auf seinen Pfad bringt. Abbildung 2 verdeutlicht den Ablauf.

Abb. 2. Bewegung des Roboters auf einer Linie zum Ziel

Als Eingang wird nur die globale Position genutzt. Die Regelung, auf einer Linie entlang zu fahren, bringt zwei wesentliche Vorteile:

1.   Der Roboter bewegt sich auf einer geraden Linie kontrolliert zum Zielpunkt.
2.   Wird die Positionsbestimmung korrigiert, etwa weil neue charakteristische Punk-te erkannt wurden, korrigiert sich auch die Bewegung des Roboters automatisch, da sofort Korrekturvektoren auf der Basis der neuen Positionsinformationen be-rechnet werden.

Damit ist der Roboter in der Lage, im globalen System Bewegungen gerichtet aus-zuführen und Zielpunkte auf einer Geraden anzufahren. Dabei reagiert er durch den Piloten auf Hindernisse, indem er vor ihnen abbremst und anhält. Erhält der Navigator einen neuen Fahrbefehl, verwirft er den alten und berechnet die virtuelle Linie erneut.

3.3 Der Pfadfinder

Der Pfadfinder realisiert das erste komplexere Verhalten des Roboters. Wenn vom Piloten über den Navigator eine Meldung zu einem Hinderniss gemacht wird, berechnet er einen Ausweichkurs um das Hindernis, ohne komplexe Informationen seiner Umgebung zu nutzen. Auch er bekommt als Befehl lediglich einen Fahrbefehl. Daraus berechnet er, wie der Navigator, eine virtuelle Linie zum Zielpunkt und gibt anschließend den Fahrbefehl weiter an den Navigator. Taucht auf dem Weg zum Ziel ein Hindernis im Sensorbereich auf und der Pilot meldet, dass er ein Hindernis in seiner Fahrtrichtung erkannt hat, sucht der Pfadfinder nach einer Lücke, durch die der Roboter passt. Aus den gefundenen Lücken wählt er diejenige mit der geringsten Abweichung vom eigentlichen Kurs und fährt dann anderthalb Roboterlängen (Dieser Wert kann auch variieren) in diese Richtung. Danach wiederholt er den Vorgang so lange bis die neue Richtung nicht mehr von der Richtung zum Hindernis abweicht.

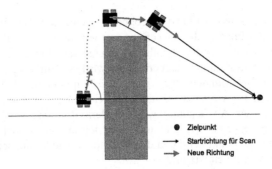

Abb. 3. kognitive Hindernisumfahrung

Dann gibt er den neuen Fahrbefehl bis zum Zielpunkt. Für den Navigator sind das stets neue Fahrbefehle.

Damit ist der Roboter zusätzlich in der Lage Hindernisse ohne Kenntnis einer globalen Karte zu umfahren und dennoch sein Ziel zu erreichen. In den darüberliegenden Schichten, können nun weitere Agenten installiert werden, die höhere Aufgaben, wie Pfadplanung oder Exploration realisieren.

4 Sensorinformationen

Jede der genannten Schichten benötigt unterschiedliche Sensorinformationen. Für den Piloten müssen die Rohdaten schnell zur Verfügung stehen, wärend der Navigator nur die Position im globalen System benötigt. Der Pfadfinder nutzt bereits vorverarbeitete Sensordaten und nutzt nur noch die Ergebnisse. Abbildung 4 zeigt wie vernetzt die einzelnen Informationen untereinander und zu den Agenten sind.

Dieser Informationsfluss ist durchaus noch ausbaufähig. Wichtig bei der Umsetzung der einzelnen Agenten ist die sichere Nutzung der Informationen. Jeder Agent muss für auf fehlende Informationen robust reagieren. So muss der Pilot zum Beispiel anhalten, wenn er keine Sensorinformationen hat. Die Position nimmt dabei eine besondere Stellung ein. Alle Agenten benötigen die Position des Roboters im globalen System und fragen diese ab. Die Positionsbestimmung oder Lokalisierung ist ein

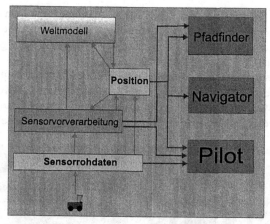

Abb. 4. Informationsfluss in der Steuerungsstruktur

eigenes Forschungsgebiet und kann hier nicht weiter betrachtet werden. Hier werden verschiedene Allgorithmen angewendet, um eine möglichst genaue Position zu bestimmen. Je genauer sich die Position bestimmen lässt, desto genauer und feinfühliger wird sich der Roboter mit dieser Struktur verhalten.

Die Daten des Weltmodells werden erst in der nächsten Ebene, zum Beispiel zur Pfadplanung genutzt. Da sie aber für die Bestimmung der Position nützlich sind, ist es bereits mit herangezogen worden.

5 Evaluierung

Die vorgestellte Steuerungsarchitektur wurde auf zwei unterschiedlichen Robotersystem in unterschiedlichen Umgebungen getestet. Abbildung 5 zeigt beide Systeme. Der rechte Roboter ist vom Typ Pioneer 2 AT der Firma Active Media [7] (HANS), der

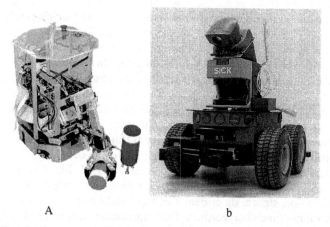

A   b

Abb. 5. Die Testplattformen: a: VIDAR, Eigenbauplattform mit holonomem Antriebssystem und Greifer; b: HANS, Pioneer 2 AT Plattform mit Sensorerweiterungen.

im Außengelände operiert und der linke Roboter ist eine Eigenkonstruktion, die ursprünglich für den Roboterwettbewerb Eurobot [6] gebaut wurde und dort unter Wettkampfbedingungen sein Verhalten erfolgreich unter Beweis gestellt hat. Sein Operationsgebiet ist der ebene Innenbereich.

Abbildung 6 zeigt die beiden Testumgebungen. Vidar fährt auf dem Spielfeld (links), auf dem auch andere Roboter agieren und in dem Objekte (farbige Kegel) zufällig herumliegen und bewegt werden können. Im Außenbereich (rechts] wurde Hans getestet. Im Gelände sind vor allem Menschen als bewegliche Objekte herumgelaufen.

a   b

Abb. 6. Die Testgebiete: a: Spielfeld für VIDAR b: Außenbereich für HANS

Auf Vidar war als oberste Ebene eine Zustandsmaschine umgesetzt, die verschiedene Aufgaben auf dem Spielfeld erfüllte. Auf Hans war eine Folge von Zielpunkten vorgegeben, die abgefahren werden sollten. Karten und eine komplexere Umweltrepräsentation wurden nicht aufgenommen. Beide Roboter zeigten ein stabiles Verhalten in ihrer Umgebung. Vidar wich stehenden Hindernissen sicher aus und fuhr sicher die von der Zustandsmaschine vorgegebenen Zielpunkte an. Als einziges dynamisches Hindernis fuhr der gegnerische Roboter herum. Diesem wich er immer sicher aus und nahm danach sofort wieder Kurs auf sein eigentliches Ziel. Er zeigte jedoch Schwierigkeiten, wenn der gegnerische Roboter auf ihn zu fuhr und ihn rammte. Da in der Architektur in den unteren Ebenen kein Zurückweichen vorgesehen ist, blieb er nur vor dem Roboter stehen und versuchte zu den Seiten auszuweichen. Wenn dergegnerische Roboter keine Kollisionsvermeidung besaß bzw. den Roboter nicht erkannt hat und weiter fuhr, kam es zur Kollision. Hier könnte man zusätzlich ein aktives Abstandhalten zum Hindernis einbauen, wodurch der Roboter auch zurückweichen würde. Eineweitere Möglichkeit wäre in einer übergeordneten Instanz ein Verhalten zu implementieren welches den Roboter einen größeren Bogen um bestimmte Hindernisse machen lässt. Hans fuhr seine Zielpunkte im Außengelände sauber ab und wich dabei zwei Felsen und einem Busch problemlos aus und erreichte seine Zielpunkte. Stellte sich ihm ein Mensch plötzlich in den Weg, wich er aus oder hielt an und nahm anschließend wieder Kurs auf sein Ziel. Ein Zielpunkt lag genau innerhalb eines Felsens, was er nicht anfahren konnte. Hier fuhr er bis an den Felsen heran und blieb dort stehen, da er dem Ziel nicht mehr näher kam. Um solche Probleme zu lösen muss eine übergeordnete Planungsinstanz entsprechend eingreifen und neue Zielpunkte vorgeben.

Literaturverzeichnis

1. Hierarchische Steuerung für einen mobilen Roboter zur autonomen Erkundung seiner Einsatzumgebung, Thomas Edlinger, 1997, Fortschr.-Ber. VDI Reihe 8 Nr. 638 Düsseldorf, VDI Verlag
2. Pioneer 2 Mobile Robot – Operations Manual.,ActivMedia Robotics, LLC, 2001
3. Intelligentes hierarchisches Regelungskonzept für autonome mobile Robotersysteme,Jürgen Guldner, 1995, Fortschr.-Ber. VDI Reihe 8 Nr. 509 Düsseldorf, VDI Verlag
4. Behavior-based Robotics, Ronald C. Arkin, 1998, ISBN 0–262-01165-4, Massechusetts Insitute of Technology
5. Konzeption eines Modells zur automatischen Navigation mit einem mobilen Roboter, Tobias Otto-Adamczak, Diplomarbeit, Chemnitz, 2002
6. Roboterwettbewerb Eurobot, www.eurobot.org
7. Activ Media, www.activmedia.com

Graphbasierte Bewegungsanalyse dynamischer Hindernisse zur Steuerung mobiler Roboter

Thorsten Rennekamp

Institut für Robotik und Prozessinformatik, Technische Universität Braunschweig,
Mühlenpfordtstraße 23, 38106 Braunschweig

E-mail: tre@rob.cs.tu-bs.de

Zusammenfassung. Diese Arbeit stellt ein kamerabasiertes Sensornetzwerk vor, welches das Bewegungsverhalten von dynamischen Hindernissen erkennt. Aus diesen Daten werden typische Bewegungsmuster als Graph mit Hilfe dynamischer Bayes-Netzwerke modelliert. Die Modellierung ermöglicht eine Vorhersage zukünftiger Hindernisbewegungen für eine geschickte Bahnplanung für mobile Roboter in dynamischen Umgebungen.

1 Einführung

Sollen mobile Roboter als dienstbare Geister in Zukunft in Bereichen eingesetzt werden, die von jedermann benutzt werden, ist es nötig, dass sich die Roboter auf das Verhalten von Personen in ihrer Umgebung einstellen und diese möglichst wenig stören. Um dieses leisten zu können, sind in den vergangenen Jahren Arbeiten entstanden, die über eine rein reaktive Behandlung dynamischer Hindernisse hinausgehen. Die vorgeschlagenen Ansätze unterscheiden sich in der Wahl der Sensoren (Kameras, Laserscanner, Ultraschall oder Infrarot), dem Ort der Sensoren (auf den Robotern oder fest in der Umgebung) und der Art der Modellierung und Vorhersage typischer Hindernisbewegungen (stochastische Raster, Graphen, Mengen von typischen Trajektorien). Ferner unterscheiden sich die Verfahren durch die Berücksichtigung der aktuellen Hindernissituation oder basieren auf einer Statistik unabhängig vom aktuellen Hindernisaufkommen. Kruse und Wahl [1] entwickelten ein Modell, das Daten aus ortsfesten Kameras für die Modellierung eines stochastischen Rasters benutzen. Bennewitz et. al. stellen in [2] ein System vor, welches Mengen von typischen Hindernisbewegungen berechnet, dessen Daten mit einem Laserscanner gewonnen werden. Vasquez et. al. [3] ermitteln ebenfalls Mengen von typischen Trajektorien und gewinnen die zugrundeliegenden Daten über eine Kamerasensorik. Liao et. al. [4] setzten ID-Sensoren ein und projizieren die Informationen über die Anwesenheit von Hindernissen auf eine Graphstruktur. Hu et. al. [5] geben eine Übersicht über verschiedenste Techniken der visuellen Überwachung von Objekten und deren Verhalten für verschiedenste Anwendungsgebiete auch außerhalb der Robotik.

Im Folgenden wird ein Verfahren vorgestellt, welches wie in [1] zwischen dynamischen und statischen Hindernissen unterscheidet. Statische Hindernisse sind ortsfeste Hindernisse, dessen Position und Orientierung konstant sind und dem System bekannt sind. Dynamische Hindernisse werden dagegen über eine globale Kamerasensorik erkannt und das typische Verhalten dieser Hindernisse mit Hilfe eines Graphen modelliert. Die Erkennung der Hindernisse basiert auf einer Differenzbildanaylse. Toyama et.al. [6] geben eine Übersicht über verschiedene Verfahren der Hintergrundmodellierung zur Differenzbildanalyse. Die Hintergrundmodellierung dieses Systems basiert auf einer Idee aus [7] und verwendet eine Generierung des Hintergrundes aus Eigenbildern, die mit Hilfe der Hauptkomponentenanalyse erzeugt werden.

Beobachtete Hindernisbewegungen werden als dynamisches Bayes-Netzwerk modelliert, dessen Topologie und Zustände mit Hilfe des Voronoigraphen des Freiraumes der statischen Hindernisse initialisiert werden. Dieses ermöglicht die Generierung eines komplexen, aussagekräftigen Modells auch auf Basis weniger Hindernisbeobachtungen. Ferner ermöglicht eine an den Voronoigraphen angelehnte Bewegungsmodellierung die effiziente Integration der Bewegungsvorhersage in ein Bahnplanungssystem für mobile Roboter und lässt sich daneben gut mit einer globalen Kamerasensorik kombinieren.

2 Systemübersicht

Die Bewegungsvorhersage dynamischer Hindernisse ist in ein verteiltes Gesamtsystem eingebettet, welches in Abb. 1 dargestellt ist. Das System ermöglicht die Steuerung zweier Mecanum-Wheel Fahrzeuge mit Hilfe eines verteilten Sensorsystems, das neben bordeigener Ultraschallsensorik vor allem auf einem Sensornetzwerk im Gebäude verteilter Kameras basiert. Auch das verteilte Softwaresystem der Steuerung ist sowohl auf den Fahrzeugen als auch ortsfest realisiert. Die Sensorik in der Umgebung (Global Sensors) als auch die Sensorik auf den Fahrzeugen (Robot Sensors) steht für alle weiteren Komponenten wie der Lokalisation der Fahrzeuge (Localisation) oder der Hindernisdetektion (Dynamic Obstacle Detection) gleichzeitig zur Verfügung. Eine zentrale Komponente (Navigation) stellt eine Schnittstelle zum Benutzer her und übernimmt die übergeordnete Steuerung der Fahrzeuge. Diese Arbeit beschreibt im Detail die Komponenten, die in Abb. 1 in der oberen rechten Ecke verzeichnet sind

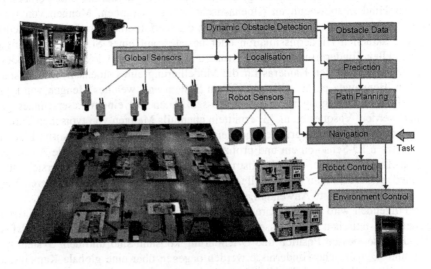

Abb. 1. Aufbau des Gesamtsystems

und Daten für das Bahnplanungssystem (Path Planning) über aktuelle Hindernisse und deren wahrscheinlicher Bewegung liefern. Im Detail sind das die Detektion dynamischer Hindernisse über die globale Sensorik, die Modellierung typischer Hindernisbewegungen und die Vorhersage aktueller Hindernisbewegungen. Des Weiteren ist eine Aktorik sowohl auf den Fahrzeuge (Robot Control) als auch in der Umgebung (Environment Control) z.B. zur Steuerung eines Fahrstuhls realisiert. Der Implementierung der Architektur liegt ein verteiltes System zugrunde, welches über die Middleware CORBA kommuniziert.

3 Sensorik

Dynamische Hindernisse werden über ein Sensornetzwerk verschiedener Kameras beobachtet, die die gesamte Umgebung überwachen. Hindernisse werden zunächst in den einzelnen Sichtbereichen der Kameras detektiert. Die Detektion in einer einzelnen Sicht basiert auf einer Differenzbildanalyse zwischen einem aktuellen Bild und einem der aktuellen Beleuchtungssituation entsprechendem Hintergrundbild.

Für die Generierung der Hintergrundbilder wird eine Menge von m Referenzbildern ohne dynamische Hindernisse vorgehalten. Diese Referenzbilder werden zuerst normiert und als Vektoren aufgefasst. Die Abhängigkeiten dieser Bilder von der Beleuchtungssituation werden mit Hilfe der Hauptkomponentenanalyse analysiert. Die Hauptkomponentenanalyse berechnet die Eigenvektoren der Kovarianzmatrix einer Matrix in der die Vektoren der Bilder spaltenweise zusammengefasst werden. Die entstandenen Eigenvektoren heißen Eigenbilder. Das aktuell aufgenommene Bild wird zu jedem Zeitschritt ebenfalls normiert und als Vektor aufgefasst und auf die n Eigenbilder mit größtem Eigenwert aus der Menge von $m > n$ Referenzbildern projiziert. Eine Rekonstruktion des aktuellen Bildes aus dieser Projektion liefert ein der entsprechenden Beleuchtungssituation angepasstes Hintergrundbild. Dieses folgt aus der Eigenschaft der Hauptkomponentenanalyse. Eigenbilder mit großem Eigenwert modellieren große korrelierte Beleuchtungsänderungen in mehreren Referenzbildern. Eigenbilder mit kleinen Eigenwerten modellieren einzelne Änderungen in einzelnen Referenzbildern.

Hindernisse im aktuellen Bild tauchen in der Rekonstruktion nicht auf, da diese nicht aus der gewichteten Summe der n Eigenbilder modelliert werden können. Diese Hindernisse werden über die Differenzbildanalyse erkannt. Das Verfahren wird über eine Ausmaskierung der Bereiche der statischen Hindernisse erweitert. Anschließend werden die Hindernisse einzelner Kamerasichten in einem globalen Koordinatensystem fusioniert. Ein letzter Schritt verknüpft die Hindernisse einzelner Zeitschritte zu Hindernistrajektorien über mehrere Zeitschritte.

4 Graphbasiertes Lernen und Vorhersagen dynamischer Bewegungen

4.1 Das Modell

Die Sensorik liefert eine Menge von beobachteten Hindernistrajektorien τ.

$$\tau_i = (x_{1i}, x_{2i}, x_{3i}, ..., x_{ni}) \quad , \quad x_{ij} \in R^2$$

Typische Hindernisbewegungen werden als probabilistisches graphisches Modell dargestellt. Die Struktur des Modells besteht aus einem geometrischen Graphen, dem Hindernisbewegungsgraph G.

$$G = (T, S)$$

Die Kanten S des Graphen werden probabilistische Trajektoriensegmente genannt. Die Knoten T des Graphen sind die Endpunkte der probabilistischen Trajektoriensegemente. Mit den Kanten des Graphen sind Zustände verknüpft, die den aktuellen Bewegungszustand dynamischer Hindernisse modellieren. Ein probabilistisches Trajektoriensegment $S_k = (T_A, T_B, M, F_A, F_B, P, P_A, P_D)$ besteht aus:

- Zwei Endpunkten T_A, $T_B \in R^2$
- Einem Bewegungszustand $M \in \{A, B\}$, der eine Bewegung in Richtung der Endpunkte T_A oder T_B modelliert
- Zwei Mengen von Folgezuständen F_A und F_B in den bei T_A und T_B anschließenden Nachbarsegmenten
- Übergangswahrscheinlichkeiten $P(F_A \mid M)$, $P(F_B \mid M)$ zu den Nachbarsegmenten und $P(M \mid M)$ zwischen den internen Zuständen
- Auftrittswahrscheinlichkeiten für neue Objekte $P_A(M)$
- und den Wahrscheinlichkeiten für das Verschwinden vorhandener Objekte $P_D(M)$

Den Aufbau eines probabilistischen Trajektoriensegments zeigt Abb. 2. Der Zustand eines beobachteten Hindernisses wird durch den Bewegungszustand auf dem aktuellen Trajektoriensegment hinreichend beschrieben. Die Verkettung probabilistischer Trajektoriensegmente mit den Zuständen M und den Übergangswahrscheinlichkeiten P modelliert ein Markov-Prozess der Ordnung 1. Es ergibt sich ein dynamisches Bayes-Netzwerk, welches typische Bewegungsmuster und deren Übergangswahrscheinlichkeiten modelliert. Ein Bewegungsmuster B ist ein Pfad im Graph G, d.h. eine Folge passierter Segmente S_k mit den dazugehörigen Bewegungszuständen M_k.

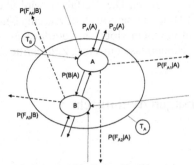

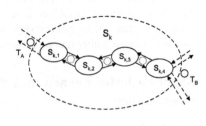

Abb. 2. Probabilistisches Trajektorien-segment S

Abb. 3. Hierarchische   Untergliederung eines Segments

Der Hindernisbewegungsgraph kann hierarchisch in verschiedener Genauigkeit modelliert werden, so dass weitreichende Bewegungsmuster und detailliertere Bewegungen gleichermaßen modelliert werden können (Abb. 3). Höhere Hierarchiestufen fassen hierbei Mengen von benachbarten Segmenten zusammen. Diese Mengen sind nur dadurch eingeschränkt, dass maximal zwei Knoten mit angrenzenden Segmenten außerhalb der Menge verknüpft sein dürfen. Diese Knoten sind die neuen Knoten T_A und T_B und die Übergangswahrscheinlichkeiten beziehen sich auf die dort angrenzenden Knoten und Segmente außerhalb dieser Menge.

4.2 Lernen der Übergangswahrscheinlichkeiten

Während einer Trainingsphase werden aus einer beobachteten Menge von Trajektorien die Übergangswahrscheinlichkeiten für das Modell geschätzt. Jede Trajektorie τ_j wird auf den Graph projiziert. Der Algorithmus basiert auf folgendem Prinzip: Für jedes benachbarte Paar $x_{i,j}$, $x_{i+1,j}$ werden die Segmente S_k und S_l mit dem kürzesten Abstand berechnet. Die Zahl der Zustandsübergänge entlang der kürzesten Pfade zwischen allen Paaren dieser Segmente liefert einen Pfad B_j im Graph, der die Trajektorie τ_j approximiert. Die Übergangswahrscheinlichkeiten $P(M_u \mid M_v)$ zweier Zustände in den Segmenten S_u und S_v berechnen sich aus der Summe aller Übergänge zwischen M_u und M_v für alle Pfade B_j und der Gesamtzahl des Auftretens des Zustands. Analog werden die Wahrscheinlichkeiten P_A und P_D für das Auftreten und Verschwinden eines dynamischen Hindernisses am Beginn und Ende einer Trajektorie berechnet. Abbildung 4 verdeutlicht die Projektion zweier Trajektorien τ_i und τ_j auf zwei Pfade B_i und B_j des Hinternisbewegungsgraphen. Übergangswahrscheinlichkeiten zwischen den der Bewegungsrichtung entsprechenden Zustände M_k der passierten Segmente S_k der Pfade werden erhöht.

4.3 Vorhersage von Hindernisbewegungen

Nach der Trainingsphase ermöglicht die Struktur die Vorhersage des Bewegungverhaltens bewegter Hindernisse. Die bisherige Bewegung eines beobachteten Objektes wird auf den Graph abgebildet. Das resultierende Segment S_k mit dem resultierenden Bewegungszustand M_k ist der aktuelle Zustand. Ausgehend von diesem

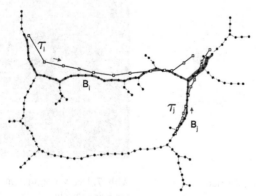

Abb. 4. Projektion zweier Trajektorien auf den Hindernisbewegungsgraph

Zustand wird der Folgepfad mit der größten Wahrscheinlichkeit als Schätzung der weiteren Bewegung des Objekts berechnet. Abbildung 5 verdeutlicht den Algorithmus. Abgebildet sind zwei gemessene Hindernistrajektorien τ_i und τ_j. Die Zustände S_{i1} und S_{j1} entsprechen der Projektion des Anfangs der Trajektorien auf den Graph. Ausgehend von diesen Startzuständen wurde die jeweils wahrscheinlichsten Pfade B_i und B_j auf dem Graph berechnet und in der Abbildung markiert.

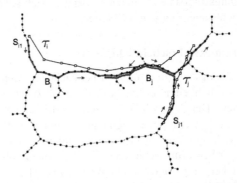

Abb. 5. Zwei Trajektorien und die Vorhersage der Bewegung ab deren Startzustand

4.4 Initialisierung der Topologie mit dem Voronoigraphen

Sowohl mobile Roboter als auch dynamische Hindernisse bewegen sich von einem Startpunkt zu einem Zielpunkt durch den Freiraum außerhalb der statischen Hindernisse. Für den diskret repräsentierten Freiraum wird das Skelett berechnet, welches den Voronoigraphen approximiert. Die Topologie des Voronoigraphen initialisiert die Struktur des Hindernisbewegungsgraphen G. Der Voronoigraph approximiert das Bewegungsverhalten von Personen in natürlicher Weise und ist zusätzlich eine geeignete Basis für die Bahnplanung mobiler Roboter. Abb. 7 zeigt den Voronoigraph auf einer Sicht der Umgebung, welche aus den Sichten der sechs einzelnen Kameras berechnet wurde.

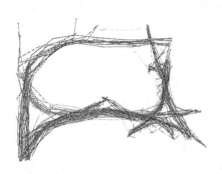

Abb. 6. Detektierte Trajektorien τ

Abb. 7. Der Voronoigraph auf einer fusionierten Kamerasicht

5 Ergebnisse und Ausblick

Mit dem Sensorssystem wurde eine Menge von 170 Beispieltrajekorien detektiert (Abb. 6) und auf den Hindernisgraphen abgebildet. Das entstandene Netzwerk erlaubte die Modellierung und Vorhersage komplexer Bewegungen dynamischer Hindernisse erstmals auf der Basis eines hierarchischen graphbasierten Modells. Es ermöglicht im Gegensatz zum Ansatz in [1] eine Vorhersage der Bewegung aktueller Beobachtungen und damit eine adaptivere Bahnplanung für mobile Roboter in dynamischen Umgebungen. Die Anlehnung der Topologie an den Voronoigraphen erlaubt im Gegensatz zu den Arbeiten in [2] eine aussagefähige Modellierung auch mit wenigen beobachteten Hindernistrajektorien. Auf größeren freien Flächen kann die Modellierung der Struktur des Graphen ungünstig sein. Hier ist eine selbstlernende Anpassung der Topologie des Graphen z.B. über die Anpassung der Lage der Trajektoriensegmente analog zu Kohonens selbstorganisierenden Merkmalskarten geplant.

Literaturverzeichnis

1. E. Kruse und F. M. Wahl: Camera-based monitoring system for mobile robot guidance. Proceedings of the IEEE/RSJ International Conference on Intelligent Robots and Systems (IROS), 1998.
2. M. Bennewitz, W. Burgard und S. Thrun: Learning Motion Patters of Persons for Mobile Service Robots. Proceedings of the International Conference on Robotics and Automation (ICRA), 2002.
3. D. Vasquez, F. Large and T. Fraichard und C. Laugier: High-Speed Autonomous Navigation with Motion Prediction for Unknown Moving Obstacles. Proceedings of the IEEE/RSJ International Conference on Intelligent Robots and Systems (IROS), Oktober 2004.
4. L. Liao, D. Fox, J. Hightower, H. Kautz und D. Schulz: Voronoi Tracking: Location Estimation Using Sparse and Noisy Sensor Data. Proceedings of the IEEE/RSJ International Conference on Intelligent Robots and Systems (IROS), Oktober 2003.
5. Weiming Hu, Tieniu Tan, Liang Wang und Steve Maybank: A Survey on Visual Surveillance of Object Motion and Behaviors. IEEE Transactions on Systems, man and cybernetics, Part C: Applications and Reviews, 34(3), August 2004.
6. K. Toyama, J. Krumm, B. Brumitt und B. Meyers: Wallflower: Principles and Practice of Background Maintenance. In Proceedings of the International Conference on Computer Vision (ICCV), September 1999.
7. N. Oliver, B. Rosario und A. Pentland: A Bayesian Computer Vision System for Modeling Human Interactions. IEEE Transactions on Pattern Analysis and Machine Intelligence, 22(8), August 2000.

Mobile Robot Motion using Neural Networks: An Overview

Mohamed Oubbati, Michael Schanz, Paul Levi

Institute of Parallel and Distributed Systems,University of Stuttgart. Universitaetsstrasse 38, D-70569 Stuttgart, Germany.

E-mail: {Mohamed.Oubbati, Michael.Schanz, Paul.Levi}@informatik.uni-stuttgart.de

Abstract. In this paper, we provide a summary of our recent results in motion control of mobile robots using recurrent neural networks. The most important asociated problems are discussed.

1 Introduction

Motion control of mobile robots has been studied by many authors in the last decade, since they are increasingly used in wide range of applications. At the beginning, the research effort was focused only on the kinematic model, assuming that there is *perfect* velocity tracking [5]. Later on, the research has been conducted to design motion controllers, including also the dynamics of the robot [3], [4]. Taking into account the specific robot dynamics is more realistic, because the assumption "perfect velocity tracking" does not hold in practice. Furthermore, during the robot motion, the robot parameters may change due to surface friction, additional load, among others. Recently, an increasing attention is being paid to recurrent neural networks (RNNs). Their ability to handle complex input-output mapping and to instantiate arbitrary temporal dynamics makes them promising for dynamical systems control. Furthermore, an interesting property of RNNs has been noticed by many authors: *Adaptive behavior with fixed weights*. It has been shown by Feldkamp et. al.[1] that a single fixed weight RNN can perform one-time-step prediction for many distinct time series. In the control domain, it is shown in [2] that a RNN can be trained to act as a stabilizing controller for three unrelated systems and to handle switch between them. More recently, we train a novel RNN called Echo State Network (ESN) to act as an adaptive identifier of non linear dynamical systems, with fixed weights [7]. This capability is acquired through prior training; instead of learning data from one system, RNNs are able to learn from different systems, or from one system in different operating conditions. The adaptive behavior of RNNs with fixed weights is named differently. It is termed "meta-learning" in [8], and "accommodative" in [6].

In this work, we provide the current state of our resarch on mobile robot control using RNNs. First, adaptive motion control using ESNs with meta-learning is described. Then, we present the real implementation of ESNs on an omnidirectional robot.

2 Control of a Nonholonomic Robot with ESN

2.1 Background

In this paper, we adopt the same robot model used by Takanori et al.[4]. In Fig. (1), P_0 is the origin of the coordinate system, where (x, y) are the coordinates of P_0. ϕ is the heading angle of the robot. The kinematic model is given by the following equation:

$$\begin{bmatrix} \dot{x} \\ \dot{y} \\ \dot{\phi} \end{bmatrix} = \begin{bmatrix} \cos(\phi) & 0 \\ \sin(\phi) & 0 \\ 0 & 1 \end{bmatrix} \begin{bmatrix} v \\ w \end{bmatrix} \tag{1}$$

We assume that the mass of the body is m_c and that of a wheel with a motor is m_w. I_c, I_w, and I_m are the moment of inertia of the body about the vertical axis through P_c, the wheel with a motor about the wheel axis, and the wheel with a motor about the wheel diameter, respectively. The dynamic model is given by:

$$M(q)\dot{v} + V(q,\dot{q})v = B(q)\tau \tag{2}$$

where $\tau = [\tau_r, \tau_l]$ are torques applied on right and left wheels, and M, V, B are

$$M = \begin{bmatrix} \frac{r^2}{4b^2}(mb^2 + I) + I_w & \frac{r^2}{4b^2}(mb^2 - I) \\ \frac{r^2}{4b^2}(mb^2 - I) & \frac{r^2}{4b^2}(mb^2 + I) + I_w \end{bmatrix}, V = \begin{bmatrix} 0 & \frac{r^2}{2b}m_c d\dot{\phi} \\ -\frac{r^2}{2b}m_c d\dot{\phi} & 0 \end{bmatrix}$$

$B = \begin{bmatrix} 1 & 0 \\ 0 & 1 \end{bmatrix}$. I and m are given by: $I = m_c d^2 + I_c + 2m_w b^2 + 2I_m$, $m = m_c + 2m_w$.

For the motion control, various feedback controllers have been proposed [5]. One of these controllers is described as following. Let x_r, y_r, and ϕ_r are the pose of a reference robot, and (v_r, w_r) are its linear and angular velocities. We define e_1, e_2, e_3 as following:

$$\begin{bmatrix} e_1 \\ e_2 \\ e_3 \end{bmatrix} = \begin{bmatrix} \cos(\phi) & \sin(\phi) & 0 \\ -\sin(\phi) & \cos(\phi) & 0 \\ 0 & 0 & 1 \end{bmatrix} \begin{bmatrix} x_r - x \\ y_r - y \\ \phi_r - \phi \end{bmatrix} \tag{3}$$

The inputs (v, w) which make e_1, e_2, e_3 converge to zero are given by [3][4]:

$$\begin{cases} v_d = v_r \cos e_3 + K_1 e_1 \\ w_d = w_r + v_r K_2 e_2 + K_3 \sin e_3 \end{cases} \tag{4}$$

where K_1, K_2, K_3 are positive constants.

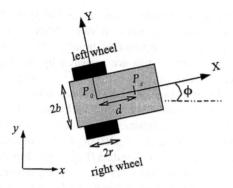

Fig. 1. Mobile robot with two actuated wheels.

2.2 Motion control using ESN

Echo state network (ESN) is formed by a "Dynamic Reservoir", which contains a large number of sparsely interconnected neurons with non-trainable weights (Fig. 2). The idea of this network is that only the weights connections from the internal neurons to the output are to be adjusted. More details on ESN training algorithm can be found in [7].

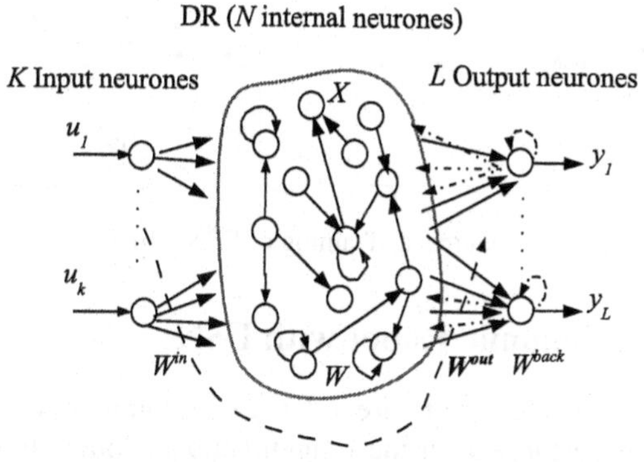

DR (N internal neurones)

Fig. 2. Basic architecture of ESN. Dotted arrows indicate connections that are possible but not required.

Control Design Here, an adaptive neurocontrol system with two levels is proposed. In the first level, a position controller (ESN_K) is designed to improve the robustness of the controller (4) and generate linear and angular velocities, necessary to track a reference trajectory. In the second level, another network (ESN_D) is designed for the model (2), which converts the desired velocities, provided by the first level, into a torque control. The advantage of the control approach proposedis that no knowledge about the dynamic model is required, since the ESN_D is designed only by learning I/O data collected from (2). Furthermore, no synaptic weight changing is needed in presence of variation of (d, m_c). This capability is acquired through prior "meta-learning".

Here, we assume that m_c vary between the values $[30,35]$ and the distance d can vary in interval of $\pm 50\%$ around its nominal value. ESN_D training involves using actual and delayed velocities of the robot as inputs, and correspondent torques as teacher signals (Fig. 3.a). Training data are prepared by driving (2) with input sequences sampled from the uniform distribution over $[-2,+2]$. 10 blocs of 1000 sequences are collected. In each bloc, m_c and d were assigned new random values taken from their respective variation intervals. The ESN_K is designed to emulate the behavior of (4), and improve its robustness in presence of noisy data. To provide ESN_K robustness against disturbances, a gaussian noise with zero mean and a variation level of $\sigma = 0.5$ was added to training data (Fig. 3.(b)). Fig. (4) shows the control results. ESN_K tracks the reference trajectory by providing desired linear and angular velocities, and ESN_D shows an excellent tracking of these velocities. Furthermore, when m_c and d values are suddenly changed, ESN_D locks on the new dynamic behavior of the robot and provides the appropriate control signal (Fig. 4 (c) and (d)), without changing any synaptic weight. We can see small adaptation periods after each variation. These are necessary for the network to switch from one familly of orbits to another. Surprisingly, the global control loop is barely affected by these switches, and the robot tracks reasonably the desired trajectory (Fig. 4 (e)).

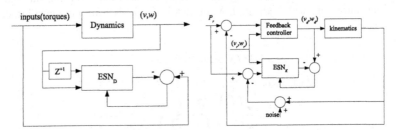

Fig 3. Training Procedures. a) Training of ESN_D. b) Training of ESN_K.

3 Control of a Holonomic Robot with ESN

Wheeled Holonomic mobile robots are often called omnidirectional robots. Here, experimentations are performed on the omnidirectional robot (Fig. 6). The robot is equipped with 3 omni-wheels equally spaced at 120 degrees from one to another.

3.1 Control Design

Here, we are interested in a problem of requiring a fixed-weights ESN to make an adaptive dynamic control for the robot, in presence of time-varying mass (a switch between $15Kg$ and $17.5Kg$). Training data were prepared by driving the robot with different velocities in different directions. 1000 I/O sequences were collected from the robot having its initial mass ($15Kg$) and other 1000 I/O sequences when the extra mass $m = 2.5Kg$ was added on the robot. Training procedure is based on the same approach presented in Fig. (3.a).

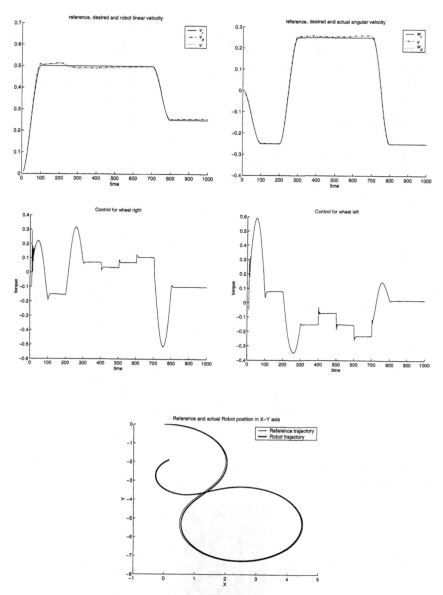

Fig 4. ESN motion control. a) linear velocity tracking. b) angular velocity tracking. c) torque for wheel right. d) torque for wheel left. e) trajectory tracking

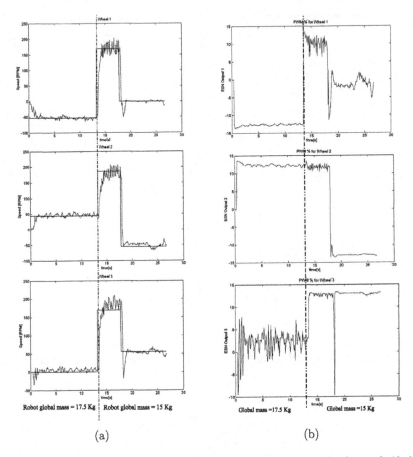

Fig 5. Control results of the case 2. a) Desired speeds(solid) and actual Wheels speeds (dashed), b) ESN control signals

Fig 6. Omnidirectional robot

Figure 5 presents the obtained results. During the first 13 seconds the ESN could successfully bring the robot to the desired velocity $(0.23m/s)$. At time $t = 13s$, the mass of the robot is reduced to $15Kg$. The controller showed a rapid adaptation to these change, and delivered the appropriate control signals to continue tracking the desired velocity.

4 Conclusion

Implementations carried out here demonstrate that a small and partially interconnected ESN can be trained to act as an adaptive controller for mobile robots, whose parameters are time varying. However, substantial investigation on ESNs and more real experiments on much larger data sets are still needed to improve the results we have achieved to date.

References

1. L. Feldkamp, G. Puskorius, and P. Moore. Adaptation from Fixed weight Dynamic Networks. In *Proc. IEEE Int. Conf. on NNs*, Washington, 1996.
2. L. Feldkamp, G.V. Puskorius, and P.C. Moore. Fixed weight Controller for Multiple Systems. In *Proc. IEEE Int. Conf. on NNs*, Texas, June 1997.
3. R. Fierro and F. L. Lewis. Control of a Nonholonomic Mobile Robot Using Neural Networks. *IEEE Trans. on NNs*, pages 589–600, July 1998.
4. T. Fukao, H. Nakagawa, and N. Adachi. Adaptive Tracking Control of a Nonholo- nomic Mobile Robot. *IEEE trans. on Robotics and Automation*, 16(5):609–615, October 2000.
5. I. kolmanovsky and N. H. McClamroch. Development in Nonholonomic Control Problems. *IEEE Control Systems*, pages 20–36, December 1995.
6. J. Lo. Adaptive vs. Accommodative Neural Networks for Adaptive System Identi-_cation. In *Proc. Int. Conf. on NNs*, 2001.
7. M. Oubbati, M. Schanz, and P. Levi. Meta-learning for Adaptive Identi_cation of Nonlinear Dynamical Systems. *In Proc. IEEE Int. Symp. on Intelligent Control*, Cyprus, June 2005.
8. D. Prokhorov, L. Feldkamp, and I. Tyukin. Adaptive Behavior with Fixed Weights in Recurrent Neural Networks: An Overview. In *Proc. Int. Conf. on NNs*, Hawaii, May 2002.

Combining Learning and Programming for High-Performance Robot Controllers

Alexandra Kirsch, Michael Beetz

Lehrstuhl für Informatik 9, Technische Universität München

Abstract. The implementation of high-performance robot controllers for complex control tasks such as playing autonomous robot soccer is tedious, error-prone, and a never ending programming task. In this paper we propose programmers to write autonomous controllers that optimize and automatically adapt themselves to changing circumstances of task execution using explicit perception, dynamics and action models. To this end we develop RoLL (Robot Learning Language), a control language allowing for model-based robot programming. RoLL provides language constructs for specifying executable code pieces of how to learn and update these models. We are currently using RoLL's mechanisms for implementing a rational reconstruction of our soccer robot controllers.

1 Introduction

The implementation of high performance robot controllers for complex control tasks such as playing autonomous robot soccer is tedious, error-prone, and a never ending programming task. Optimizing robot behavior requires parameter tweaking, situation specific behavior modifications, and proper synchronizations between concurrent threads of control. The mutual interactions between parameter settings, synchronizations, and situation-specific hooks imply that even tiny changes to the program may have large and unpredictable effects on the robot behavior.

Even worse, not only changes to the program cause drastic behavior changes, but even changes to the environment such as switching from one soccer field to another, changing lighting conditions, or simply playing against another team might cause the same controller to produce very different behavior. In one case we played on a softer carpet which caused the robot to sink deeper and the ball to get stuck between the guide rail for dribbling and the floor. As a result, the dribbling skill and the action selection had to be substantially modified. The robots had to be equipped with mechanisms to detect situations where the ball is stuck in order to avoid overheating the controller board. Or, in another case the playing strategy had to be extended substantially when we played against a team with bigger robots. The bigger robots occluded landmarks on the field and the ball. As a consequence the robots needed much more sophistication in situations where they couldn't localize themselves and the ball [2].

In order to enable programmers to implement controllers that can adapt themselves quickly to such changing circumstances we propose to specify the code pieces requiring adaptation or optimization based on experience explicitly as executable learning problems. For this purpose we develop a language RoLL (Robot Learning Language) that integrates learning mechanisms into the existing robot control and plan language

RPL [4]. Using ROLL programmers can specify robot controllers that represent, acquire, and reason about models of their control routines. In this paper we focus on the acquisition of appropriate models by automatic learning. Besides models we also learn routines that execute the robot's desired actions using the automatically acquired models.

With this approach the learning process is executable and can be performed automatically. Therefore the models and parts of the control program can be relearned without human interaction whenever the environment changes substantially. But the integration of learning and programming is not only a convenience to the developper, it also enhances the learning performance. Robot learning is a complex task that requires the adjustment of parameters like the experiences and the learning bias. On the one hand, these parameters can be found empirically in a convenient way, as they are represented explicitly and the learning process is performed automatically. On the other hand, these parameters could be acquired by a meta-learning process.

Still, an explicit representation of the learning process does not suffice to make a robot learning task reproducible, because there is a huge uncertainty in the training experiences. To get the experiences under control we explore the use of datamining techniques. The field of datamining has yielded strong mechanisms for producing clean, understandable data. We found that the learning performance can be enhanced signifi- cantly by applying data mining techniques.

In order to underline our arguments, Fig 1(a) shows a code fragment of a simple control program for an autonomous soccer robot. The parts in italics denote the functions or routines that require an adaptation to the environment.We see that in this case all the routines needed to fulfill the goals could be learned. Also, the choice of the next goal could be made by a learned decision tree depending on the opponent team.

```
repeat
  cond
    opponent-attack? → defend-own-goal
    near-ball?       → score-goal
    supporting-mate? → support-team-mate
    else             → annoy-opponent
```

```
with-models
  action-model, perception-model, ...
seq
  tagged learning-phase
    to-be-learned ← extract-non-operational-routines()
    for-every problem in to-be-learned
      learn(problem)
  tagged execution-phase
  repeat
    current-goal ← choose-goal(current-situation)
    try-all
      wait-for tasks-changed?
      achieve current-goal
```

(a) Simple robot controller.   (b) Controller with learned models.

Fig. 1. Example controller for an autonomous soccer robot with and without learned models.

In contrast, Fig 1(b) shows a control program that uses explicit, learned models. The first step in the execution is to learn all the routines that have not been learned yet. After that the robot starts its normal course of action. Here the choice of the next goal is performed by a learned function and also the execution routines are learned. In the rest of the paper we glimpse at the main concepts of our learning language ROLL.

Then we will briefly go into how we tested parts of this language and how we plan to evaluate it in the future. We will conclude with section 4.

2 Learning Mechanisms

Figure 2 shows the parts of a learning agent. Every aspect of a learning problem is represented explicitly within ROLL.

The *performance element* realizes the mapping from percepts into the actions that should be performed next. The control procedures therein might not yet be executable or optimized. These are the procedures we want to learn.

The *critic* is best thought of as a learning task specific abstract sensor that transforms raw sensor data into information relevant for the learning element. To do so the critic monitors the collection of experiences and abstracts them into a feature representation that facilitates learning. The experiences are stored in a *database*, which allows us to employ datamining mechanisms for data cleaning.

The *learning element* uses experiences made by the robot in order to learn the routine for the given control task. It can choose a learning algorithm from a library of so-called *learning systems*. The *problem generator* generates goals that are achieved by the performance element in order to gather useful experiences for a given learning problem. The problems are generated according to a probability distribution as given in the learning problem specification.

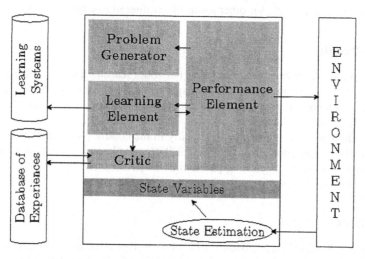

Fig. 2. Parts of a learning agent after Russell and Norvig.

3 Progress and Further Work

We are currently using ROLL's mechanisms for implementing a rational reconstruction of our autonomous soccer robot controllers and for the control of an autonomous household robot (Fig. 3). The salient features of the code are that it is self-optimizing and adaptive.

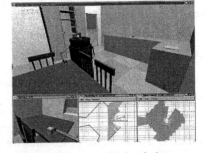

(a) Autonomous soccer robots.   (b) Realistic household simulation.

Fig. 3. Evaluation testbeds for RoLL: robotic soccer and a simulated household robot.

We used the learning mechanisms described in section 2 for learning navigation tasks in the domain of robot soccer. The learned routines were successfully applied in the RoboCup world championship 2004 in Lisbon [3,1]. The model-based paradigms were also employed on that occasion, as well as in the domain of the household robot.

The next step in our research is to combine the underlying model-based language with the learning concepts, so that the scenario in figure 1(b) can be implemented as shown. We will then be able to perform extensive testing on how the overall performance changes when the learning parameters are changed. After that we intend to extend the learning constructs for other forms of learning like reinforcement learning.

4 Conclusion

We have given a very brief overview over our robot learning language RoLL, which combines learning and programming in a synergetic way. By first learning appropriate models (perception, actions, dynamics), we then proceed to learning control routines and decision functions.

The learning process is described explicitly in the control program and is executed automatically by the controller. This makes it repeatable and reproducible. Besides, the parameterization for a learning problem on a special robot can be carried over to different robot platforms and similar learning problems.

References

1. M. Beetz, A. Kirsch, and A. Müller: RPL-LEARN: Extending an autonomous robot control language to perform experience-based learning. 3rd International Joint Conference on Autonomous Agents & Multi Agent Systems (AAMAS), 2004.
2. M. Beetz, T. Schmitt, R. Hanek, et al.: The AGILO robot soccer team experience-based learning and probabilistic reasoning in autonomous robot control. Autonomous Robots, 17(1):55–77, 2004.
3. M. Beetz, F. Stulp, A. Kirsch, et al.: Autonomous robot controllers capable of acquiring repertoires of complex skills. RoboCup International Symposium 2003, Padova, July 2003.
4. D. McDermott: A Reactive Plan Language. Research Report YALEU/DCS/RR-864, Yale University, 1991.

Automatic Neural Robot Controller Design using Evolutionary Acquisition of Neural Topologies

Yohannes Kassahun, Gerald Sommer

Christian Albrechts University, Institute of Computer Science and Applied Mathematics, Department of Cognitive Systems, Olshausenstr. 40, D-24098, Kiel, Germany

Abstract. In this paper we present an automatic design of neural controllers for robots using a method called Evolutionary Acquisition of Neural Topologies (EANT). The method evolves both the structure and weights of neural networks. It starts with networks of minimal structures determined by the domain expert and increases their complexity along the evolution path. It introduces an efficient and compact genetic encoding of neural networks onto a linear genome that enables one to evaluate the network without decoding it. The method uses a meta-level evolutionary process where new structures are explored at larger time-scale and existing structures are exploited at smaller time-scale. We demonstrate the method by designing a neural controller for a real robot whichshould be able to move continously in a given environment cluttered with obstacles. We first give an introduction to the evolutionary method and then describe the experiments and results obtained.

1 Evolutionary Acquisition of Neural Topologies

Evolutionary Acquisition of Neural Topologies (ENAT) [6,7] is an evolutionary reinforcement learning system that is suitable for learning and adaptation to the environment through interaction. It combines meaningfully the principles of neural networks, reinforcement learning and evolutionary methods.

The method introduces a novel genetic encoding that uses a linear genome of genes (nodes) that can take different forms. The forms that can be taken by a gene can either be a neuron, or an input to the neural network, or a jumper connecting two neurons. The jumper genes are introduced by the structural mutation along the evolution path. They encode either forward or recurrent connections.

Figure 1 shows an example of encoding a neural network using a linear genome. As can be seen in the figure, a linear genome can be interpreted as a tree based program if one considers all the inputs to the network and all jumper connections as terminals.

The linear genome has some interesting properties that makes it useful for evolution of neural controllers. It encodes the topology of the neural controller implicitly in the ordering of the elements of the linear genome. This enables one to evaluate a neural controller represented by it without decoding the neural controller. The evaluation of a linear genome is closely related to executing a linear program using a postfix notation. In the genetic encoding the operands (inputs and jumper connections) come before the operator (a neuron) if one goes from right to left along the linear genome. The linear genome is *complete* in that it can represent any type of neural network. It is also a *compact* encoding of neural networks since the length of the linear genome is

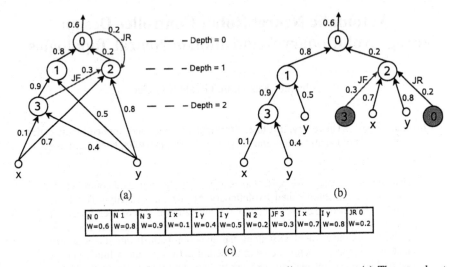

N 0	N 1	N 3	I x	I y	I y	N 2	JF 3	I x	I y	JR 0
W=0.6	W=0.8	W=0.9	W=0.1	W=0.4	W=0.5	W=0.2	W=0.3	W=0.7	W=0.8	W=0.2

(c)

Fig. 1. An example of encoding a neural network using a linear genome. (a) The neural network to be encoded. It has one forward and one recurrent jumper connection. (b) The neural network interpreted as a tree structure, where the jumper connections are considered as terminals. (c) The linear genome encoding the neural network shown in (a). In the linear genome, pcrN stands for a neuron, pcr I for an input to the neural network, pcr JF for a forward jumper connection, and pcrJR for a recurrent jumper connection. The numbers beside pcr N represent the global identification numbers of the neurons, and pcrx or pcry represent the inputs coded by the input gene (node).

the same as the number of synaptic weights in the neural network. It is *closed* under structural mutation and under a specially designed crossover operator. An encoding scheme is said to be closed if all genotypes produced are mapped into a valid set of phenotype networks [5]. The crossover operator exploits the fact that structures originating from some initial structure have some parts in common. By aligning the common parts of two randomly selected structures, it is possible to generate a third structure which contains the common and disjoint parts of the two mother structures. This type of crossover is introduced by Stanley [10]. An example of the crossover operator under which the linear genome is closed is shown in Fig. 2.

If one assigns integer values to the nodes of a linear genome such that the integer values show the difference between the number of outputs and number of inputs to the nodes, one obtains the following rules useful in the evolution of the neural controllers. The first is that the sum of integer values is the same as the number of outputs of the neural controller encoded by the linear genome. The second is that a sub-network (sub-linear genome) is a collection of nodes starting from a neuron node and ending at a node where the sum of integer values assigned to the nodes between and including the start neuron node and the end node is *one*. Figure 3 illustrates the concept.

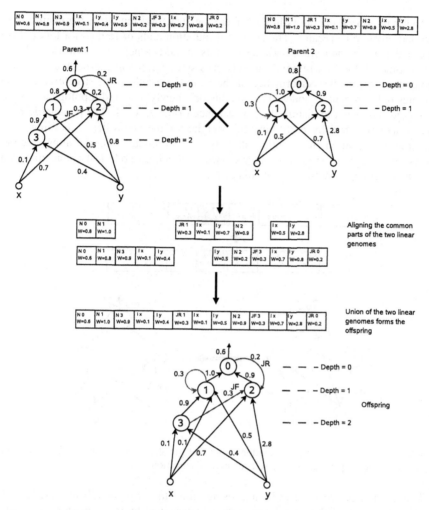

Fig. 2. Performing crossover between two linear genomes. The genetic encoding is closed under this type of crossover operator since the resulting linear genome maps to a valid phenotype network. The weights of the nodes of the resulting linear genomesare inherited randomly from both parents.

N 0	N 1	N 3	I x	I y	I y	N 2	JF 3	I x	I y	JR 0
W=0.6	W=0.8	W=0.9	W=0.1	W=0.4	W=0.5	W=0.2	W=0.3	W=0.7	W=0.8	W=0.2
[-1]	[-1]	[-1]	[1]	[1]	[1]	[-3]	[1]	[1]	[1]	[1]

Fig. 3. An example of the use of assigning integer values to the nodes of the linear genome. The linear genome encodes the neural network shown in Fig. 1. The numbers in the square brackets below the linear genome show the integer values assigned to the nodes of the linear genome. Note that the sum of the integer values is *one* showing that the neural network encoded by the linear genome has only *one* output. The shaded nodes form a sub-network. Note also that the sum of the integer values assigned to a sub-network is always *one*.

The main search operators in EANT are the structural mutation, parametric mutation and crossover operator. The structural mutation adds or removes a forward or a
recurrent jumper connection between neurons, or adds a new sub-network to the linear genome. It does not remove sub-networks since removing sub-networks causes a
tremendous loss of the performance of the neural controller. The structural mutation
operates only on neuron nodes. The weights of a newly acquired topology are initialized to zero so asnot to disturb the performance of the network. The parametric mutation is accomplished by perturbing the weights of the controllers according to the
uncorrelated mutation in evolution strategy or evolutionary programming. Figure 4
shows an example of structural mutation where a neuron node lost connection to an
input and received a self-recurrent connection.

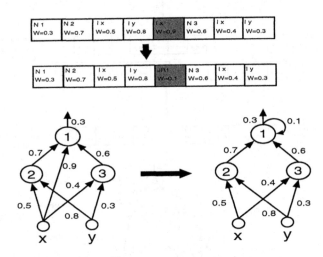

Fig. 4. An example of structural mutation. Note that the structural mutation deleted the input
connection to N1 in pcr and added a self-recurrent connection to it.

The initial structures are generated using either the grow or full method [2]. The
initial complexity of the neural structures is determined by the domain expert and is
specified by the maximum depth that can be assumed by the initial structures. The
depth of a neuron node in a linear genome is the minimal number of neuron nodes
that must be traversed to get from the output neuron to the neuron node, where the
output neuron and the neuron node lie within the same sub-network that starts from
the output neuron. The system starts with exploitation of structures that are already in
the system. By exploitation, we mean optimization of the weights of the structures.
This is accomplished by an evolutionary process that occurs at smaller time-scale.
The evolutionary process at smaller time-scale uses parametric mutation and recom-
bination operators as a search operator. Exploration of structures is done through
structural mutation and crossover operator. The structural selection operator that occurs at larger time-scale selects the first half of the population to form the next generation. In order to protect the structural innovations or discoveries of the evolution,
young structures that are less than few generations old with respect to the larger time-
scale are carried over along the evolution regardless of the results of the selection

operator. New structures that are introduced through structural mutation and which are better according to the fitness evaluations survive and continue to exist.Since sub-networks that are introduced are not removed, there is a gradual increase in the complexity of structures along the evolution. This allows EANT to search for a solution starting from a minimum structural complexity specified by the domain expert. The search stops when a neural controller with the necessary optimal structure that solves a given task is obtained.

2 Evolving Neural Controller for Navigation

The aim of this experiment is to demonstrate the automatic design of neural controllers for robots using EANT. We start with sonar based robot navigation for developing the controllers in simulation and then transfer the developed controllers to real robot. We evolved the structure and weights of the neural controller which enables B21 [9] robots to autonomously explore the environment and avoid obstacles. The controller is expected to avoid dead lock situations where Braitenberg-like controllers [3] have difficulties of escaping them. In these situations, they either come to a rest or start to oscillate left to right.

We used the sonar sensors of the B21 robot for detecting the obstacles. The B21 robot has 24 sonar sensors which are symmetrically distributed around its cylindrical body. We used the 8 in front and 2 in the rear sonar sensors as inputs to the neural controller. The sonar sensors give the distance of obstacles in millimeters measured from the center of the robot. The values returned by the sonar sensors are transformed using equation (1) before feeding them to the neural controller.

$$V_n = \begin{cases} \dfrac{-V_s + 2000}{2000} & \text{if } V_s < 1000 \\ 0 & \text{otherwise} \end{cases} . \tag{1}$$

In the equation, V_n is the transformed and normalized sonar reading and V_s is the actual reading returned by a particular sonar sensor. The value V_n lies between 0 and 1 for obstacles which are located at a distance less than 2 m from the center of the robot.

The initial controller has two output neurons and each neuron is connected to all sensors. The outputs of the neurons are connected to the motor apparatus of the robot. In addition to the sensor inputs, each neuron has a constant bias input connected to it. The forward translational velocity and rotational velocity of the robot are given by $V_t = 0{,}5(O_1 + O_2)$ and $R_t = O_1 + O_2$, respectively. The quantities O_1 and O_2 are the outputs of the neural network. Since the output of the neurons is between −1 and 1, the maximum and minimum forward velocity of the robot is 1 m/s and −1 m/s, respectively. The rotational velocity is bounded between 2 rad/s and −2 rad/s.

The initial controller is similar to Braitenberg-like controller and is not capable of avoiding dead lock situations. The algorithm is expected to find a controller which is complex enough for solving the navigation problem with the ability of avoiding dead lock situations. The fitness function used to evaluate the controllers is given by

$$F = \sum_{t=1}^{T} D(t) e^{-100(H(t)-H(t-1))^2} (1 - S_{max}(t)), \tag{2}$$

where $D(t)$, $H(t)$ and $S_{max}(t)$ are the distance traveled, the heading of the robot, and the maximum value of the currently perceived normalized sonar readings respectively. The fitness function favors controllers that move straight as long and as fast as possible and controllers that give the robot the maximum distance from the obstacles. Figure 5 shows the initial neural controller and the final controller obtained by our algorithm. The ability of avoiding the deal lock situations comes because of the recurrent connections. The result is similar to that obtained by Nolfi and Floreano [8] and Hülse and Pasemann [4] but in both cases the structure of the neural controller is determined manually beforehand. Ahrns et.al [1] designed a fuzzy-neuro controller for solving the robot navigation with obstacle avoidance. They solved the dead lock situations problem by designing a feature extraction mechanism that extracts afree space direction closest to the heading of the vehicle. They further stored the sonar readings in a short time memory to extract the coarse model for the direct robot surroundings. They used the feature extraction mechanism since the fuzzy-neuro controller does not have recurrent connections.

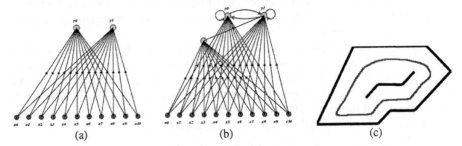

(a)   (b)   (c)

Fig. 5. (a) The initial Braitenberg-like controller (b) The best neural controller found by EANT that is capable of avoiding dead lock situations. (c) Trajectory of the robot controlled by the best neural controller.

3 Conclusion

In this paper we have demonstrated the automatic design of controllers for real robots using EANT taking as an example robot navigation with reactive obstacle avoidance. EANT found a clever solution that gives the robot the ability to explore the environment without being trapped in dead-lock situations, where simple designs like Braitenberg-like controllers have difficulties of escaping them.

4 Acknowledgment

This work is sponsored by the German Academic Exchange Service (DAAD) under grant code R-413-A-01-46029 and the COSPAL project under the EC contract no. FP6-2003-IST-2004176, which are duly acknowledged.

References

1. I. Ahrns, J. Bruske, G. Hailu, and G. Sommer. Neural fuzzy techniques in sonarbased collision avoidance. In L.C. Jain and T. Fukuda, editors, *Soft Computing for Intelligent Robotic Systems*, Studies in Fuzziness and Soft Computing, pages 185–214. Physica-Verlag (Springer), 1998.
2. W. Banzhaf, P. Nordin, R. E. Keller, and F. D. Francone. *Genetic Programming: An Introduction on the Automatic Evolution of Computer Programs and Its Applications*. Morgan Kaufmann, San Francisco, CA, 1998.
3. V. Braitenberg. *Vehicles. Experiments in Synthetic Psychology*. The MIT Press, Massachusetts, London, 1994.
4. M. Hülse and F. Pasemann. Dynamical neural schmitt trigger for robot control. In *Proceedings of International Conference on Artificial Neural Networks (ICANN 2002)*, pages 783–788. Springer-Verlag, 2002.
5. J. Jung and J. Reggia. A descriptive encoding language for evolving modular neural networks. In *Proceedings of the Genetic and Evolutionary Computation Conference (GECCO)*, pages 519–530. Springer-Verlag, 2004.
6. Y. Kassahun and G. Sommer. Eficient reinforcement learning through evolutionary acquisition of neural topologies. In *Proceedings of the 13th European Symposium on Artificial Neural Networks (ESANN)*, pages 259–266, Bruges, Belgium, April 2005. d-side publications.
7. Y. Kassahun and G. Sommer. Evolution of neural networks through incremental acquisition of neural structures. Technical Report 0508, Institute of Computer Science and Applied Mathematics, Christian-Albrechts University, Kiel, Germany, June 2005.
8. S. Nolfi and D. Floreano. *Evolutionary Robotics. The Biology, Intelligence, and Technology of Self-Organizing Machines*. The MIT Press, Massachusetts, London, 2000.
9. RWI. *B21 Users Guide*. Real World Interface, Inc., 1997.
10. K. O. Stanley. *Efficient Evolution of Neural Networks through Complexification*. PhD thesis, Artificial Intelligence Laboratory. The University of Texas at Austin., Austin, USA, August 2004.

KAWA-I krabbelt!
Entwurf, Aufbau und Steuerungsarchitektur des Colani-Babys

Tilo Gockel, Tamim Asfour, Joachim Schröder, Luigi Colani, Rüdiger Dillmann

Institut für Technische Informatik (ITEC), Universität Karlsruhe,
Technologiefabrik, Haid-und-Neu-Straße 7, D-76131 Karlsruhe

E-mail: {gockel, asfour, schroede, dillmann}@ira.uka.de

Zusammenfassung. Im April dieses Jahres trat Prof. Colani an unser Institut heran mit der Bitte, den Prototyp seines Roboterbabys *KAWA-I* mit Aktuatorik auszustatten. Hieraus ist ein Blockpraktikum für acht Studenten des Fachbereichs Informatik an der Universität Karlsruhe entstanden. Die Aufgabe umfasste mechanische, elektromechanische und konstruktive Arbeiten, Steuerungstechnik und Programmierung des dezentralen Mikrocontrollers im Roboter sowie des PC-seitigen Benutzer-Interfaces.

1 Die Aufgabe

KAWA-I ist das japanische Wort für „niedlich". Dieses Adjektiv hat Prof. Colani inspiriert, eine Babypuppe zu designen, welche als Spielgefährte, ähnlich dem Roboterhund *AIBO* oder der Robotrobbe *PARO*, beim Menschen positive Emotionen weckt (Abb. 1, [13, 11, 5]). Nach der Designphase und dem Aufbau eines Musters aus 2-Komponenten-Polyesterkunststoff trat Herr Prof. Colani an unser Institut heran mit der Bitte, diesen Prototypen mit Motoren auszustatten und die zugehörige Steuerung zu entwerfen. Entstanden ist hieraus ein Blockpraktikum für Informatikstudenten der Universität Karlsruhe mit den Inhalten Mechanik, Elektronik, Steuerungstechnik, Robotik, µC- und PC-Programmierung. Realisiert werden sollte eine krabbelnde Bewegung des Babys mit einer möglichst geringen Anzahl von Aktuatoren, eine schnelle

Abb. 1. Prof. Colani mit *KAWA-I*-Abformpositiv

anschauliche Umsetzung, um mit einfachen Mitteln möglichst schnell und auch kostengünstig einen Demonstrator zur Verfügung stellen zu können. Die langfristige Planung umfasst aber auch die Herstellung einer ultraleichten und auf die Unterbringung der Aktuatoren optimierten Abformung, sowie eine Steuerung, welche dem Baby in der Zukunft ermöglichen soll, selbstständig zu lernen und hiermit die Krabbelbewegung später in eine Gehbewegung zu wandeln.

2 Systementwurf

2.1 Spezifikation

Während das Baby vermessen wurde und als Volumenmodell im CAD-Programm ProE modelliert wurde, um hiermit eine Simulation mit MCA2 ([MCA2-05], vgl. Abschnitt 3.3) aufsetzen zu können, erfolgte parallel hierzu bereits eine Einpassung der Antriebs- und Lagerungskomponenten. Nach einer Auftrennung des Korpus und einer Volumenbetrachtung wurden für das vorliegende Projekt Servomotoren, also Stell-Aktuatoren aus dem Modellbaubereich, gewählt. Die Festlegung der notwendigen Drehmomente erfolgte nach Experimenten an Korpus und Segmenten mit einer Federwaage. Der verwendete Servo-Typ HITEC HS-5945MG ist digital geregelt und besitzt ein Stellmoment von 130 Ncm. Die interne Untersetzung erfolgt durch ein stabiles Metallgetriebe. Da in den Segmenten des Babys durch die dicke Kunststoffwandung nur sehr wenig Platz zur Verfügung stand, wurde entschieden, nur die für eine Krabbelbewegung absolut notwendige Aktuatorik einzubauen. Die Anordung der aktiven Freiheitsgrade ist in Abb. 2 dargestellt, zusätzlich wurden auch passive, federgestützte Freiheitsgrade in den Kniegelenken, in den Ellenbogengelenken und im Hals eingeführt.

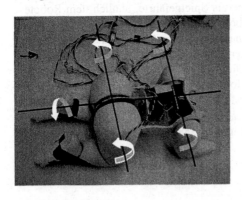

Abb. 2. Aktive Freiheitsgrade von KAWA-I

2.2 Architektur und Schnittstellen

Die in KAWA-I umgesetzte Steuerungsarchitektur ist in Abb. 3 zu sehen. In KAWA-Is Schädel wird ein Mikrocontrollermodul mit dem Controller AT89S8252 zur Servoansteuerung verwendet [2]. Trajektorien werden entweder PC-seitig interaktiv vom Benutzer vorgegeben, PC-seitig in Form von Makros abgespielt oder für einen *Stand-Alone*-Betrieb lokal im Flash des μCs abgelegt.

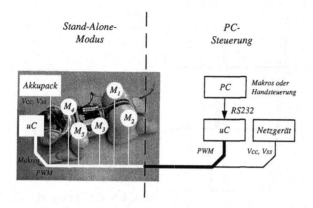

Abb. 3. Aufbau der Steuerung

Die Programmierung des Flash-Controllers und auch die Kommunikation mit dem Controllerboard erfolgen via RS232, die Geschwindigkeit beträgt 19.200 bps. Hier kann auch mittels Easy-Radio-Funkmodulen eine Funkkommunikation zur Bewegungsvorgabe oder zum erneuten Flashen des Controllers realisiert werden. Das sehr einfache serielle Protokoll ist folgendermaßen aufgebaut:

$$Datenpaket = „S" + (Stellwert\ Motor1) + .. (Stellwert\ Motor\ n)$$

Dabei ist *Stellwert* $\in$ *[0..255]*. Es entstehen also beim Einsatz von fünf Motoren Datenpakete der Länge sechs Bytes, das Zeichen „S" dient hier auch als Stopp-Byte zur Fehlererkennung. Andere Zeichen des Protokolls sind *I* für *Init* und „*M*" für „*Max*".

3 Umsetzung

3.1 Hardware

Die interne Hardware des Roboters umfasst die Motoren sowie ein in der Größe auf die Schädelinnenmaße angepasstes µC-Board [6,9]. Ein Akkupack für den autonomen Betrieb konnte nicht mehr im Inneren des Babys untergebracht werden und ist in Form eines kleinen Rucksacks außen auf dem Rücken des Babys angebracht. Für die Steuerung per PC steht ein zweites, externes und entsprechend vorverdrahtetes Board zur Verfügung. Weiterhin wird das Baby dann per Nabelschnur durch ein Labornetzgerät versorgt. Die Motoren sind als Direktantriebe ohne weitere Untersetzungen in der Brust und in den Oberschenkeln montiert, der Hüftantrieb ist über eine zusätzliche Lagerung mechanisch entkoppelt. (Abbildungen 3 und 7)

3.2 Software und Algorithmik

3.2.1 Servo-Timing

Die verwendeten Modellbauservomotoren werden durch ein pulsbreitenmoduliertes Signal angesteuert ($f = 50\ Hz$, $T_{Puls} \approx 1,0 .. 2,0\ ms$, $\theta \approx -90° .. +90°$, vgl. Timing-Diagramm in Abb. 4). Die PWM-Generierung erfolgt im Atmel-Controller durch

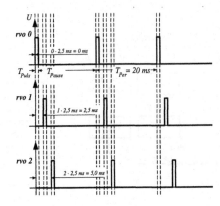

Abb. 4. Servo-Timing, vgl. auch [9]

einen hochprioren Timer. Wie im Timing-Diagramm gut zu erkennen ist, lassen sich auf diese Art und Weise durch einen einzelnen Timer bis zu acht Servomotoren synchron ansteuern.

Sollten in einer späteren Ausbaustufe mehr als acht Motoren verwendet werden, so ist die einfachste Möglichkeit der Einsatz eines weiteren µC-Boards, dessen RS232-Leitung RXD parallel zum ersten Board angeschlossen wird (der zweite MAX232-Baustein entfällt). Dann können beide Boards an nur einer Schnittstelle auf unterschiedliche Datenpakete reagieren.

3.2.2 Adaptierter Bresenham-Algorithmus zur Bewegungssteuerung

Als Bewegungsart für den Roboter wurde die klassische Punkt-zu-Punkt-Steuerung gewählt. Hierbei sind zwei Varianten möglich: Gleichzeitige Bewegung aller Achsen mit maximaler Geschwindigkeit, bzw. Beschleunigung oder Beendigung der Bewegung aller beteiligten Achsen zur gleichen Zeit [1, Abschnitt 4.4.1]. Für die vorliegende Anwendung wurde Variante Zwei gewählt, wobei in unserem Falle nicht direkt Geschwindigkeiten vorgegeben werden, sondern Gelenkwinkel über diskrete Zeitabschnitte. Es wird also der größtmögliche Gelenkwinkel entsprechend der gegebenen Auflösung (hier ca.: $180° / 256 \approx 0{,}7°$-Schritte) als Vorgabe für die anderen Motoren gewählt und diese vollziehen dann die $0{,}7°$-Schritte nicht nach jedem Zeitabschnitt t_i, sondern nur nach $n \cdot t_i$ mit $n = \theta_k / 256$.

Der Roboter kann in zwei Modi angesteuert werden: Im ersten Modus erfolgt die Ansteuerung über den Leitstands-PC und es können die einzelnen Gelenke via Slidern in der GUI bewegt werden oder auch abgespeicherte Sequenzen über die serielle Schnittstelle geschickt werden. Im zweiten Modus – dem *Stand-Alone*-Modus – erfolgt die Bewegung autonom entsprechend der im Flash-Speicher des Controllers abgelegten Sequenz. Für den *Stand-Alone*-Modus wurde obenstehende Gleitkommadivision für n vermieden, um Controller-Ressourcen zu sparen. Die Berechnung erfolgt hier nach dem Bresenham-Algorithmus, der für die vorliegende Anwendung leicht verändert wurde. In Abb. 5 ist der Pseudo-Code hierzu gelistet. [3,7]

```
// pseudo code for one movement (S teta_1 teta_2 teta_3..teta_n)
// similar to Bresenham, but with const. predefined MAX
signed counter[1..n] = 0;// n motors have n counter variables
signed teta[1..n];      // these are the given angles (constants)
lauf = 0;
const MAX = 256;
for (i = 1..n){
    counter[i] = MAX / 2;
}
while(lauf < MAX){
    // all motors have to reach End Position in MAX TimeSteps
    for (i = 1..n){
        counter[i] = counter[i] - teta[i];
        if (counter[i] < 0){
            // move Motor[i] one step (approx. 180°/MAX):
            STEP[i];
            counter[i] = counter[i] + MAX - teta[i];
        }
    }
    lauf++;
}
```

Abb. 5. Pseudo-Code zum angepassten Bresenham-Algorithmus

3.3 Das Software-Framework MCA

MCA (Modular Controller Architecture, [12]) ist ein Software-Framework, welches am Forschungszentrum Informatik (FZI) in der Gruppe Interaktive Diagnose- und Servicesysteme (IDS) entwickelt wurde und unter der GNU Public Licence (GPL) steht. Es handelt sich um ein modulares, netzwerktransparentes und betriebssystemsübergreifendes C++-Framework zur Unterstützung der Implementierung von komplexen, echtzeitfähigen Steuerungen. MCA2 bietet ein sog. Modul-Framework mit standardisierten Schnittstellen. Die elementaren Einheiten (Module) können zu komplexeren Gruppen zusammengefasst werden, sodass komplexe Sachverhalte realisiert werden können. Dabei liegen die Vorteile in der hohen Wiederverwendbarkeit der Komponenten und in der Möglichkeit der Parameteränderung zur Programmlaufzeit. MCA2 kommt bereits seit Jahren an den Instituten IDS und ITEC zur Steuerung von komplexen Laufmaschinen und humanoiden Robotern zum Einsatz. Die Vorteile dieses offenen Steuerungskonzeptes sollen zukünftig auch in die Trajektoriengenerierung für das Roboterbaby einfließen. Hierfür wurde, wie oben bereits angesprochen, ein Volumenmodell der Segmente von KAWA-I erstellt und eine einfache Visualisierung von Bewegunsgabläufen des Roboters realisiert. Weiterhin erfolgt die Ansteuerung der seriellen Schnittstelle über die graphische Schnittstelle MCAGUI. Die Synchronisation der Sollwertvorgabe für die einzelnen Antriebe übernimmt ein Trajektorien-Manager. (Abb. 6).

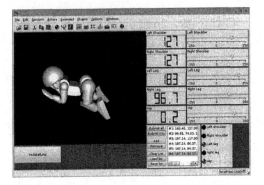

Abb. 6. Screenshot aus MCA2

4 Ergebnisse und Ausblick

In Abb. 7 ist der fertiggestellte Roboter einmal in geöffnetem Zustand und einmal im geschlossenen Zustand mit Akkupackrucksack zu sehen. Aus dem Korpus herausgeführte Dip-Schalter ermöglichen die Wahl unterschiedlicher Bewegungsmuster. In Abb. 8 ist ein Bewegungszyklus des Babys dargestellt. Es konnte gezeigt werden, dass eine anthropomorphe Krabbelbewegung auch mit wenigen aktiven Gelenken realisiert werden kann. Auch die Umsetzung der einzelnen Arm- und Beinbewegungen in eine resultierende Vorwärtsbewegung gelingt gut. Bisher wurden die Bewegungen im Gelenkwinkelraum des Roboters realisiert. Hierzu dienten Videoaufzeichnungen von krabbelnden Babys. Zukünftig werden Bewegungen eines krabbelnden Babys, die mithilfe des VICON-Motion-Capture-System erfasst wurden, auf das Baby übertragen. Dadurch sollen realistische, babyähnliche Bewgungen generiert werden. Eine sensorische Erweiterung des Babyroboters ist jedoch notwendig, um Bewegungen des Babyroboters aus einer vorgegebenen Körpertrajektorie zu erzeugen. Ein weiterer Hardware-Ausbau des Roboters durch zusätzliche Aktuatoren wird im Moment durch das hohe Gewicht der teilweise sehr dickwandigen Kunststoffkarosserie verhindert. In der nächsten Generation wird eine Abformung des ursprünglichen Babykörpers (vgl. Abb. 1) mit laminiertem Glas- oder Kohlefasergeflecht erfolgen, um somit eine wesentlich leichtere Ausgestaltung zu erreichen. Geplant ist auch der Einsatz einer digitalen Kamera, um auf Umweltereignisse entsprechend reagieren zu können [4], eine Erweiterung der aktiven Freiheitsgrade (Hände, Füße und Hals) und eine Sprachausgabe.

Abb. 7a. AWA-I ist

Abb. 7b. ... fertiggestellt

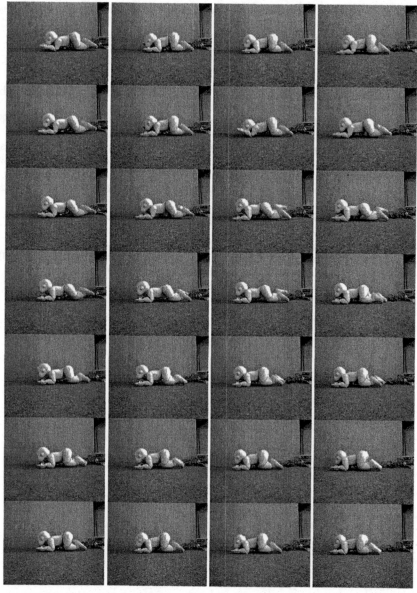

Abb. 8. Bewegungsphasen von KAWA-I beim Vorwärtskrabbeln (Reihenfolge: ↓↓↓↓)

Literatur

1. T. Asfour: Sensomotorische Bewegungskoordination zur Handlungsausführung eines humanoiden Roboters. Dissertation, Universität Karlsruhe (TH), FB Informatik, GCA-Verlag, Herdecke, 2003.
2. Homepage des IC-Herstellers Atmel http://www.atmel.com

3. Jack E. Bresenham: Algorithm for computer control of a digital plotter. *IBM Systems Journal* Vol. 4, No. 1, 1965. Vgl. auch:
http://www.research.ibm.com/journal/sj/041/ibmsjIVRIC.pdf

4. Homepage der Carnegie Mellon Univ., Entwickler der CMUcam
http://www-2.cs.cmu.edu/~cmucam/

5. Homepage von Prof. Luigi Colanihttp://www.colani.de

6. Elektor, Zeitschrift für Elektronik, Artikel: AT89S8252-Flashboard, Ausgabe Dezember 2001, Elektor-Verlag, Aachen, 2001.

7. Elektor, Zeitschrift für Elektronik, Artikel: Synchrone Servosteuerung, Ausgabe März 2005. Elektor-Verlag, Aachen, 2005.

8. T. Gockel, O. Taminé, R. Dillmann: EduKaBot – Aufbau eines edukativen Roboterbaukastensystems. *Tagungsband zum 18. Fachgespräch Autonome Mobile Systeme (AMS 2003)*, Karlsruhe, Dezember 2003.

9. T. Gockel, R. Dillmann, A. Bierbaum, A. Piaseczki, J. Schröder, P. Azad: *Embedded Robotics – Das Praxisbuch*. Elektor-Verlag, Aachen, 2005.

10. Homepage der FZI-Gruppe IDS, hier: Veröffentlichungen zu MCA2
http://www.fzi.de/ids/publikationen.php

11. Homepage mit Informationen zur Roboterrobbe PARO http://paro.jp/english/

12. K.U. Scholl, J. Albiez und B. Gassmann: MCA – An Expandable Modular Controller Architecture. In: *3rd Real-Time Linux Workshop*. Milano, Italy, 2001.

13. Homepage mit Informationen zum Roboterhund AIBO http://www.aibo-europe.com

Autorenverzeichnis